Illuminate
Publishing

OCR
Geology
for A Level & AS

Stephen Davies, Frank Mugglestone, Ruth Richards and Tony Shelton
with Vince Williams, Malcolm Fry and Debbie Armstrong.

Consultant Editor: Ruth Richards

Dedication

This book is dedicated to Frances Stratton. Without her vision and inspiration, this book would not have been possible.

Published in 2017 by Illuminate Publishing Limited, an imprint of Hodder Education, an Hachette UK Company, Carmelite House, 50 Victoria Embankment, London EC4Y 0DZ

Orders: please contact Hachette UK Distribution, Hely Hutchinson Centre, Milton Road, Didcot, Oxfordshire, OX11 7HH. Telephone: +44 (0)1235 827827. Email: education@hachette. co.uk. Lines are open from 9 a.m. to 5 p.m., Monday to Friday. You can also order through our website: www.hoddereducation.co.uk

British Library Cataloguing-in-Publication Data

A catalogue record for this book is available from the British Library ISBN 978-1-911208-14-3

Printed by Ashford Colour Press, UK

Impression 4
Year 2023

Hachette UK's policy is to use papers that are natural, renewable and recyclable products and made from wood grown in well-managed forests and other controlled sources. The logging and manufacturing processes are expected to conform to the environment regulations of the country of origin.

Editor: Geoff Tuttle
Design and layout: Kamae Design

Acknowledgements

Frank Mugglestone would like to thank Amy and Charlotte Mugglestone, Brian and Dorothy Harmer for their invaluable help

Stephen Davies acknowledges the support given by his late wife, Ann, to his geological pursuits over a period of 50 years.

Ruth Richards acknowledges the outstanding support from Graeme and Isaac Richards, Gren Ireson, Philippa Baker, Geoff Tuttle, Ian Smith, David Siveter and Derek Siveter.

Photographs

p24 Obsidian – www.sandatlas.org; p86 Mississippi delta – NASA; p122 Crater – USGS; p146 San Andreas – Thule Scientific; p160 Ocean map – NASA/USGS; p162 Crysotile – Eurico Zimbres; p259 Brachiopods – Wooster College Geologists; p265 Archaeopteryx – Ann Musser; p277 Archaeopteryx feather – NATURKINDER MUSEUM BERLIN; p269 Gravity map – Canadian Geological Survey; p269 Shocked quartz – USGS; **IStock** p191 Copper mine – Pavliha; p210 Landfill – bokan76; p211 Sheep – IG Royal; p211 Radiation sign – BackYard Production; p234 Subsidence – Stephen Clarke; p236 Ropoto village – siete_vidas; p237 Rockfallsign – tinieder; p242 Pisa – musat; p243 Core box – eredre; p243 Drilling – fbxx; p253 DNA – pagadesign; p277 Archaeopteryx – demarfa; **Shutterstock** p42 Lava fissure – Photovolcanica. com; p116 Alpine lake – Bjoern Alberts; p122 Protoplanetary disc – Mopic; p192 Placer ore – oneword; p207 Dam – tishmir; p246 Stromatolites – John Carnemolla; p248 Sand dunes – David Steele; p250 Nuclear explosion – Everett Historical; p260 Lungfish – sciencepics; p260 Coelacanth – Catmando; p287 Pterodactytus – Valentyna Chukhlyebova

Cover Alex Fieldhouse / Alamy Stock Photo

All other photos: Frances Stratton, Ruth Richards, Frank Mugglestone, Tony Shelton, Stephen Davies, Malcolm Fry, Vince Williams, Debbie Armstrong.

FSC
www.fsc.org

MIX
Paper | Supporting responsible forestry
FSC™ C104740

OCR
Geology
for A Level & AS

Contents

How to use this book

In this book, you will find a number of features to help you.

This book is organised as double-page spreads, and each of these has relevant and related material arranged together. All of the related content are arranged in chapters, regardless of their position in the specification. The AS topics appear first, followed by the A Level content; all positioned in the same chapter. For example, all of the igneous petrology is in Chapter 2 and all of the sedimentology is in Chapter 3; allowing a coherent journey through the material.

The pages are colour coded, shown as a strap line at the top of the page. The AS level content has a pale coloured strap line, whilst the A level has a darker coloured strap line. This is to ensure that AS students can identify the relevant content to study, whilst not interfering with the 'journey' of the A level students.

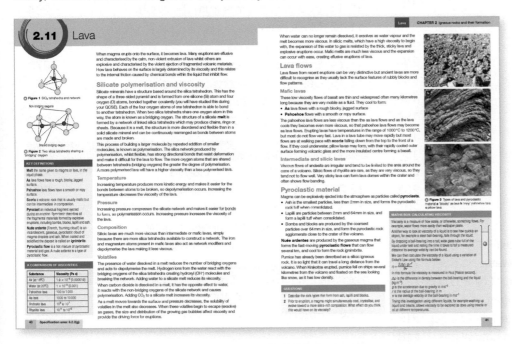

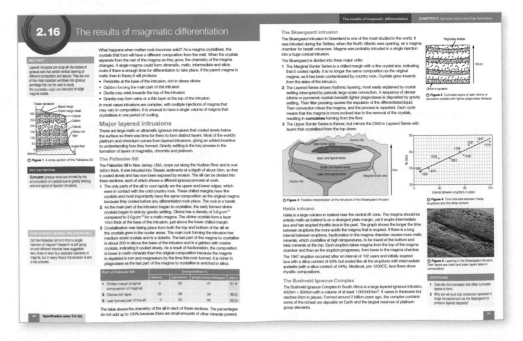

The page layout

Specification references

Specification references

These are given at the foot of the left-hand page of a double spread, and link to the content of the OCR specification. This is also coupled with the colour of the strap line at the top of the page. The same chapter will have the same main colour for the strapline; with the AS content paler, and the A level content darker.

KEY DEFINITIONS

These are terms that appear in the specification, or need to be understood to explore the content fully. These terms are given in bold in the text, where appropriate.

KEY FACTS

Some spreads have summarised key facts for part of the topic.

HOW SCIENCE WORKS

Aspects that relate to everyday life, or where there is ongoing scientific work in a field; these have been identified as 'How science works' in the text. These are designed to allow a fuller understanding of the topic, by giving it everyday context.

MATHS BOXES

These show you how calculations should be done and are often completed as worked examples or given full explanations, so that students can access the material. Where appropriate they include formulae that are specific to the specification.

STUDY TIPS

Where students make common errors, there are selected points exemplified to draw your attention to them. They may also be tips to help you remember the content.

CASE STUDIES

In some spreads, there are case studies describing specific examples, directly related to the contents of the double-page spread. These are designed to give a more 'in-depth' understanding of the topic.

QUESTIONS

At the end of each double-page spread, there are questions that relate to the content and challenge your understanding of the topic.

Practice questions

These are given in a separate section, towards the end of the book, and provide integrated questions. We would also recommend that you access any available past papers and mark schemes; plus any sample assessment material that may be available.

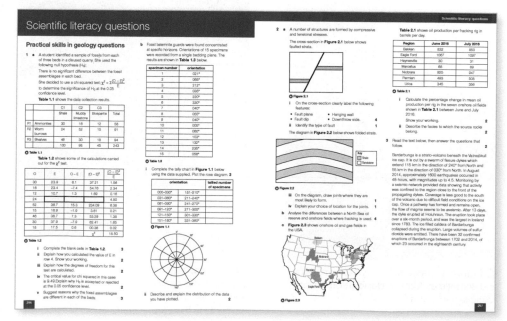

A **mineral** is a naturally occurring chemical substance having a definite composition and crystalline structure. The most important **rock**-forming minerals are silicates, which means that they contain silicon (Si) and oxygen (O) in different proportions, and they form most of the rocks on Earth. There are other minerals, such as the metallic minerals, which are important for their economic value. Most minerals are compounds, but some contain one element only. Examples of these are native copper (Cu) and native sulfur (S).

Elements and minerals

Most minerals are compounds of the most abundant elements in the Earth's crust. More than 98% of the crust consists of eight elements, so it's not surprising that they occur often in the chemical formulae of many minerals. Quartz is one of the most common minerals in the crust. It is composed of the two most abundant elements. Its formula is SiO_2. Common in limestones, the mineral calcite has the formula $CaCO_3$. Some elements are compounds of sulfur rather than oxygen. They include galena (PbS) and pyrite (FeS_2).

Abundance of elements in Earth's crust		
Element	Mass per cent	Chemical symbol
Oxygen	46.6%	O
Silicon	27.7%	Si
Aluminium	8.1%	Al
Iron	5.0%	Fe
Calcium	3.6%	Ca
Sodium	2.8%	Na
Potassium	2.6%	K
Magnesium	2.1%	Mg
All others	<1.5%	

▲ **Figure 1** Native sulfur crystals are yellow and sometimes form by **deposition** around volcanic vents and fumaroles

▲ **Figure 2** Well-formed crystals of quartz. It is composed of silicon and oxygen, the two most abundant elements in the Earth's crust

▲ **Figure 3** Crystals of galena. Composed of lead and sulfur, it is the principal ore of lead and has metallic **lustre**

Minerals and crystals

Minerals are solid **crystals**, but they often grow from liquids. For example, the elements that form the mineral quartz (Si and O) are present in magma, a liquid from which crystals grow and combine to make rocks. Ions of these, and all other elements, behave differently when liquid or solid. They move at high velocity in a liquid, because they have greater thermal energy. They are disorganised and do not form a pattern. Slow cooling of a liquid reduces the thermal energy and gives ions time to organise into patterns. These organised patterns of elements give the internal structure and external appearance to crystals. In the case of Si and O, they would form crystals of the mineral quartz. If liquids are cooled rapidly, they freeze the disorganised arrangement of a liquid, and a natural **glass** is produced. In a silicon- and oxygen-rich liquid the glass is called obsidian. Glass is not a mineral.

Mineral a naturally occurring chemical substance having a definite composition and crystalline structure.

Rock an aggregate or mixture of one or more minerals.

Sublimation is the transition of a substance directly from the solid phase to the gas phase, without passing through the intermediate liquid phase. The reverse of this process is known as **deposition** (or de-sublimation).

Lustre the surface appearance of a mineral, as it interacts with light.

Crystal a solid with plane faces formed when atoms are arranged in a structurally ordered pattern.

Glass an amorphous solid with no crystalline structure.

Grain boundary the line of contact between mineral crystals in a rock.

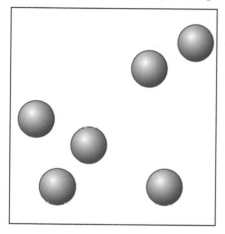

a Mobile atoms in a liquid

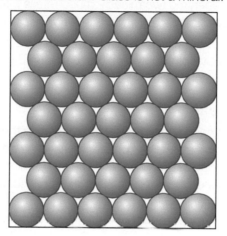

b Atoms ordered in a crystalline solid

◄ **Figure 4** Diagrams of atoms arranged in a liquid, and in a solid that formed from a slowly cooled liquid. Red colour represents moving atoms with high thermal energy in a liquid. Blue colour represents atoms in an ordered structure in a solid with low thermal energy.

Size and shape of crystals

Mineral crystals grow from the centre outwards by adding to their faces. Although the faces may be of different sizes, the angle between them is always constant for any given mineral. This is due to the regular arrangement of the atoms in their structure. If a liquid is cooled slowly, fewer crystals begin to form than if it is cooled more quickly. Slower cooling gives this smaller number of crystals more space into which they can grow, so slowly cooled liquids produce larger crystals. Eventually, the crystals will meet each other, and the **grain boundaries** between them will be irregular. If they grow into a cavity they will then have good crystal shape. Interestingly, crystals of a particular mineral do not always grow into exactly the same shape. Calcite ($CaCO_3$) can grow to form two crystal shapes. They look quite different, but both have the same symmetry and other properties of calcite.

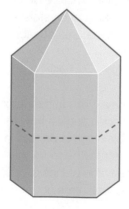

a Diagram of a quartz crystal.

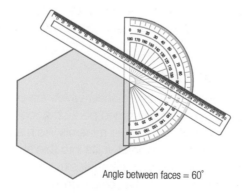

Angle between faces = 60°

b Cross-section of quartz crystal, showing that angles between faces are constant even though faces may not be uniform size.

◄ **Figure 5** Measuring the angle between crystal faces

1 Describe one difference between rocks and minerals.

2 Explain why a naturally formed glass is not a mineral.

3 Why are silicates the most important rock-forming minerals?

4 Research the seven crystal systems and compare their shape and symmetry.

△ **Figure 1** Special properties.
The mineral magnetite is magnetic and can be identified in this rock specimen because it attracts a small magnet.

Characteristics of rock-forming minerals

Most rock-forming minerals are silicates, the main exception being calcite, which forms the sedimentary rock limestone and the metamorphic rock marble.

Minerals have physical characteristics that can help geologists to recognise them.

Colour

Colour can be used to identify some minerals, but this is not very reliable as some minerals can occur in several different colours. Colour is used to distinguish between the two micas – muscovite is pale grey and biotite is black. Quartz can be purple, pink, grey, yellow, white or clear, so colour is not diagnostic.

White or colourless minerals contain metallic elements largely from groups one and two of the periodic table. Coloured minerals tend to contain elements from the transition metals.

Streak

A mineral specimen is scraped across the surface of a piece of unglazed porcelain (known as a streak plate). Streak is the colour of the powder left by the mineral on the plate. Most silicates have a white streak, which does not help to distinguish between them. Galena is a lead-grey colour and the streak is the same colour, while iron pyrite is brassy yellow but the streak is black. Minerals harder than the streak plate will scratch the streak plate without leaving a streak.

Lustre

The surface appearance of a mineral is described as its lustre and depends on its ability to reflect light. Minerals that are shiny like metal have metallic lustre. Minerals that are dull are described as earthy. Minerals that shine like diamonds are described as adamantine. There are other descriptive terms such as pearly, silky and resinous. Most rock-forming minerals have vitreous lustre, like glass.

Shape

The shape or habit of some mineral crystals can be very distinctive. Cubic crystals such as halite and fluorite are easy to recognise. Garnet is a complex shape based on a cube forming a dodecahedron. The hexagonal shape of quartz can be seen in Spread 1.1. Some mineral crystals grow in pairs and are said to show twinning. This can be a feature of feldspars.

There are actually seven crystal systems, mimicking the internal structure of the minerals themselves. An awareness of this is helpful when looking at crystals in hand specimen without the need to know the characteristic features of each system.

Cleavage and fracture

Some minerals have cleavage planes, which are planes of weakness in their atomic structure. Cleavage may be termed perfect, good or poor depending on how easily it splits along a cleavage plane. In minerals like muscovite, cleavage is perfect in one direction so that the mineral splits into thin sheets. There may be more than one direction of cleavage and the angles made between cleavage planes can also help in identification. Calcite has cleavage in three directions which are not at right angles.

Minerals that break along an irregular surface tend to fracture. They do not have cleavage because the bonds between atoms are strong. This is fracture. Sometimes the fracture is a series of concentric curved cracks, rather like a broken glass bottle, which is called a conchoidal fracture. A mineral with cleavage planes may also fracture if it breaks along a direction that is not a cleavage plane.

Hardness

The hardness of minerals is measured on Mohs' scale of hardness. Friedrich Mohs, a German mineralogist, used a set of ten reference minerals in his ten-point scale, showing distinctly different hardness. Hard minerals scratch softer ones. Soft minerals write on harder ones. The streak plate has a hardness of around 6.5 and so will be scratched by harder minerals. Hardness is a non-linear scale.

Hardness	Reference mineral
1 (soft)	Talc
2	Gypsum
3	Calcite
4	Fluorite
5	Apatite
6	Feldspar
7	Quartz
8	Topaz
9	Corundum
10 (hard)	Diamond

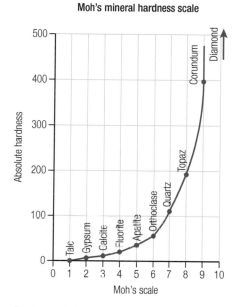

Mica has perfect cleavage in one direction

Calcite has perfect cleavage in three directions not at right angles

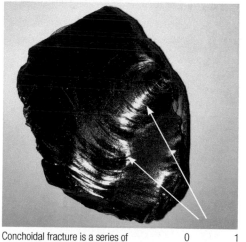

Conchoidal fracture is a series of concentric curved cracks in obsidian.

▲ **Figure 2** Graph showing Mohs scale and absolute hardness, shown as non-linear. Most naturally occurring minerals have a hardness of 7 or below.

▲ **Figure 3** Cleavage and fracture

Density

Mineral density and specific gravity are similar properties. Density is mass per unit volume and is measured in grams per cubic centimetre ($g\,cm^{-3}$). Provided density is expressed in these standard units, the figure obtained is the same as the one for specific gravity. Specific gravity is the ratio of the mass of a mineral compared with the mass of an equal volume of water. Because it is a ratio, specific gravity has no units.

Quartz has a density of $2.65\ g\,cm^{-3}$, and a specific gravity of 2.65.

Reaction with acid

Calcite reacts strongly with dilute HCl. It fizzes as it gives off CO_2 gas. This is a very good way to distinguish it from other pale-coloured vitreous minerals.

MATHS BOX: SPECIFIC GRAVITY

If a mineral specimen with a mass of 150 g displaces $20\ cm^3$ of water, its density is $7.5\ g\,cm^{-3}$.
The density of water is $1\ g\,cm^{-3}$. So, $20\ cm^3$ of water has a mass of 20 g. To calculate the specific gravity of a mineral, divide the mass of the specimen by the mass of the water it displaces.
In this example:
$$\frac{150}{20} = 7.5$$

QUESTIONS

1 How can calcite be distinguished from quartz?

2 Comment on the chemical formulae of native sulfur, native copper, calcite, quartz, pyrite and galena.

3 Why is streak a more reliable diagnostic property than colour?

Minerals are used to help classify sedimentary, igneous and metamorphic rocks. So, before dealing with the classification of rocks, we need to be able to identify the minerals they are likely to contain. The diagnostic properties we looked at in the last spread are the key to doing this.

Diagnostic properties of some rock-forming minerals

Mineral	Shape	Colour	Hardness	Cleavage/fracture	Lustre	Density (g cm⁻³)	Streak
Quartz	Irregular	Varies; often white or grey	7	None/irregular	Vitreous / shiny	2.7	None
Feldspar (orthoclase/plagioclase)	Approximately rectangular	Pink, white or grey	6	Two directions	Vitreous / dull	2.6	White
Mica	Sheets / flakes	White or grey (muscovite) Black (biotite)	2.5–3	Good in one direction	Vitreous	2.8	White/grey
Augite (most common pyroxene)	Square or eight sided in cross-section	Dark green to black	5.5–6.5	Two directions almost at right angles (93°)	Vitreous	3.2–3.6	White/grey
Hornblende (most common amphibole)	Diamond shaped in cross-section	Green to black	5–6	Two directions at about 60° (56°)	Vitreous / shiny	3.0–3.5	White/grey
Olivine	Rounded	Yellow green	6.5–7	Poor/conchoidal	Vitreous	3.2–4.4	White/grey
Calcite	Irregular / rhombic	White, grey or transparent	3	Three directions rhombohedral	Vitreous	2.7	White
Garnet	Rounded / six sided in cross-section	Deep red to brown	6–7.5	None /poor conchoidal	Vitreous	4	White
Kyanite	Flat / blades	Blue, white or grey	5.5–7	Two directions	Vitreous	3.5–3.7	White

Several minerals have special properties that are diagnostic.
Calcite reacts strongly with dilute HCl. Magnetite, which is found in some igneous rocks, is magnetic.

Making observations

'You see, but you do not observe' is what Sherlock Holmes said to Dr Watson. Being observant is important in science. It means being able to notice significant details. Many people see the beauty of minerals, but far fewer notice the details that allow us to do more than just admire them. Minerals are the basic components of geology because all rocks contain one or more of them. You can improve your skills of observation through practice. Make observations of minerals in hand specimen. This can't be done by reading a book. What can be done in this book is to see some observations of diagnostic properties that can be made from photographs and thin section diagrams, to see how minerals can be identified.

Minerals in photographs

🔻 **Figure 1** Feldspar, quartz and mica

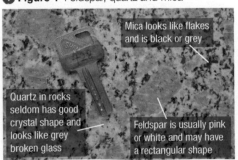

Mica looks like flakes and is black or grey

Quartz in rocks seldom has good crystal shape and looks like grey broken glass

Feldspar is usually pink or white and may have a rectangular shape

Quartz has good crystal shape if it has grown into a cavity

Citrine crystal

0 1
cm

Quartz can have a range of colours produced by impurities. Amethyst is purple and smoky quartz is brown.

Lines going through crystals are cleavage

Crystals are roughly rectangular shape

▲ **Figure 2** Feldspar. White feldspars in dark-coloured rocks are usually plagioclase.

Dark crystals could be augite or homblende

▲ **Figure 3** Augite

A crystal weathered from the rock has eight sides, which is diagnostic of augite, not homblende.

0   0.5
cm

QUESTION

1 Describe how you could distinguish hornblende from augite in hand specimen.

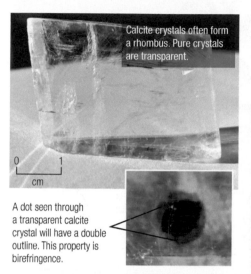

Calcite crystals often form a rhombus. Pure crystals are transparent.

0   1
cm

A dot seen through a transparent calcite crystal will have a double outline. This property is birefringence.

Fossils are often composed of calcite

◀ ▲ **Figure 4** Calcite

Cleavage direction

Cleavage direction

cm 1   2   3   4

▲ **Figure 5** Kyanite

STUDY TIP

You need to be able to classify mineral specimens using photographs that show their physical properties. A good way to improve your skill in this respect is to take your own photographs and look for the diagnostic properties yourself. It also gives you the opportunity to make and record observations and measurements, and present your data in appropriate ways. Figure 6 shows how to do it, and Figure 7 is an example of the results. Why not have your labelled photographs mounted and used as displays in your Geology Lab, giving you the chance to become familiar with all the minerals needed for the specification?

Although pyrite has a brassy yellow colour, its streak is black.

Streak

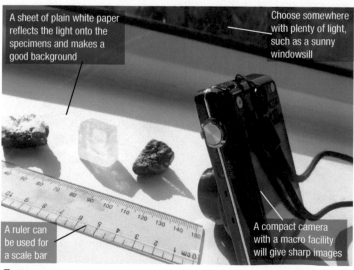

A sheet of plain white paper reflects the light onto the specimens and makes a good background

Choose somewhere with plenty of light, such as a sunny windowsill

A ruler can be used for a scale bar

A compact camera with a macro facility will give sharp images

▲ **Figure 6** Recording the properties of minerals on photographs

Crystals of iron pyrite are usually cubic and brassy yellow in colour. Surfaces of the cubes often have striations.

Striations

cm

▲ **Figure 7** Physical properties of iron pyrite.

Minerals in thin section diagrams

A thin section is a very thin slice of rock, mounted on a glass slide and viewed using a petrological microscope. Diagrams can be drawn (and photographs taken) of what is seen.

Twinning is a feature of feldspars that is most easily seen in thin section. It occurs when two or more crystals grow with different orientations from a common plane. One is usually a mirror image of the other.

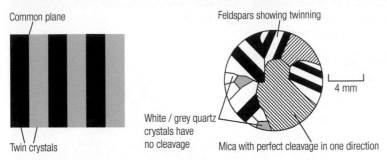

Common plane

Twin crystals

Feldspars showing twinning

White / grey quartz crystals have no cleavage

Mica with perfect cleavage in one direction

4 mm

△ Figure 1 Thin section diagrams of feldspar, quartz and mica

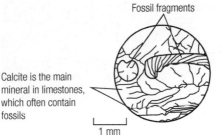

Fossil fragments

Calcite is the main mineral in limestones, which often contain fossils

1 mm

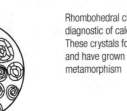

Calcite cement

Spherical grains with an internal concentric structure are ooliths which are composed of calcite

Pattern of radiating lines shows growth of calcite crystals

1 mm

Rhombohedral cleavage is diagnostic of calcite. These crystals form marble and have grown during metamorphism

2 mm

△ Figure 2 Thin section diagrams of calcite

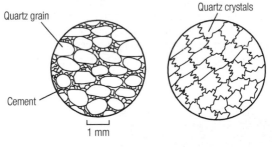

Quartz grain

Cement

Quartz crystals

Both diagrams show quartz: Quartz grains are found in sandstones but metamorphosed, new crystals grow and replace the cement

1 mm

△ Figure 3 Thin section diagrams of quartz

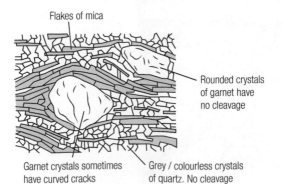

Flakes of mica

Rounded crystals of garnet have no cleavage

Garnet crystals sometimes have curved cracks

Grey / colourless crystals of quartz. No cleavage

Andalusite has the same formula as Kyanite, but has a different crystal form

△ Figure 4 Thin section diagrams of garnet, mica and andalusite

Testing for mineral hardness

Mohs' hardness test compares the resistance of a mineral to being scratched by ten reference minerals. If the reference minerals are not available, some common objects can be used instead.

Mohs' scale of hardness		
Hardness	**Reference mineral**	**Common objects**
1 (soft)	Talc	Scratched by fingernail
2	Gypsum	
3	Calcite	Scratched by 2p coin
4	Fluorite	Scratched by steel nail
5	Apatite	
6	Feldspar	Can sometimes be scratched by steel nail
7	Quartz	Scratch glass
8	Topaz	
9	Corundum	
10 (hard)	Diamond	

STUDY TIP

A useful way to remember the reference minerals in the correct order of increasing hardness in Mohs' scale is:

The **G**eologists **C**limbed **F**ast **A**nd **F**elt **Q**uite **T**ired **C**oming **D**own

Carry out the test by placing a fresh-looking pointed surface of one specimen on a flat surface of another specimen and trying to produce a scratch. Take a reference specimen, or one of the common objects, and drag it across the surface of the unknown mineral. If it produces a scratch, the unknown mineral is softer than the reference mineral. If it doesn't, or if it writes on the unknown mineral, then the unknown mineral is harder than the reference mineral. Always do the test twice to confirm your result.

Most common minerals have a uniform hardness. Some vary a little because their chemical composition can vary, within limits. Kyanite has different hardness in different directions (along or across the crystal).

Measuring mineral density

- Use a balance to find the mass of the mineral specimen and record the result. Use a balance that can measure up to two decimal places if possible.
- Fill a measuring cylinder (graduated in ml) with water up to a given level, which you need to record.
- Place the mineral into the cylinder then observe and record the new level.
- Subtract the value of the original level from the new level, which will give an answer in ml.
- The volume of 1ml of water is 1cm^3, so the result is suitable for calculating density.

    $$\text{Density } (\rho) = \frac{\text{mass (g)}}{\text{volume (cm}^3)}$$

- Divide your mass result by your volume result to give density in g cm^{-3}

Alternatively, you could use a displacement (eureka) can for this density experiment.

QUESTIONS

1 Explain why quartz does not produce a streak.
2 Explain why some properties are likely to be less useful than others in identifying rock-forming minerals.
3 Describe how you could discover the hardness of an unknown mineral.
4 Identify minerals A to E shown in Figure 5.

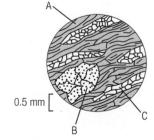
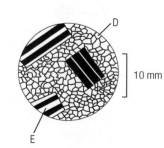

▲ **Figure 5** Thin section diagrams for use with question 4

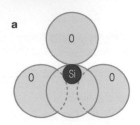

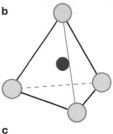

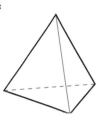

Figure 1 a The small radius of the silicon atom allows only four oxygen atoms to surround it in the form of a tetrahedron.
b Oxygen atoms shown at the corners of a tetrahedron with a small silicon at the centre.
c Silica tetrahedron as shown in diagrams. The oxygen atoms are at the four points of the tetrahedron.

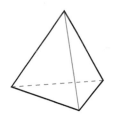

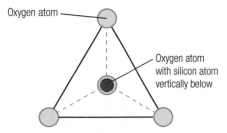

Oxygen atom

Oxygen atom with silicon atom vertically below

Figure 2 Single tetrahedra in elevation and plan view.
Olivine and garnet are formed of single tetrahedra.

The atomic structure of minerals influences many of their physical properties.

Silicon atoms are **cations** with a positive charge of 4 (4+). Oxygen atoms are **anions** with a negative charge of 2 (2−). They are attracted to each other because they can share electrons (one from each atom) and form **covalent bonds**.

The atomic radius of a silicon atom is smaller than the atomic radius of oxygen. This allows each silicon to be surrounded by just four oxygen atoms that are in contact, in the form of a **tetrahedron**. Each tetrahedron has a charge of 4− because one electron from each of the oxygen atoms is shared with the silicon atom. This leaves the other four electrons available to form bonds with other cations. In the case of quartz (SiO_2) they bond with silicon. The structures of silicate minerals are made up of tetrahedra arranged in different ways. Cations such as sodium (Na^+) potassium (K^+) calcium (Ca^{2+}) magnesium (Mg^{2+}) iron (Fe^{2+}) aluminium (Al^{3+}) fill the spaces in between.

Single tetrahedra

In the simplest silicate structures, single tetrahedra are bonded to cations. They have a silicon to oxygen ratio of 1:4.

Olivine [$(Mg,Fe)_2SiO_4$] has either two Mg^{2+} ions, two Fe^{2+} ions, or one of each, to balance the 4− charge of the silica tetrahedron. Many of the cations in silicate minerals have similar atomic radii (like Mg and Fe) and so can substitute for one another in a process called **solid solution**. Olivine forms at high temperatures, earlier than minerals that have more complex structures.

Chains

Silicate tetrahedra form chains when each one shares two of its oxygen atoms (known as **bridging oxygens**) with adjacent tetrahedra, giving a silicon to oxygen ratio of 1:3. Each chain of two tetrahedra has a 2− charge, which is balanced by cations such as Mg^{2+} and Fe^{2+} that link the chains together.

Pyroxenes [$(Mg,Fe)_2 Si_2O_6$] are an example. Weaker bonds between the chains result in two cleavage directions, roughly at 90°.

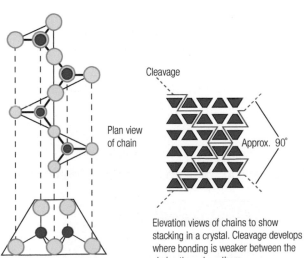

Cleavage

Plan view of chain

Approx. 90°

Elevation views of chains to show stacking in a crystal. Cleavage develops where bonding is weaker between the chains than along them.

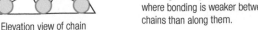

Elevation view of chain

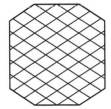

Viewed across the end of the chains pyroxene crystals have eight sides and two cleavages almost at 90°.

Figure 3 Chains in elevation and plan view. Pyroxenes have chain structure.

Double chains

Two single chains are combined in some minerals to form double chains. Alternate tetrahedra in two parallel chains are linked, giving a silicon oxygen ratio of 4:11, so each double chain has a 6– charge. However, holes between the double chains accommodate OH^- ions resulting in $[Si_4O_{11}(OH)]^{7-}$ found in amphiboles. Double chain structures are less compact than single chains, which reduces their density. Mg^{2+}, Fe^{2+}, and Al^{2+} often link the chains together.

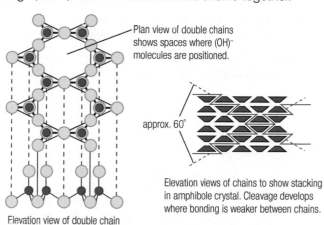
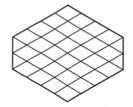

Plan view of double chains shows spaces where (OH)⁻ molecules are positioned.

approx. 60°

Elevation views of chains to show stacking in amphibole crystal. Cleavage develops where bonding is weaker between chains.

Viewed across the end of the double chains, amphibole crystals have six sides and two cleavages almost at 60°.

Elevation view of double chain

⬆ **Figure 4** Double chains in elevation and plan view. Amphiboles have double chain structure.

Sheets

Sheet structures form when each tetrahedron shares three of its oxygen atoms with adjacent tetrahedra, giving a silicon to oxygen ratio of 2:5. The spaces in the sheets accommodate OH^- ions and account for low densities. Each sheet has a 4– charge $(Si_4O_{10})^{4-}$. Between each of the sheets there are weak K–O bonds in micas, which explains their good cleavage.

Frameworks

Framework silicates form when all four oxygens of the tetrahedron are shared by adjacent tetrahedra. They have a silicon to oxygen ratio of 1:2 and so carry no charge. Quartz is a framework silicate and has strong Si–O bonds only. Feldspars are also frameworks but they have some weaker bonds between oxygen and cations (usually K^+, Ca^{2+} and Al^{3+}). This gives them cleavage in two directions almost at 90°.

Polymerisation

Some small molecules can join up to make larger ones. This process is called polymerisation. Silicate tetrahedra link to form **polymers** by oxygen sharing. Resistance to flow is affected in liquids where polymers are formed. You will discover more about this when you explore the flow rates of lavas later.

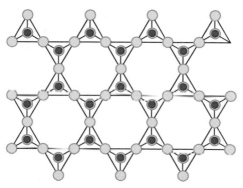

⬆ **Figure 5** Sheet silicate structure in plan view. Micas and clay minerals are sheet silicates. They are tabular and have low densities.

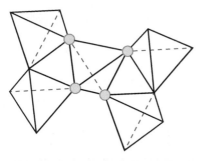

⬆ **Figure 6** Framework silicates have all four oxygens of the tetrahedron shared by adjacent tetrahedra, as shown by the middle tetrahedron. Quartz and feldspars have framework structure.

QUESTIONS

1. Explain why the angles between the cleavage in augite and hornblende are different.
2. What characteristics of atoms determine whether they will form part of silicate structures?
3. Why does quartz lack cleavage?
4. Explain why density varies between minerals having different silicate structures.

The classification of rocks

Pressure, temperature and rocks

Use Figure 1 to remind yourself about the rock cycle and the three classes of rocks (introduced in KS3 Science).

- The boundary between the **diagenesis** of sediments and **metamorphism**, which occurs under conditions of higher temperature and pressure, is gradational. For example, a sedimentary shale grades into a metamorphic slate.
- The boundary between **igneous** and **metamorphic rocks** is where **partial melting** occurs. Metamorphic rocks **recrystallise** in the solid state but igneous rocks **crystallise** from magma.

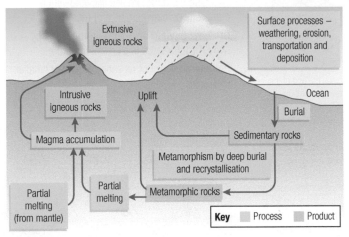

Figure 1 The rock cycle: a sequence of processes that links all three rock classes

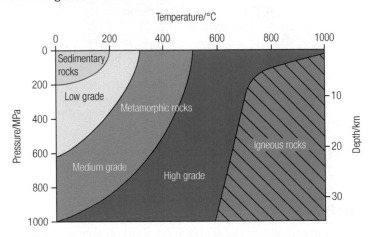

Figure 2 Pressure, temperature and classes of rocks

KEY DEFINITIONS

Diagenesis includes all processes that occur in sediments at low temperature and pressure at or near the Earth's surface.

Metamorphism the changing of rocks in the Earth's crust by heat and/or pressure and/or volatile content. It is isochemical and occurs in the solid state.

Igneous rock a rock that has crystallised from a magma.

Metamorphic rock is formed by the recrystallisation of other rocks in the solid state due to pressure, temperature or both.

Partial melting occurs when only a portion of rock in the lower crust or upper mantle is melted.

Recrystallisation is the solid state process that changes minerals into new crystalline metamorphic minerals.

Crystallisation occurs during the cooling of magma or lava so that solid mineral crystals form.

Burial occurs when sediment is covered by younger layers of sediment.

Sedimentary rock a rock composed of fragments that have been deposited compacted and cemented.

Clast a fragment of broken rock produced by mechanical weathering and erosion.

Igneous, sedimentary and metamorphic rocks

Rocks are classified according to the way they are formed.

Sedimentary rocks

These rocks form at relatively low pressure and temperature at shallow depths below the Earth's surface. Existing rocks exposed at the Earth's surface are broken down by weathering and erosion. The rock cycle diagram (Figure 1) shows that broken fragments of these rocks and ions in solution are transported usually to the oceans by rivers. The energy of the water decreases when it spreads out, where rivers enter the oceans, and the fragments of rock are deposited on the ocean floor.

Over millions of years more and more fragments are deposited. **Burial** causes compaction into layers to make **sedimentary rocks**, for example sandstone. Many sedimentary rocks are made up of **clasts** (fragments or grains) held together by cement, rather than interlocking crystals. The grain size of clasts can vary considerably. Other sedimentary rocks are biologically formed by the precipitation of ions from sea water. Sedimentary rocks may contain fossils of plants, animals or the traces of animal activity. They are usually built up in layers or beds.

Igneous rocks

Igneous rocks are formed from magma (liquid rock) within the Earth. They may form at high or low pressures depending on depth, but they all form at high temperatures. Magma eventually cools to form igneous rocks composed of interlocking crystals. The mineral crystals do not usually have any preferred alignment. There are two main types of igneous rocks.

- **Intrusive** (hypabyssal or plutonic) igneous rocks – these have medium crystal size if the magma cooled fairly slowly at intermediate depths, for example dolerite, or coarse crystal size if the magma cooled slowly at greater depth inside the Earth's crust, for example granite. The minerals are hard silicates, such as quartz and feldspar, and separate mineral crystals can be seen in the rock.
- **Extrusive** (volcanic) igneous rocks – these have a fine crystal size because the magma reached the Earth's surface. Erupted as lava or ejected as fragments, they cooled rapidly, producing rocks such as basalt or pumice. Individual crystals cannot usually be seen.

Metamorphic rocks

Within the Earth, rocks may be subjected to heat or pressure or both. This causes their texture and mineral composition to change. Minerals in the original or parent rock recrystallise in the solid state to form new minerals. This means that metamorphic rocks are composed of interlocking crystals, but they often show preferred alignment if they have been affected by pressure. There are three main types of metamorphism: burial, contact and regional. If they contain flat crystals, rocks affected by pressure as part of burial or regional metamorphism show **foliation**. Foliated layers commonly contain mica. Metamorphic minerals tend to be hard silicates but may be composed of calcite, a carbonate mineral.

KEY DEFINITIONS

Magma accumulation is magma collecting within a magma chamber.

Intrusions composed of igneous rock formed below the Earth's surface, where magma is forced into pre-existing rocks.

Extrusion the emission of magma onto the Earth's surface where it forms a lava flow.

Foliation is a texture in metamorphic rocks formed by the preferred alignment of flat/tabular minerals.

Deposition the laying down of sediment that occurs when a transporting agent loses energy.

Weathering the breakdown of rocks in situ.

Erosion the removal of weathered material, usually by the physical action of transported fragments.

Transport the means by which weathered material is moved from one place to another by water, wind, ice or gravity.

Uplift the return of buried rocks to the Earth's surface by tectonic forces.

	Igneous	Sedimentary	Metamorphic
Texture	Crystalline Minerals interlocked in a mosaic	Fragmental Made of grains/fragments or fossils cemented together	Crystalline Minerals interlocked in a mosaic
Mineral alignment	Grains not usually lined up, unfoliated	Grains often lined up /aligned, foliated if affected by pressure	
Main minerals	Quartz Plagioclase and K feldspar Hornblende (amphibole) Augite (pyroxene) Olivine Micas, muscovite and biotite	Quartz K feldspar Calcite Mica, mainly muscovite Clay minerals Rock fragments	Quartz Plagioclase and K feldspar Micas, muscovite and biotite Garnet Calcite Al_2SiO_5 polymorphs
Formation	Crystallised from magma	Deposition of rock and mineral particles	Recrystallisation of other rocks
Features	As igneous intrusions – dyke, sill, batholith or lava flow, no beds	In beds	Large areas or zones – no beds
Fossils	None	May be present	Usually none, deformed if present
Ease of breaking	Hard and does not split	May be soft and crumbly but some are difficult to break if well cemented	Hard but may split in layers
Thin sections			

QUESTION

1   Describe how rock classes can be controlled by the pressure and temperature regimes during their formation.

The classification of igneous rocks

The internationally agreed classification of **igneous rocks** is based upon chemical analyses or proportions of the minerals present. These methods require expensive measurement or time-consuming microscopic analyses of thin sections. In the field, neither of these methods help identify igneous rocks, instead we look to the physical characteristics and observe the crystal grain size, composition and colour.

Crystal size

For the purpose of classification, igneous rocks are divided on the basis of the average crystal, or grain size. This means looking at the crystals in a specimen and ignoring any very large crystals called **phenocrysts**. If no crystals can be seen, even using a hand lens, the rock has a glassy texture showing it cooled too quickly for any crystals to form.

- Fine grained crystals are less than 1 mm in size, so that individual crystals cannot be identified by eye.
- Medium grained crystals are 1 to 5 mm in size and can be seen by eye.
- Coarse grained crystals are greater than 5 mm in size and can be seen easily and identified by eye.

Mineral composition

Igneous rocks usually contain many different rock-forming minerals but only a few **essential minerals** are used in classification.

Felsic minerals

The **felsic minerals** are quartz and feldspar, which are both rich in silica. The term felsic comes from the words feldspar and silica. Quartz is only found in silicic or intermediate rocks. The **magma** must be oversaturated with silica so that excess silica is left after the other rock-forming minerals crystallise.

Recognising quartz in igneous rocks	Mineral characteristics of quartz
• It is very different from looking at individual crystals of quartz. • Crystals are grey in colour, with the appearance of broken bottle glass and lacking cleavage. • Crystals have an amorphous shape as they crystallised in an enclosed space.	• Hardness 7 (quartz crystals can't be scratched with a steel nail) • Colour white or grey or transparent • No cleavage • Vitreous lustre (glassy) • Composition SiO_2

The feldspars are the most common rock-forming mineral in igneous rocks. They form about 60% of the minerals in these rocks so are essential to igneous classification. The main types are potassium (K) feldspar and plagioclase feldspar. K feldspar (orthoclase) is only found in silicic and intermediate igneous rocks, and is a good mineral to use for classification. They often form large phenocrysts as well as the majority of the groundmass of granite.

Recognising feldspars in igneous rocks	Mineral characteristics of feldspars

• The pink colour of most K feldspar makes it very easy to identify.
• Grey or white feldspar is usually plagioclase.
• If you look at it with a hand lens you will see the glassy appearance and smooth, straight cleavage surfaces.
	• Hardness 6 (can be scratched with a steel nail)
	• Colour pink for K feldspar, white or grey for plagioclase feldspar
	• Good cleavage in two directions
	• Vitreous lustre (glassy)
	• Composition:
	  K feldspar – $KAlSi_3O_8$
	  Plagioclase feldspar – $NaAlSi_3O_8$ to
	  $CaAl_2Si_2O_8$

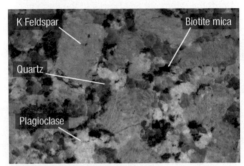

🔺 **Figure 1** Photo of granite labelled to show grey quartz, pink K feldspar, white plagioclase and black biotite mica

0   1
cm

🔺 **Figure 2** Photo of gabbro labelled to show grey plagioclase and brown/green augite

0   1
cm

KEY FACTS: SILICIC, INTERMEDIATE, MAFIC AND ULTRAMAFIC ROCKS

Silicic igneous rocks have a composition rich in silica, of more than 66% and are light coloured (leucocratic).

Intermediate igneous rocks have a silica content of 66 to 52% and are grey (mesocratic).

Mafic igneous rocks have a silica content of 52 to 45% and are dark (melanocratic).

Ultramafic igneous rocks have a silica content of less than 45%.

Mafic minerals

These are minerals that contain both magnesium and iron, so the term **mafic** comes from the words magnesium and ferric (iron). These minerals are also referred to as ferromagnesian. The micas are biotite, found in silicic or intermediate rocks, and muscovite found in silicic rocks only.

Hornblende is one of a group of minerals called *amphiboles*, which are particularly common in intermediate rocks. They are not easy to identify in hand specimens and can easily be confused with augite. Augite belongs to the *pyroxene* group of minerals and is one of the main minerals in mafic and ultramafic rocks. Olivine is found in some mafic and ultramafic rocks. It will form only when the magma is under-saturated in silica.

Mineral characteristics of hornblende (amphibole)	Mineral characteristics of augite (pyroxene)	Mineral characteristics of olivine
• Hardness 5.5 • Colour black • Two cleavages at 60° • Forms 6-sided crystals • Vitreous lustre (glassy)	• Hardness 5.5 • Colour black to dark green • Two cleavages at 90° • Forms 8-sided crystals • Vitreous lustre (glassy)	• Hardness 6.5 • Colour light green • No cleavage • Vitreous lustre (glassy)

Silica percentage

Chemical analyses give the overall chemical composition of a rock, rather than which minerals are present. The majority of SiO_2 in igneous rock is not present as free quartz (SiO_2), but is 'locked up' within other silicate minerals, such as feldspar and olivine.

It is easier to link the SiO_2 percentage to colour as a short cut to identification. The mafic rocks are rich in dark-coloured ferromagnesian minerals such as olivine and augite. Silicic (felsic) rocks are rich in the light-coloured minerals quartz and feldspars. But colour can be misleading in rocks such as obsidian, which looks black but is clear in thin section and is a silicic rock.

Recognising biotite and muscovite mica in igneous rocks	Mineral characteristics of biotite and muscovite mica
• Shiny black crystals of biotite • Silvery flakes of muscovite	• Hardness 2.5 (scratched with a copper coin) • Colour of biotite is black and muscovite silvery • One perfect cleavage that forms flakes • Vitreous lustre (glassy)

KEY FACT

Silica (SiO_2) percent is not the same thing as the mineral quartz. Quartz is made entirely of silica, but the silica percent includes the silica in all of the rock-forming minerals. A rock with no quartz will still contain silica in silicate minerals such as olivine. You cannot see the silica content in the field, and measuring SiO_2 percentages requires laboratory work. However, colour can be used as an indicator:
• Silicic (felsic) rocks are rich in light-coloured minerals;
• Mafic rocks are rich in dark-coloured minerals.

QUESTIONS

1 Four plutonic igneous rocks have been chemically analysed and found to contain SiO_2 % of 71.2, 54.5, 48.1 and 41.7 respectively. What are the four rock types based upon these compositions?

2 Explain why SiO_2 % is not used to identify igneous rocks in the field.

		SILICA CONTENT			
		>66%	52–66%	45–52%	<45%
		Silicic	**Intermediate**	**Mafic**	**Ultramafic**
	Colour	Light	Medium →		Dark
Crystal size	Mode of origin				
Glassy	Volcanic extrusive	Obsidian**		Scoria	
Fine <1mm	Volcanic extrusive	Pumice Rhyolite	Andesite	Basalt	
Medium 1-5mm	Minor intrusion (Hypabyssal)	Microgranite	Microdiorite	Dolerite	
Coarse >5mm	Major intrusion (Plutonic)	Granite Granodiorite	Diorite	Gabbro	Peridotite
		Quartz	Quartz (rare)		
		K feldspar	K feldspar		
		Plagioclase feldspar (Na)	Plagioclase feldspar	Plagioclase feldspar (Ca)	
		Biotite	Biotite	Augite	Augite
		Muscovite	Hornblende	Olivine	Olivine

** NB Obsidian is an exception to the rule, being silicic but dark in colour

Using all of the methods described, an igneous classification table (above) can be created that makes the classification and description of igneous rocks possible.

Look for the following characteristics to identify any igneous rock (in the field, drawings, photos or photomicrographs):

- Crystal size, which indicates the rate of cooling
- Mineral composition and colour to help estimate the silica content
- Igneous textures, which indicate the origin.

Silicic igneous rocks

High-silica magma contains 66 to 75% silica and is mainly generated by melting of the Earth's crust at convergent plate margins. Rocks formed from high-silica magma tend to have a lot of quartz and other high-silica minerals such as feldspar, muscovite and biotite micas. They will be light in colour.

Silica-rich magmas are very viscous, much stiffer than toothpaste, so only a small amount of magma reaches the surface. Granite is the most common silicic rock and these magmas solidify at relatively low temperatures, typically 600 to 900°C. Rapid cooling produces obsidian, a volcanic glass. The explosive eruption of gas-rich, viscous magma produces pumice, which has a vesicular, glassy texture. The surface feels rough. Rhyolite can show a banded, streaky appearance (**flow banding**) due to the friction within the viscous, magma as it flows.

Once you have decided that a rock belongs to the silicic group of rocks, look at the crystal size and texture to identify the rock.

	Obsidian	Pumice	Rhyolite	Granite and Granodiorite
Crystal size	Glassy	Glassy shards and fine crystals	Fine crystals	Coarse crystals
Texture	Glassy with **conchoidal** fracture	**Vesicular**	Flow banded	**Porphyritic** or **equigranular**

Identifying specific feldspars in hand samples can be difficult. Whilst potassium feldspar is often pink, it can be white or grey just like plagioclase feldspar and quartz. Using a petrological microscope, in cross-polarised light (XPL), K feldspars and plagioclase can have distinctly different twinning characteristics (the dark and light bands aligned parallel within the crystal).

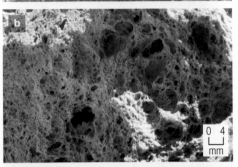

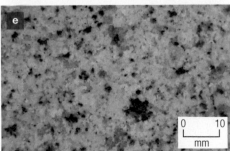

🔺 **Figure 1** Photos showing characteristic minerals, crystal grain size and textures for: **a** obsidian; **b** pumice; **c** rhyolite; **d** granite; **e** granodiorite

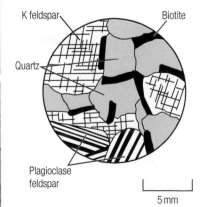

🔺 **Figure 2** XPL photomicrograph showing potassium (K) feldspar with characteristic 'crosshatched' twinning and plagioclase feldspar showing characteristic parallel twinning

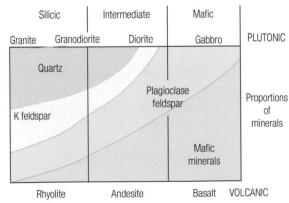

🔺 **Figure 3** The distribution of rock-forming minerals in igneous rock

Intermediate igneous rocks

Intermediate magma contains 52 to 66% silica. The most common igneous rock formed from this type of magma is andesite, a fine grained volcanic rock. The coarse grained plutonic rock is diorite. They contain abundant plagioclase feldspar, together with dark ferromagnesian minerals such as the pyroxene, augite and the amphibole, hornblende. They contain little or no quartz. The magma is less viscous than a silicic magma but still very sticky (viscous). It solidifies at a higher temperature than a silicic magma. Andesite is common at convergent plate margins.

	Andesite	Diorite
Crystal grain size	Fine	Coarse
Texture	Vesicular, **amygdaloidal**, porphyritic or equigranular	Equigranular or porphyritic

Mafic igneous rocks

The first thing to look for to identify a mafic rock is the colour, which should be dark due to the abundant ferromagnesian minerals. It will contain pyroxene, augite, and sometimes olivine with Ca-rich plagioclase, but no quartz. Low-silica magma contains only 45 to 52% silica. This type of magma is easy to recognise as it contains low amounts of silica and large amounts of MgO and FeO. The magma has low viscosity, so it easily reaches the surface as **lava**, making basalt the most common mafic igneous rock. It solidifies at high temperatures above 1000°C. Most low-silica magma is produced by melting of the Earth's mantle.

Once you have identified that a rock belongs to the group of mafic rocks, then look at the crystal grain size to identify the actual rock.

	Basalt	Dolerite	Gabbro
Crystal grain size	Fine	Medium	Coarse
Texture	Vesicular or amygdaloidal or porphyritic or equigranular	Porphyritic or equigranular	Equigranular

Olivine and pyroxene can both be green in hand specimen. In thin section, olivine has a distinctive irregular fracture pattern which can distinguish it from pyroxene (augite). Both olivine and pyroxene show high birefringence colours in XPL which means that they stand out from the black and white stripes of the plagioclase crystals.

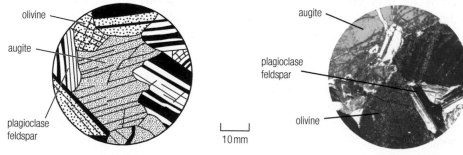

Ultramafic rocks

These rocks contain less than 45% silica and consist almost entirely of ferromagnesian minerals. The most common ultramafic rock is peridotite, which forms much of the Upper Mantle. This is mainly olivine with some pyroxene and a small amount of Ca-rich feldspar. Some ultramafic rocks are monomineralic, meaning they consist of only one mineral.

Ultramafic rocks have extremely high melting points. We can find ultramafic rocks from the mantle, where fragments have been brought up to the surface by volcanoes. The most famous of these are the kimberlite pipes of South Africa. Others are found as **ophiolites** in mountain ranges where a slab of the oceanic crust has broken off at a subduction zone.

▲ **Figure 5** Photos showing the differences between the mafic rocks **a** porphyritic dolerite; **b** equigranular gabbro

◄ **Figure 4** XPL photomicrographs showing olivine, pyroxene (augite) and plagioclase feldspar

▲ **Figure 1** Obsidian – silica-rich volcanic glass, showing conchoidal fracture

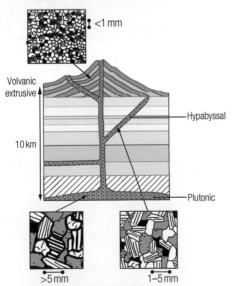

▲ **Figure 2** Cross-section through volcanic, hypabyssal and plutonic rocks with photomicrograph drawings of fine, medium and coarse crystal size

Igneous rocks consist of interlocking crystals formed from the cooling and solidification of magma or lava, which may be at the surface or at depth. Crystallisation begins with microscopic nuclei that have planar natural surfaces called crystal faces. These are the external expression of the atomic structure of the mineral. Examining crystals in igneous rocks is one of the quickest ways to identify rocks in the field or from samples.

Crystal grain size

The rate of cooling is the main factor that controls crystal size and this is determined largely by depth below the Earth's surface. The deeper the intrusion, the more slowly it will cool. The general rule is the slower the rate of cooling, the larger the crystal size.

However, there are others factors that that can be important. Where an intrusion is large, the inside of the intrusion will cool more slowly than the outer part. This is because it is better insulated and will lose heat more slowly. If a large intrusion cooled at a depth of 1 km, it could have a range of crystal grain sizes from fine at the edges to coarse in the centre.

Glassy

Lava **extruded** onto the surface may cool so quickly there is no internal atomic order and no crystals with planar faces. The result is volcanic glass. Obsidian forms in this way. It looks black because light will not pass through it but in thin section it is a clear glass.

The conchoidal fracture we see when volcanic glass breaks is a result of curved irregular faces. Cooling usually occurs within hours. If lava flows into the sea, it can cool in minutes. This is why some pillow lavas show a layered structure with glass on the outside and fine crystals inside (as the inside cooled more slowly).

Fine

Fine crystals are less than 1 mm in size, so individual crystals cannot be seen by eye. They can form in lava extruded onto the surface that cools rapidly over weeks or months. Although still described as rapid cooling, it is slower than the cooling rate that produces glass. They can also form below the surface, as chilled margins at the edges of minor **intrusions** such as sills and dykes.

Medium

Medium crystals are between 1 and 5 mm in size, so individual crystals can be seen by eye, but are difficult to identify. They form below the surface, in minor intrusions (**hypabyssal**), at a depth of approximately 1 km. They cool slowly over a few thousand years.

Coarse

Coarse crystals are over 5 mm in size, so can be seen easily and identified in hand specimens. Major intrusions form 10 km or more beneath the surface, as batholiths or **plutonic** rocks, cooling very slowly over millions of years.

Texture

The crystal grain size, crystal shape and mutual relationship between crystals is known as the texture of an igneous rock. It is determined by the nucleation rate of new crystals, the growth rate and the diffusion rate (the rate at which atoms can move through the molten magma to form new crystal faces). How slowly the magma cools and how long it stays at a particular temperature have a significant influence on texture.

The nucleation of crystals is normally an endothermic process, requiring energy from the magma. Conversely, crystal growth is an exothermic process, putting energy into the magma, keeping it hot. The insulating effect of rock surrounding deep intrusions acts to keep the nucleation rate low. Because fewer crystals are able to nucleate, those that do are free to grow to a larger size. However, the process is poorly understood and there are other models of nucleation.

Crystal grain shape

The crystal shape is strongly affected by the rate of cooling. Very slow cooling, in deep, plutonic intrusions tends to produce **euhedral** tabulate plagioclase, **sub-equant** olivine and **prismatic** pyroxene crystals. For as long as they are free to grow, crystals will add material to each face and maintain their shape.

When a crystal grows unimpeded by the growth of neighbouring crystals, it is able to assume its equilibrium habit with sharp, well-formed and fully developed crystal faces. This is known as a euhedral crystal. Early formed crystals are generally euhedral.

As magma cools, the growing crystals touch each other and this prevents crystal faces forming properly. Crystals that have no well-formed crystal faces are known as **anhedral** and those that have some well-formed faces are called **subhedral**. In the more rapid cooling of shallow and hypabyssal intrusions, crystals are usually subhedral to anhedral; the emergent crystals tend to crowd each other and no single crystal can evolve as euhedral.

Pegmatite

A pegmatite is an igneous rock that has exceptionally large crystals. The crystals are normally larger than a few centimetres and can often be tens of centimetres, or even metres long. However, in this case the large crystal size is not a product of slow growth. Pegmatites can produce large crystals in a short (geologically) period of time.

Many magmas contain a significant amount of dissolved water and this is not removed during early crystallisation. Rare Earth Elements (REE), that are not readily incorporated into essential minerals, become concentrated in the magma and the water becomes rich in their dissolved ions. Pegmatites form in the last stages of crystallisation, from this water-rich residual magma, where the low viscosity enables crystals to grow relatively quickly due to the high diffusion rate (the high mobility of the ions). The concentration of REE means that pegmatites can be of economic importance.

a   b
c   d

▲ **Figure 3** Typical crystalline habits of common rock-forming minerals when allowed to grow freely: **a** sub-equant crystal of olivine and euhedral crystals of **b** plagioclase feldspar; **c** pyroxene (augite) and **d** K feldspar

▲ **Figure 4** Microgranite formed from anhedral and subhedral crystals that have crowded each other preventing euhedral growth

0   1
cm

QUESTIONS

1 Explain why igneous rocks have crystal grains with angular interlocking shapes.

2 Explain how you would distinguish between a hypabyssal and a plutonic rock.

3 How does a glassy texture form?

Igneous textures

Igneous rocks generally consist of interlocking crystals, which gives them a crystalline texture. Pyroclastic rocks will have a fragmental texture as they consist of fragments of glass, crystals, ash and rock, as a result of explosive volcanic activity.

Glassy texture

A glassy or vitreous texture means that there are no crystals and the rock resembles a block of (coloured) glass. This is a result of very rapid cooling where crystals had no time to form.

Equicrystalline (equigranular) texture

An equicrystalline (equigranular) texture simply means that all the crystals are of equal size but may be fine, medium or coarse. The relationship of crystal size to cooling is the larger the crystal size the longer the time of cooling:

- Fine crystals form by rapid cooling at the surface and are described as extrusive.
- Medium grained crystals formed by slow cooling below the surface are hypabyssal.
- Coarse crystals formed by very slow cooling, deep below the surface are plutonic.

Both hypabyssal and plutonic rocks are also known as intrusive.

Usually the crystal size can be related to the depth of cooling but there are exceptions:

- An intrusion that cooled 2 km below the surface will have medium crystals but will also have fine crystals forming chilled margins at the edges which cooled rapidly.
- A lava flow that cooled at the surface will have fine crystals but if it is very thick then the centre of the flow may contain medium-sized crystals.

Vesicular and amygdaloidal

Vesicular texture is where gas bubbles are trapped in lava as it cools rapidly, leaving holes where the gas was present. The holes, called vesicles, are usually oval or ellipsoid in shape and elongated parallel to the direction of flow. The bubbles are found near the top of the lava flow or sometimes at the edge of an intrusion. Vesicular texture is common in basalt and pumice.

Amygdaloidal texture is where vesicles are later infilled by minerals deposited from percolating groundwater. A vesicular rock will be very porous so that groundwater can flow through it relatively easily. The groundwater contains dissolved minerals such as calcium carbonate, so calcite will be precipitated into the holes. The most common minerals are calcite or quartz; but sometimes rarer minerals are precipitated, and are known as zeolites. This can happen thousands to millions of years after the vesicular rock formed. Each infilled hole is called an amygdale, though large ones, partially filled with crystals growing in towards the centre, are called geodes. It is most common in basalt.

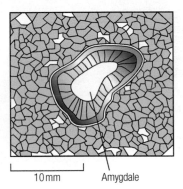

10 mm   Vesicle

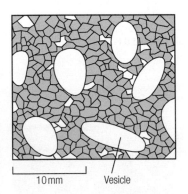

10 mm   Amygdale

▲ **Figure 1** Photomicrographs of vesicular and amygdaloidal texture

▶ **Figure 2** Basalt showing **a** vesicular texture and **b** amygdaloidal texture, with white calcite crystals infilling the amygdales

0   10
mm

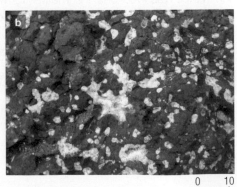

0   10
mm

Flow banding

Flow banding occurs where layers of dark and light minerals form due to the separation of minerals within a lava flow. They will be aligned parallel to the flow direction, though they are often contorted because the lava was viscous so flowed slowly. Flow banding is usually seen in **rhyolite**.

Porphyritic texture

A porphyritic texture forms where a rock has two stages of cooling resulting in two distinct sizes of crystals. Large crystals are called phenocrysts and form first by cooling slowly. They are surrounded by a finer groundmass, which cooled more quickly. Porphyritic texture can be found in any igneous rock but it is common in porphyritic basalt and porphyritic granite, though each has formed at different depths.

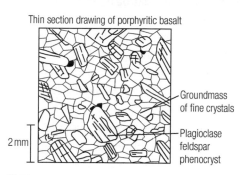
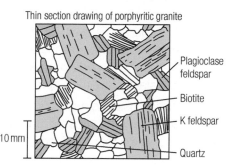

▲ **Figure 4** Thin section drawings of a porphyritic basalt and a porphyritic granite

In a porphyritic basalt, plagioclase feldspar phenocrysts start to form in the magma chamber below the volcano. When an eruption occurs, magma and plagioclase crystals are erupted to the surface where rapid cooling forms the fine crystals of the groundmass. Sometimes the phenocrysts will be aligned parallel to the direction of flow of the lava.

The phenocrysts of K feldspar in a porphyritic granite start to form very slowly, at a great depth, perhaps around 30 km. The magma and feldspar crystals move up within the crust to form a batholith at a depth of 15 km. The rest of the magma then cools very slowly to form the coarse crystals of the groundmass.

Ophitic texture

An ophitic texture forms when an elongate crystal is enclosed by another mineral. This is common in dolerite and gabbro, where plagioclase 'laths' are enclosed by augite.

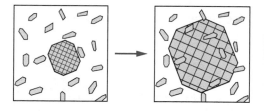

▲ **Figure 7** Ophitic texture: an augite crystal grows around and encloses plagioclase crystals

Cumulate texture

A cumulate texture forms when crystals settle out of the magma, typically on the floor of the magma chamber, but can be on the walls or roof, and accumulate in mutual contact. The crystals continue growing after they settle.

Part of crystal formed after settling

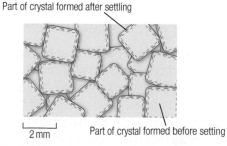

Part of crystal formed before setting

▲ **Figure 8** Cumulate texture, dashed lines show continued layers of growth of the crystal after settling

▲ **Figure 3** Flow banding in a silicic rock, with dark layers that are glassy with no crystals and lighter layers with very fine crystals.

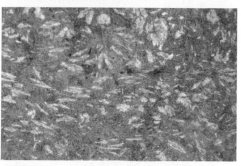

▲ **Figure 5** Porphyritic texture in basalt (with phenocrysts aligned in direction of flow). Phenocrysts approximately 4 mm long.

▲ **Figure 6** Porphyritic texture in granite, with large K feldspar phenocrysts approximately 2 cm long.

HOW SCIENCE WORKS: MICROSCOPY

It can be difficult to see individual minerals in a hand specimen. so a very thin slice 0.03 mm of a rock is sawn, ground and glued to a microscope slide. Using an ordinary microscope, just the outline of the crystals can be seen but in polarised light the minerals have very distinctive colours. Photos taken through a microscope are called photomicrographs. Often drawings of photomicrographs are used to illustrate rock texture.

QUESTIONS

1 Explain why the large white crystals in a porphyritic basalt will have a different origin and composition to the large white crystals in an amygdaloidal basalt.

2 Describe the similarities and differences between an amygdaloidal basalt and a vesicular basalt.

2.5 Formation of magmas

The behaviour of seismic waves tells us that the Earth's crust and mantle are almost entirely solid rock, yet evidence from volcanoes shows that magma must originate somewhere. If the average geothermal gradient is 30 °C per kilometre near the surface, then you might expect that the increase in temperature with depth would produce molten rock at little more than 40 km. However, this does not happen as pressure also increases with depth and the rock then requires a higher temperature to melt.

Convection within the Earth's mantle allows hot, less dense rock to rise and cool, more dense rock to sink without any melting. This happens because the geothermal gradient of the mantle approaches that of an **adiabatic** system. As hot mantle rock rises, it does not gain or lose heat to the surrounding mantle but remains in virtual equilibrium with it. The reduction in pressure allows the rising rock to expand; however, the increase in volume reduces the overall temperature in a process known as **adiabatic cooling** (the heat energy within the rising rock remains constant, it is simply spread over a greater volume). No melting occurs.

The mantle is closest to melting in the seismic low velocity zone of the asthenosphere. Whilst still solid, it behaves in a more ductile manner and it is this that allows the motion of the lithospheric plates above. Here any increase in temperature, reduction in pressure or lowering of the rock melting point will generate magma. Since some rock-forming minerals have a lower melting point than others, it is normal for any melting to be partial.

Magma at divergent plate margins

At mid-ocean ridges (MOR) and continental rift valleys, the Earth's outer layer, the lithosphere, is being pulled apart. These are **divergent plate margins**. As it stretches and thins, the asthenosphere upwells closer to the surface and pressure is reduced. At MORs this can be as little as 5 km below the surface. Decompression of the ultramafic peridotite causes **partial melting** and produces mafic magmas. Basalts are produced, which mostly erupt as pillow lavas on the sea floor, with dolerite dykes and sills intruded below. In continental settings, magma rising through the thicker continental crust is modified and can result in a wide range of volcanic products.

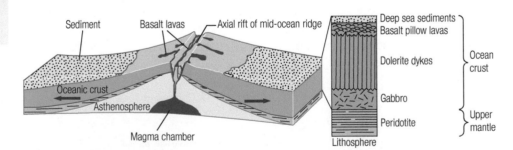

Magma at hot spots

A similar process is operating at **hot spots** where mantle plumes rise from deep in the mantle and partial melting of ultramafic material results. Convection in the mantle slowly transports heat from the core to the Earth's surface. Mantle plumes carry the heat upwards in narrow, rising columns of hot material, which spreads out when the plume head meets the base of the rigid lithosphere. Lower pressures allow decompression and partial melting of mantle peridotite, in a concentrated zone of the asthenosphere, forming enormous volumes of basaltic magma. This basalt may erupt onto the surface over very short time scales (less than 1 million years) to form flood basalts.

If the plume provides a continuous supply of magma in a fixed location, it forms a hot spot. As the lithosphere moves over this fixed hot spot due to plate tectonics, the eruption of magma onto the surface forms a chain of volcanoes parallel to the movement of the plate. A good example of this is the Hawaiian island chain in the Pacific Ocean.

Magma at convergent plate margins

When two plates collide, this is known as a **convergent plate margin** and magma will be generated. As a plate is forced down, the geothermal gradient increases and partial melting results.

Magma at subduction zones

Where plates converge and an oceanic plate is subducted, water in the rocks of the oceanic crust is carried down into the hot mantle. As the descending plate heats up, this water is released into the mantle rock above the plate. The water lowers the melting points of minerals in the mantle rock and partial melting of the mantle produces magma. This process is known as flux melting.

Magma at an oceanic-oceanic plate margin

Where the overriding plate is oceanic, the erupting magma will be mafic to intermediate in composition, as initially the magma has only thin oceanic crust to rise through. If magma rises quickly at shallow depths then it will be basalt. In larger, older island arc systems such as Japan, the crust is thicker and intermediate (andesitic) magmas can evolve.

Magma at a continental-oceanic plate margin

Where the overriding plate is continental, the magma must rise through thicker silicic crust and this may partially melt due to the increasing temperature. The rising mafic magma may be at a temperature in excess of 1000 °C and silicic material melts at just 800 °C. The result is mixing of the mafic and silicic melts to give intermediate to silicic magma. The two components have different viscosities, so mixing is difficult. Some of the magma will reach the surface to form intermediate volcanoes. Most will be intruded to form granite **batholiths**.

Magma generated at a continental-continental plate margin

Where two continental plates converge, neither continent will be subducted. However, the high pressures and the mass of sediments, which have been deformed to form fold mountains, combine to force the base of the crust down. Partial melting at the base of the continental crust will produce silicic magma. As continental crust is silicic and starts to melt at 800 °C, this is the temperature below the mountains due to the geothermal gradient and the presence of water. As the magma rises it intrudes to form granite batholiths. There are no volcanoes present as the silicic magma is too viscous to rise to the surface.

HOW SCIENCE WORKS: CONVERGING PLATES

Where plates converge and an oceanic plate is subducted, hydrated minerals (such as clays) in the oceanic crust are carried down into the hot mantle. As the descending plate heats up these crustal minerals dehydrate releasing water into the mantle rock above the plate.

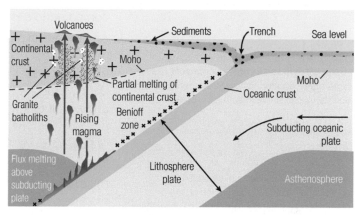

▲ **Figure 2** Magma forming at a continental-oceanic plate margin, shown in cross-section.

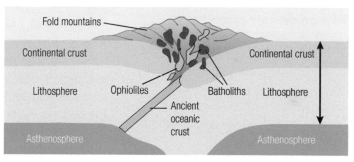

▲ **Figure 3** Magma forming at a continental-continental plate margin, shown in cross-section.

HOW SCIENCE WORKS: HEAT FLOW

Heat flow near the Earth's surface, in boreholes and in mines, can be measured directly using thermometers and thermocouples.

Heat flow at depth is determined by the effect it has on seismic waves. Hot areas slow down seismic waves.

QUESTIONS

1 Summarise the tectonic settings and processes that lead to melting of the crust.
2 What is an adiabatic process?
3 Describe and explain the effect of adding sediments, and so water, at subduction zones.

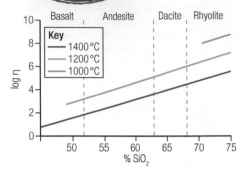

Figure 1 Viscosity of silicate magmas. Note: for any given SiO_2 content, magma will be more viscous at low temperature than at high temperature, and for any given temperature, the viscosity (η) will increase with SiO_2 content

Magma is a mixture of molten rock, solid material (crystals) and volatiles (dissolved gases). The driving force that causes magma to rise is its buoyancy. Magma migrates toward the surface because it is less dense than the surrounding rock. Whether a body of magma can rise buoyantly depends on its viscosity, the ease with which it flows.

Density of magma

The density of magma is determined by its composition, pressure and temperature. Composition is the most important factor. Mafic magma is relatively rich in iron, compared to silicic magma, and this makes it more dense than silicic magma.

An increase in pressure causes magma to compress, which increases its density making it less buoyant. An increase in temperature causes magma to expand, which decreases its density making it more buoyant. The average density of the Earth's crust, at $2.75\ \mathrm{g\,cm^{-3}}$, is generally higher than that of most basaltic magmas and basalt tends to erupt in almost all geological settings.

Viscosity of magma

The viscosity of magma is determined by its composition and temperature. Mafic magma has a low viscosity and flows easily because its silica content is low. Silicic magma has a high viscosity; it is sticky and resists flow because it has a high silica content. The higher the silica content, the more viscous the magma and the slower it will move through conduits. However, for any given composition of magma, an increase in temperature reduces viscosity. The hotter a magma is, the more easily it will flow.

Magma recharge

After an eruption, volcanoes can recharge with new magma from deep within the Earth. This can be monitored using geophysical techniques. Microgravity sensors can detect local changes in subsurface density produced by magma movement, and seismometers can track the migration of small earthquakes from depth toward the surface. As ground water is heated, the increase in pressure can cause crustal deformation and the release of gases, which can also be measured.

How frequently new magma is injected into a chamber can be seen in the crystals within the erupted lava. Zoned plagioclase and olivine crystals are often used for this purpose as each zone shows the composition of the magma at the time that zone was formed.

A 'high level' magma chamber, in which magma is crystallising, can be periodically or continually fed with fresh magma from a lower chamber. If a more evolved magma, richer in silica, is recharged by a mafic magma, this is often the trigger for an eruption. The magmas can mix to form an intermediate eruption, or there can be a simultaneous eruption of two magmas of different composition.

HOW SCIENCE WORKS: MAGMA COMPOSITION

By measuring the chemical composition of zones around crystals such as olivine or plagioclase, geologists can determine how the composition of the magma changed over time. A little like reading tree-rings.

The history of magma recharge can be tracked as each injection of fresh magma changed the composition in the magma chamber.

It is also possible to determine if a fresh injection of magma triggered an eruption or if further growth (a new zone) occurred after the injection.

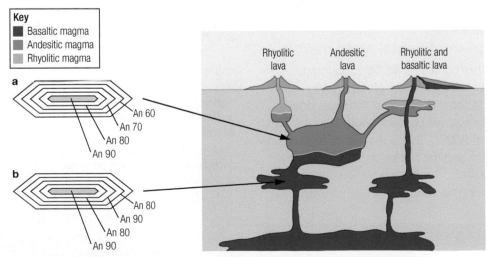

Figure 2 Crystals in erupted lava can show the history of recharge. The zoned plagioclase crystal **a** shows continuous reduction in % anorthite (An) during crystallisation and so no recharge, whereas crystal **b** shows an injection of fresh magma increasing the % anorthite part way through crystallisation. Note: recharge can cause the same volcano to erupt magma of different compositions

Magma mixing

When two magmas of different compositions are combined, there are a number of factors that act to prevent mixing. In the case of a silica-rich rhyolitic magma and a mafic basaltic magma, the density contrast would cause the less dense rhyolite to simply float on top of the basalt. The viscosity contrast would require some vigorous stirring to mix the thick, sticky, rhyolite with a runny, basalt and the temperature contrast would create a thermal barrier. Where the two magmas meet, the hotter, mafic basalt will cool and crystallise. The rhyolite will heat up causing its early formed crystals to dissolve back into the magma.

Magma mixing does occur, but is often incomplete. Igneous rocks show evidence that early formed crystals from one magma, became incorporated into magma of a distinctly different composition before eruption. We also find enclaves (or blobs) of one magma completely surrounded by magma of a different composition. The enclaves tend to be of the host magma, surrounded by the recharge magma. This is thought to be an intermediate step before full mixing, with convection within the magma chamber providing the necessary stirring.

Figure 3 Magma mixing: Igneous rock showing enclaves of one magma with early formed crystals, incorporated within another magma of distinctly different colour and composition. Finger for scale

Ground water and magma

When rising magma encounters groundwater, it can turn the water to steam and cause explosive eruptions.

When groundwater percolates down through faults, fractures and porous rock in the proximity of magma chambers, geysers and hot springs can result. Geysers are hot springs from which a column of hot water and steam is explosively discharged at intervals. Hot springs are common in volcanic areas as the groundwater is heated by the magma at depth. Convection causes the gas-rich, superheated water to rise. As it rises, the pressure decreases and 'flash' boiling into steam results. The column of cooler water above it is thrust up explosively to form the geyser. The water then drains back into the ground and is heated for the cycle to start again.

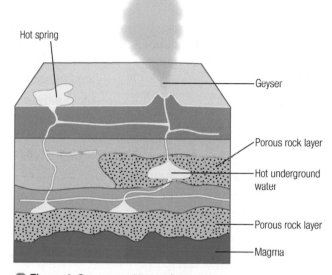

Figure 4 Geysers and hot springs

Exsolution of volatiles

As magma rises toward the surface, the reduction in pressure allows dissolved gases to exsolve, or come out of solution, forming bubbles. This happens because there is less magma above it pressing down. Bubbles can also form in a magma chamber due to recharge and mixing, as rhyolitic magmas have a higher gas content than basaltic magmas. The bubbles increase the pressure in the chamber and this can drive the magma up toward the surface, triggering an eruption.

Note: Volatility is quantified by the tendency of a substance to vaporise - referring to its ability to move through the environment in a dissolved or gaseous form.

MATHS BOX: HOW FAST IS MAGMA EJECTED OUT OF A VOLCANO?

The volume of magma thrust up a volcanic conduit and erupted out of a vent each second is known as the **discharge rate (Q)**. It depends upon the dimensions of the conduit, the properties of the magma and the conditions in the chamber.

Given the **depth (d)** of the magma chamber, the **radius (r)** of the volcanic conduit, the **viscosity (μ)** of the magma and the **pressure (P)** within the magma chamber, the discharge rate can be calculated using the following simplified formula:

		d km	**P** Pa	**r** m	**μ** Pa s
$Q = \dfrac{\pi r^4 P}{8\mu d}$	Iceland	5.4	3.6×10^6	2.0	1000
	Hawaii	5.0	5.0×10^6	1.0	100
	Mt Etna	7.4	7.0×10^6	2.5	450
	Mt St Helens	7.0	3.0×10^7	50.0	2.0×10^6

Calculate the volumetric discharge rate Q in $m^3 \, s^{-1}$ for each volcano in the table above.

Which volcano will be the most hazardous, which the least and why?

QUESTION

1 Describe how the zoning of plagioclase crystals in erupted lava can be used to determine past conditions in the magma chamber.

Intrusive igneous rocks are those that crystallise below the Earth's surface, so intrusion is a process that cannot be observed. You can only see intrusive rocks after the rocks that previously overlaid them have been eroded. Most of the intrusive rocks that we see are millions of years old.

Intrusions are described as **concordant** where they are intruded along a bedding plane and **discordant** where they cross cut a bedding plane. When magma is moving up through **country rocks**, the intrusions will follow any pre-existing line of weakness, such as a fault, joint or bedding plane. The intrusions will also create new channels, as new fractures are created by the pressure of magma injection. Igneous intrusions are often found below areas of volcanism. Dykes extend from the magma chamber to the vents, acting as feeders for the overlying lavas.

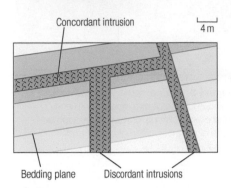

△ **Figure 1** Concordant and discordant intrusions

KEY DEFINITIONS

Concordant intrusions are parallel to the existing beds.

Discordant intrusions cut across the existing beds.

Country rock any rock into which an igneous rock intrudes.

Dyke a discordant, sheet-like intrusion.

Minor intrusions cool at hypabyssal depth below the surface and include sills and dykes.

Sill a concordant, sheet-like intrusion.

Minor intrusions

The word minor suggests that these intrusions are small, but many are not. Certainly some are small with **dykes** that are just a few centimetres wide but hundreds of metres long. Many **minor intrusions** are huge, such as the Palisade Sill, which is 300 m thick, cropping out for 60 km. The 100 m thick Great Whin Sill underlies a large area of north-east England and crops out along the line of Hadrian's Wall.

Most **sills** and dykes are made of the medium grained, mafic rock dolerite, except along the margins where the chilled margins are basalt. Some sills and dykes have granite as a source, these will be made of porphyry or microgranite, which are both medium grained silicic rocks.

Sills

Sills are sheet-like intrusions that are concordant and parallel to the beds. They may have a sinuous outcrop pattern as they are usually intruded along bedding planes. Some sills occasionally cut across the beds in steps from one bed to another, to form transgressive **sills**.

Sills are intruded when the fluid pressure is so great that the underlying magma actually lifts the overlying rocks. This means that sills usually form at shallow depths in the Earth where the mass of overlying rocks is not too great.

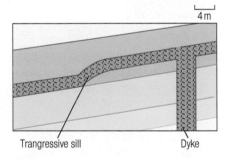

△ **Figure 2** Transgressive sill and dyke

Dykes

Dykes are sheet-like intrusions which are discordant and often vertical or near vertical. They range in size from a few centimetres to more than 100 m thick, but are commonly 1 to 2 m across. As most dykes are vertical they form straight lines in outcrop, as long, thin features. They are often intruded along zones of weakness in the country rock, such as faults or joints.

Dykes are common as feeder vents under volcanoes, so may have a radial pattern.

△ **Figure 3** Photo of a dyke with chilled margins and baked margins visible. Foot for scale.

Ring complexes

Dykes can form curved, ring-like structures on the surface, known as ring complexes. There are two types:

1 Ring dykes: form because of roof collapse when the magma pressure decreases in the underlying pluton or magma chamber. Magma oozes out into the fractures surrounding the collapse and produces dykes that dip away from the centre of the underlying pluton or magma chamber.

2 Cone sheets: form due to magma pressure pushing up and fracturing the overlying roof rocks. Magma is forcibly injected into the fractures producing a conical emplacement with the dykes dipping toward the buried centre of the magma source.

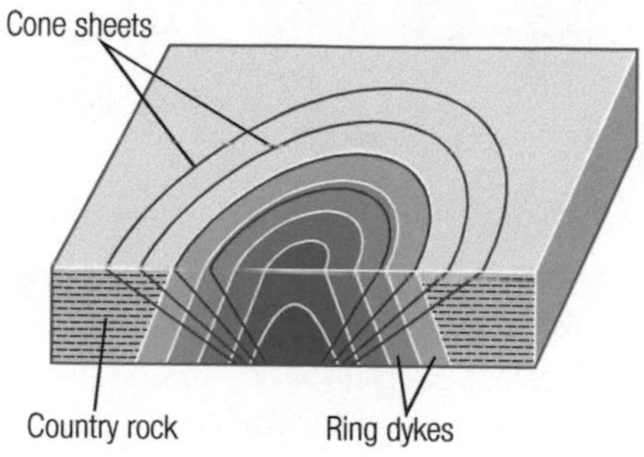

Figure 4 Ring complexes, ring dykes and cone sheets

Volcanic plugs

A volcanic plug is an intrusion formed when magma crystallises in the conduits or vent of a volcano. Over time, weathering and erosion removes the less resistant lava flows and volcaniclastic material making up the flanks of the volcano leaving behind the more resistant plug that formed.

The processes of intrusion

Magma commonly moves as **diapirs** since the process of partial melting produces a melt phase that occupies a larger volume than the solid phase it replaces.

With minor intrusions, there is a large temperature contrast between the country rock and the magma, metamorphosing the country rock in the process. The magma, in turn, is cooled by the country rock, along the **contact** between them. **Baked margins** are produced in the country rock and **chilled margins** are produced at the edges of the intrusion.

KEY DEFINITIONS

Diapir a body of relatively low density material that pierces and rises up through overlying material of a higher density.

Contact where the igneous rock meets the country rock.

Baked margin in the country rock where it was heated by the intrusion and altered.

Chilled margin where the igneous rock has cooled rapidly so it has fine crystals.

Chilled margins

The edge of the igneous intrusion, which is closest to the cold country rock, cools more rapidly than the rest of the intrusion and forms the small crystals of the chilled margin.

The chilled margin may be only a few centimetres to several metres wide.

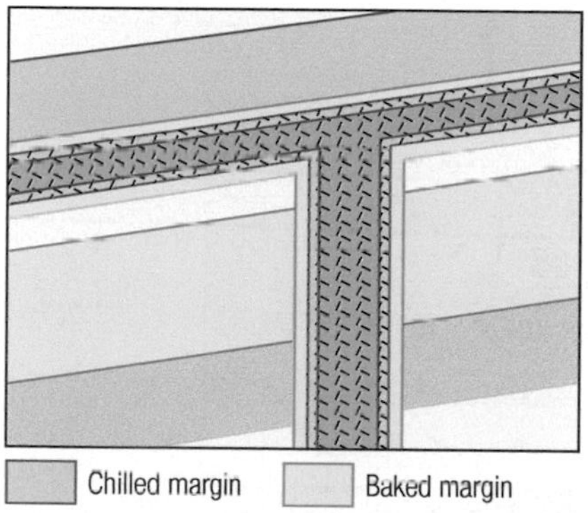

Figure 5 Chilled and baked margins of a dyke and sill

Baked margins

The area of the country rock adjacent to a minor intrusion gets altered by the heat from the intrusion and is called the baked margin. The country rock is recrystallised by the heat by contact metamorphism. The rocks in the baked margin have a sugary texture and become harder and lighter in colour. The baked margin may be a few centimetres to a few metres wide, depending on the size of the intrusion. The larger the intrusion, the wider the baked margin.

QUESTIONS

1 Describe how you would identify a sill from a dyke on a map.
2 Explain why a sill might become transgressive.
3 Describe the changes that occur to both intrusion and country rock along the contact.
4 Explain why magma rises as a diapir.

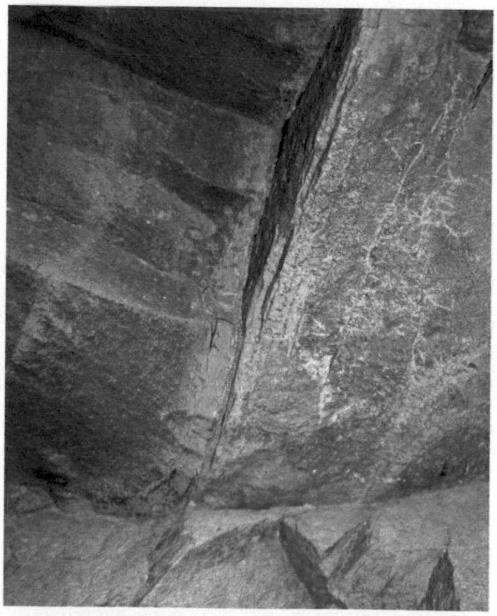

Figure 6 Chilled margin at the edge of a dyke, picked out by iron staining. The dyke is to the left and the country rock to the right.

0 10 cm

Major intrusions are plutonic and solidify deep below the surface. If you find a coarse grained igneous rock in the field, then it is likely to be from a major intrusion that crystallised several km below the Earth's surface. A **pluton** represents one magma body. A **batholith** is much larger (a hundred km or more across). Batholiths represent a long period of repeated igneous intrusions and may consist of multiple plutons of similar composition.

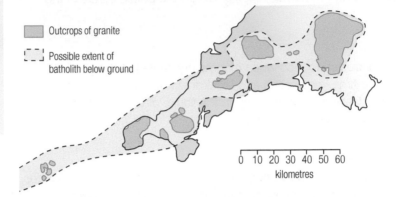

☐ Outcrops of granite

☐ Possible extent of batholith below ground

0 10 20 30 40 50 60 kilometres

🔺 **Figure 1** The granite batholith of south-west England

Batholiths

The most common form of major intrusion is a batholith, which can be very large. Most batholiths are discordant, where they are intruded into the country rock. The outcrop pattern for many batholiths is roughly circular in shape, with steep sides. They are normally composed of granite, although they can be composed of granodiorite or even diorite, all of which are coarse grained. They solidify very slowly at depths of 5 to 30 km as plutonic rocks.

- The Peru–Chile batholith in the Andes of South America is 4500 km long.
- The Cornubian batholith that extends from Dartmoor in Devon to Lands End in Cornwall and to the Isles of Scilly is over 235 km long.

Metamorphic aureole

Batholiths, as major intrusions, heat a large zone of country rock so that a **metamorphic aureole** can be hundreds of metres to several kilometres wide. The rocks in the aureole are altered at the **contact** by contact metamorphism to produce new metamorphic rocks. The metamorphic grade of an aureole zone is quantified by which minerals occur within the zone. If the batholith is intruded into a country rock of shale, then a clear sequence of changes takes place as follows:

- Closest to the intrusion, where the heat is higher for the longest time, the rocks will be totally recrystallised to form hornfels.
- Further away from the intrusion the rocks will be recrystallised to form a rock containing the metamorphic mineral andalusite.
- Towards the edge of the metamorphic aureole, where the heat was least, recrystallisation is partial, so that a spotted rock forms.

The width of a metamorphic aureole will depend on: the size and temperature of the intrusion at the time of emplacement, the dip of the sides of the intrusion and the composition of the surrounding rocks. For example, limestones are altered much more easily than sandstones.

Plan-view

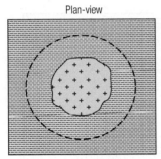

Cross-section

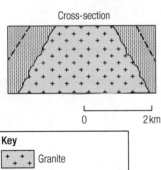

0   2km

Key
+ + Granite
     Metamorphic aureole
     Country rock

🔺 **Figure 2** A map and cross-section of a granite pluton

Formation of granite magmas

Most batholiths are intruded at convergent plate margins, where they form the core of fold mountains, which have resulted from plate collision. High temperature mafic magmas will be rising up through the crust and the heat from these will partially melt continental crust. **Partial melting** produces silicic magma, which rises because it is less dense than the rock from which it was melted. Silicic magma is viscous and rises through the crust slowly. It is difficult for it to mix with other magmas because the viscosities are so different. Batholiths are huge bodies of magma usually produced by repeated intrusion of magma in the same area. As the magma moves up it generates more magma by **stoping** and assimilation.

Stoping and assimilation

At shallow depths of about 10 km, the crust is cold and brittle and batholiths may intrude by stoping. Magma moves upwards along joints, faults and bedding planes, separating masses of country rock. Eventually, pieces of country rock are detached and settle into the magma where they are known as **xenoliths**. No new space is created during stoping, as the magma simply fills the space formerly occupied by country rock. As the xenoliths fall into the magma, some may be preserved but most will be **assimilated** as they gradually melt to become more magma.

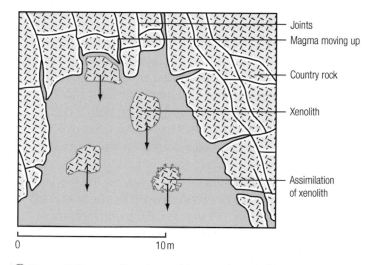

Figure 3 Cross-section of part of the top of a batholith showing stoping and formation of xenoliths

The composition of the crust is commonly different from the composition of the magma trying to pass through it. This means that assimilation is always a possibility, but there is a temperature balance involved. In order to assimilate country rock, enough heat must be provided by the magma to melt the country rock. Granitic magma is cooler than mafic magma and is therefore less able to melt and assimilate country rock.

Assimilation can significantly alter the bulk composition of a magma body, as the magma becomes 'contaminated' by the country rock. There are other magmatic processes, such as magma mixing and magmatic differentiation, that can produce a daughter magma of distinctly different composition from the parent magma. Often the only clear evidence of stoping and assimilation is the presence of partially assimilated xenoliths preserved within the plutonic rock.

KEY DEFINITIONS

Partial melting occurs when only a portion of a rock is melted. When a rock is heated, those minerals with lower melting temperatures will melt. Those with higher melting temperatures will remain solid.

Stoping the process that accommodates the magma, as it moves upwards into the country rock, by the mechanical fracturing of the surrounding country rock.

Xenoliths clasts or blocks of pre-existing rock contained within an igneous rock.

Assimilation the melting process that incorporates blocks of country rock, freed by stoping, into the magma.

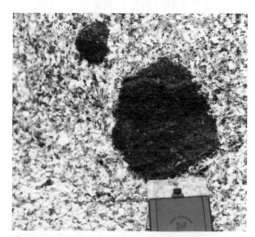

Figure 4 Photo to show xenoliths in granodiorite. Compass clinometer for scale.

CASE STUDY: VARISCAN OROGENY

The granites of the Cornubian batholith were intruded towards the close of a major mountain building event, called the Variscan Orogeny (300 to 275 Ma). The main exposures can be seen on Dartmoor, Bodmin Moor, St Austell, Carnmenellis, Lands End and the Isles of Scilly. Most of the individual plutons forming the batholith have their own distictive composition. They are largely coarse grained but are, in places, pegmatitic. K-feldspars of several cm long are typical. Metal ore mineralisation of tin, copper and tungsten is common.

QUESTIONS

1 A silicic magma is at a temperature of 850°C when it is intruded. What will be the difference in temperature between the country rock and an intrusion at a depth of 10 km?

2 Describe how the process of stoping and assimilation can change the composition of a magma.

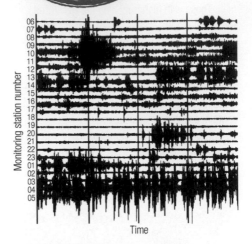

▲ **Figure 1** Seismogram showing four hours of high intensity harmonic tremors, July 6 2013, Popocatepetl, Mexico

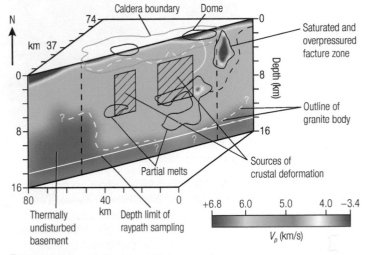

▲ **Figure 2** Interpretive geologic cross-section from P-wave model of Yellowstone caldera

▲ **Figure 3** Global navigation satellite system (GNSS) used to monitor ground movement via the GPS network

Earthquakes commonly provide the earliest warning of magma moving and a volcano preparing to erupt, and earthquake swarms immediately precede most volcanic eruptions. Seismicity is of three types:

- Short-period earthquakes including the microseismics, caused by the fracturing of brittle rock as magma forces its way upwards.
- Long-period earthquakes, believed to indicate increased gas pressure in the magma chamber of a volcano.
- Harmonic tremors, sustained vibrations in the range of 0.5–5 Hz, the result of magma vibrating in the vent as it moves upwards.

Where is the magma chamber?

Most magma chambers are areas where the crust or mantle is partially melted rather than a great lake of magma below the surface. One way to think about a magma chamber is to picture a water-filled sponge. The sponge would represent the solid rock and the water the magma. Beneath Yellowstone, on average, the magma chamber is about 90% solid rock containing 10% liquid in its pores.

Seismic survey

One way to locate a magma chamber is to put seismometers all around a volcano and record the seismic waves generated by earthquakes. When seismic waves encounter molten material they either slow down (P waves) or are stopped (S waves). If material is partially melted, then both P and S waves will slow down. Some of the waves travel directly under the volcano and the seismometer on the opposite side of the volcano from where the earthquake originated will not receive or will get delayed P and S waves. The magma chamber creates a shadow zone allowing its size and depth to be calculated. Below Yellowstone, the magma chamber has a volume of approximately $15\,000\,km^3$ and is about 80 km long and up to 40 km wide. The top of the magma chamber is about 8 km deep.

Ground movement

Swelling of a volcano suggests that magma is moving up under the volcano. The rate of swelling is measured using **tiltmeters** and **GPS** that can accurately measure both vertical and horizontal movements to an accuracy of one millimetre. This is often coupled with satellite-borne radar (interferometric synthetic aperture radar) and thermal monitoring (infrared band satellite imagery) to detect magma movement just below the volcano's surface.

Gas emissions

As magma nears the surface and the confining pressure decreases, gases escape. It is similar to what happens when you open a bottle of fizzy drink and carbon dioxide escapes. A change in gas emissions can indicate magma movement. Sulfur dioxide is one of the main components of volcanic gases, and increasing amounts being released suggests that magma is nearing the surface. There may also be changes in gas composition in the ratio of CO_2/SO_2. Changes in **fumarole** gas composition, or in the emission rate of SO_2 and other gases, may indicate variation in magma supply rate or a change in magma type.

Changes in ground water

Borehole and well measurements are increasingly used to monitor changes in a volcano's subsurface gas pressure and thermal regime. Increased gas pressure will cause water levels to rise and suddenly drop right before an eruption, and the increased local heat flow can reduce flow in aquifers.

Dyke orientation

The propagation of dykes is controlled by the underlying stress field in the country rock. Dykes that are parallel to each other tend to form in response to a regional stress field, whereas radial dykes tend to form due to stress around a pluton or volcano. The orientation of dykes can therefore be used as an indicator of the stress field in volcanic areas and to identify volcanic centres.

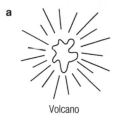
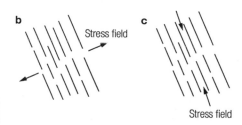

🔺 **Figure 4** Dyke orientations, **a** radial around a volcanic centre, **b** parallel in an extensional regime, **c** parallel in a compressional regime (arrows show direction of plate movement)

MATHS BOX: MEASURING DYKE ORIENTATIONS

The stretching of the crust as the Atlantic Ocean opened up created weaknesses in the surface of the rock, at right angles to the direction of tension. Magma rising up through the cracks formed dyke swarms across northern England, southern Scotland and the Inner Hebrides, forming the British Tertiary Volcanic Province.

The data below, dyke orientations in degrees from North, was obtained from the southern coastline of Arran over a distance of 5 km:

142, 121, 122, 130, 128, 128, 234, 127, 122, 224, 132, 148,

133, 212, 146, 220, 219, 136, 229, 206, 293, 296, 209, 225,

263, 264, 246, 282, 281, 290, 291, 296, 328, 235, 265, 302,

293, 312, 224, 320, 305, 309, 288, 325, 301, 312, 295, 351,

220, 321, 330, 350, 320, 320, 327, 201, 297, 297, 294, 290.

Plot the dyke orientations on a rose diagram with segments at 15 degree intervals.

1 What was the direction of the extensional forces?

2 What might explain the NE/SW trending dykes?

3 Given a mean average dyke thickness of 72 cm, use the above data to calculate the total crustal extension in metres and the % of crustal extension for this area of southern Arran.

Guidance: dyke orientations are trends not directions so opposite segments of a rose diagram have the same trend. Tally all orientations between 1–15 degrees and 181–195 degrees, add them together and plot the total in both the 0–15 and the 181–195 segments of the rose diagram. Then repeat for successive segments.

🔺 **Figure 5** Major intrusive centres of the British Tertiary Volcanic Province

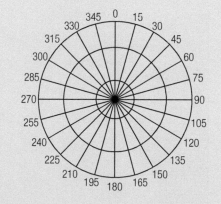

KEY DEFINITIONS

Tiltmeters designed to measure very small changes in vertical level.

GPS stands for Global Positioning System and is the radio navigation system that allows the determination of an exact position.

Fumaroles openings in or near a volcano, through which hot gases emerge.

CASE STUDY: EARTHQUAKES AND ERUPTIONS

Mount St Helens

On March 20 1980 a magnitude (M) 4.2 earthquake occurred beneath Mount St Helens. Three days later there was another of 4.0M. Harmonic tremors and minor eruptions occurred throughout March and April. By mid-April the frequency of minor eruptions had slowed but earthquake monitoring showed no such decline. Earthquakes greater than 3.0 M continued to average more than 30 a day. The deformation of Mount St Helens prior to the May 18 1980 eruption was also continuously monitored and this showed the North side of the volcano bulging upwards at 2m per day as magma built up beneath.

Popocatépetl

In December 2000, scientists predicted an eruption within two days at Popocatépetl, on the outskirts of Mexico City. Their prediction used an increase in the numbers of long-period earthquakes, as an indicator of an imminent eruption. The government evacuated tens of thousands of people; 48 hours later, the volcano erupted as predicted. It was Popocatépetl's largest eruption for 1000 years, yet no one was injured.

QUESTIONS

1 What are harmonic tremors and what do they indicate?

2 How does the movement of magma affect the level of ground water in wells?

How do you distinguish sills from lava flows?

In the field, it can be very difficult to be sure if a mass of rock is an ancient lava flow, now covered in more recent sediments, or a sill intruded between two beds of rock.

Lava flows cool at the Earth's surface and are extrusive igneous rocks, whilst sills cool below the surface and are intrusive igneous rocks. To distinguish between these two methods of formation, you will need to make observations of the characteristics of the rock, particularly crystal grain size. Looking at the field relationships between igneous rocks and the country rock is another key method. Chilled and baked margins are good indicators of intrusive igneous rocks. Most igneous textures can be found in any rocks, but flow banding and vesicular texture nearly always indicate an extrusive origin.

Crystal size

In general, lava flows cool rapidly and have fine crystals compared to the medium or coarse crystals of sills. The interior of minor intrusions like sills, typically cool more slowly and produce a coarser crystal size. If a sill cools slowly enough, magmatic differentiation can produce **cumulate** layers.

Baked and chilled margins

With both sills and lava flows, there is a large temperature contrast between the country rock and the magma or lava. As a result, the magma or lava cools more rapidly and forms a fine or glassy **chilled margin** along the contact with the country rock. The country rock, on the other hand, is baked by heat. As a result, the country rock is altered and recrystallised along the contact, often developing a sugary texture and becoming harder and lighter in colour, forming a **baked margin**.

One key difference between sills and lava flows is that a sill will produce two baked margins in the country rock, one above and one below the sill, whereas a lava flow will only produce one baked margin, in the country rock below the flow.

▲ **Figure 1** Cross-section through a 200 m thick sill showing variation in crystal size, coarser in the centre due to slower cooling. The bulk of the sill has medium crystal size, fine crystal size at the chilled margins

Labels (top to bottom): Sandstone; Baked margin; Fine crystal size; Medium crystal size; Coarser crystal size in centre; Sill; Cumulate layer; Fine crystal size; Baked margin; Sandstone

KEY DEFINITIONS

Cumulate an igneous rock produced by gravity settling of crystals in a magma body.

Chilled margin where the igneous rock has cooled rapidly so it has fine crystals.

Baked margin in the country rock where it was heated by the intrusion and altered.

Palaeosol a soil horizon that was formed in a past geological age.

KEY FACT

Magma and lava are terms which are often confused, so it is worth remembering that molten rock is magma when it is below the surface and lava at the surface once it has erupted.

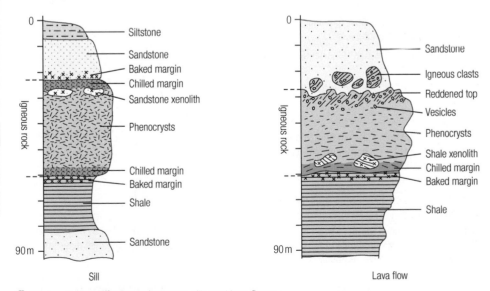

Sill (left) labels: Siltstone; Sandstone; Baked margin; Chilled margin; Sandstone xenolith; Phenocrysts; Chilled margin; Baked margin; Shale; Sandstone

Lava flow (right) labels: Sandstone; Igneous clasts; Reddened top; Vesicles; Phenocrysts; Shale xenolith; Chilled margin; Baked margin; Shale

▲ **Figure 2** The differences between sills and lava flows

Xenoliths

As a sill is being intruded, it can incorporate xenoliths (pieces of country rock) ripped from above and below it. A lava flow can only rip up rock from below.

Upper surface

Lava flows have an irregular upper surface, which can consist of scoria (a highly vesicular, mafic or intermediate volcanic rock) and or rubble. If the flow is not rapidly buried by subsequent flows, it can be exposed to weathering and develop a reddened surface. Prolonged exposure to weathering can allow a **palaeosol** (a soil horizon) to develop. Sills are protected by the country rock into which they intruded so there will be no weathering of the upper surface.

Igneous texture

A vesicular texture is common in the upper part of lava flows. When magma erupts onto the surface and becomes lava, the dissolved gases are able to exsolve, due to changes in pressure, and produce vesicles. The confining pressure below ground makes vesicles rare in sills.

Tabular phenocrysts can become aligned parallel to the flow direction.

Friction within silicic lava flows can cause flow banding, the separation of minerals such that they form light and dark layers.

If a sill is thick enough to cool very slowly, early formed dense crystals (such as olivine) can sink through the melt. These crystals can settle to the bottom of the sill producing a different compositional layer.

Using all of the methods described, a table can be created to help with identification:

 Figure 3 Aa lava flow with an irregular upper surface
`0  25 cm`

Figure 4 Dolerite sill with sharp, regular upper surface
`0  1 m`

Figure 5 Scoria (vesicular, mafic, volcanic rock) reddened by weathering
`0  1 cm`

Figure 6 Flow banding in rhyolite lava
`0  5 cm`

Sills have:	Lava flows have:
Two chilled margins due to contact with cool country rock, both above and below the intrusion	Only one clear chilled margin below, while the top may be chilled or made of scoria
Two baked margins, both above and below, because the sill heats all the country rock around it	One baked margin below the lava flow and none above, as it was exposed to air
Xenoliths from rocks above and below the sill as the sill rips up material during emplacement	No xenoliths from above the sill, but possibly some from below
May show differentiation of magma, or gravity settling, if the sill is thick and cools slowly	May have flow banding or phenocrysts, may be aligned parallel to the direction of flow
Medium average crystal size	Fine average crystal size
Dolerite as the most common rock	Basalt as the most common rock
Rare vesicles	Vesicles common in the upper part, may become amygdales when infilled by minerals
No fragments of the sill in overlying rocks	Lava fragments in overlying sedimentary rocks as a result of erosion and re-deposition
A regular upper surface	Irregular upper surface made up of scoria or rubble
No weathering of the upper surface	A reddened surface or even an ancient soil if the flow was chemically weathered

QUESTIONS

1 Describe the relationship between basalt and dolerite, and rhyolite and microgranite. Summarise the distribution of these rock types in sills and lava flows.

2 Describe and explain three methods of distinguishing a sill from a lava flow in the field.

3 Research the Carboniferous Great Whin Sill and write an account of its emplacement, composition and size.

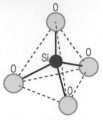

Figure 1 SiO_4 tetrahedra and network

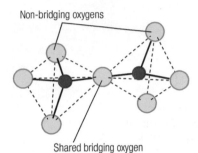

Non-bridging oxygens

Shared bridging oxygen

Figure 2 Two silica tetrahedra sharing a 'bridging' oxygen

KEY DEFINITIONS

Melt the name given to magma or lava, in the liquid phase.

Aa lava flows have a rough, blocky, jagged surface.

Pahoehoe lava flows have a smooth or ropy surface.

Scoria a volcanic rock that is usually mafic but can be intermediate in composition.

Pyroclast an individual fragment ejected during an eruption. Pyroclastic describes all the fragmental materials formed by explosive eruptions, including bombs, blocks, lapilli and ash.

Nuée ardente (French, 'burning cloud') is an incandescent, gaseous, pyroclastic cloud of magma droplets and ash. When cooled and solidified the deposit is called an **ignimbrite**.

Pyroclastic flow is a hot mixture of pyroclastic material and gas. A nuée ardente is a type of pyroclastic flow.

A COMPARISON OF VISCOSITIES

Substance	Viscosity (Pa s)
Air (at 18°C)	1.9×10^{-5} (0.000019)
Water (at 20°C)	1×10^{-3} (0.001)
Pahoehoe lava	100 to 1 000
Aa lava	1000 to 10 000
Andesite lava	10^6 to 10^7
Rhyolite lava	10^{11} to 10^{12}

When magma erupts onto the surface, it becomes lava. Many eruptions are effusive and characterised by the calm, non-violent extrusion of lava whilst others are explosive and characterised by the violent ejection of fragmented volcanic materials. How lava behaves on the surface is largely determined by its viscosity and this relates to the internal friction caused by chemical bonds within the liquid that inhibit flow.

Silicate polymerisation and viscosity

Silicate minerals have a structure based around the silica tetrahedron. This has the shape of a three-sided pyramid and is formed from one silicone (Si) atom and four oxygen (O) atoms, bonded together covalently (you will have studied this during your GCSE). Each of the four oxygen atoms of one tetrahedron is able to bond to another tetrahedron. When two silica tetrahedra share one oxygen atom in this way, the atom is known as a bridging oxygen. The structure of a silicate **melt** is formed by a network of linked silica tetrahedra which may produce chains, rings or sheets. Because it is a melt, the structure is more disordered and flexible than in a solid silicate mineral and can be continuously rearranged as bonds between atoms are made and broken.

This process of building a larger molecule by repeated addition of smaller molecules, is known as polymerisation. The silica network produced by polymerisation, whilst flexible, has strong directional bonds that resist deformation and make it difficult for the lava to flow. The more oxygen atoms that are shared between tetrahedra (bridging oxygens) the greater the degree of polymerisation. A more polymerised lava will have a higher viscosity than a less polymerised lava.

Temperature

Increasing temperature produces more kinetic energy and makes it easier for the bonds between atoms to be broken, so depolymerisation occurs. Increasing the temperature decreases the viscosity of the lava.

Pressure

Increasing pressure can compress the silicate network and make it easier for bonds to form. However, it can also reduce the number of bonds depending on other compositional and temperature factors. So increasing pressure can increase or decrease the viscosity of a magma.

Composition

Silicic lavas are much more viscous than intermediate or mafic lavas, simply because there are more silica tetrahedra available to construct a network. The iron and magnesium atoms present in mafic lavas also act as network modifiers and depolymerise the lava making it less viscous.

Volatiles

The presence of water dissolved in a melt reduces the number of bridging oxygens and acts to depolymerise the melt. Hydrogen ions from the water react with the bridging oxygens of the silica tetrahedra creating hydroxyl (OH^-) molecules and breaking the network. Adding water to a silicate melt reduces its viscosity.

When carbon dioxide is dissolved in a melt, it has the opposite effect to water, it reacts with the non-bridging oxygens of the silicate network and causes polymerisation. Adding CO_2 to a silicate melt increases its viscosity.

As a melt moves towards the surface and pressure decreases, the solubility of volatiles in the melt also decreases. When these volatiles begin to escape (exsolve) as gases, the size and distribution of the growing gas bubbles affect viscosity and provide the driving force for eruptions.

When water can no longer remain dissolved, it exsolves as water vapour and the melt becomes more viscous. In silicic melts, which have a high viscosity to begin with, the expansion of this water to gas is resisted by the thick, sticky lava and explosive eruptions occur. Mafic melts are much less viscous and the expansion can occur with ease, creating effusive eruptions of lava.

Lava flows

Lava flows from recent eruptions can be very distinctive but ancient lavas are more difficult to recognise as they usually lack the surface features of rubbly blocks and flow patterns.

Mafic lavas

These low viscosity flows of basalt are thin and widespread often many kilometres long because they are very mobile as a fluid. They cool to form:

- **Aa** lava flows with a rough blocky, jagged surface
- **Pahoehoe** flows with a smooth or ropy surface.

The pahoehoe lava flows are less viscous than the aa lava flows and as the lava cools they becomes even more viscous, so that pahoehoe lava flows may become aa lava flows. Erupting lavas have temperatures in the range of 1000 °C to 1200 °C, but most do not flow very fast. Lava in a lava tube may move rapidly but most flows are at walking pace with **scoria** falling down from the top to the front of a flow. If they cool underwater, pillow lavas may form, with their rapidly cooled outer surface forming volcanic glass and the more insulated centre forming a basalt.

Intermediate and silicic lavas

Viscous flows of andesite are irregular and tend to be limited to the area around the cone of a volcano. Silicic flows of rhyolite are rare, as they are very viscous, so they tend not to flow well. Very sticky lava can form lava domes within the crater and often shows flow banding.

Pyroclastic material

Magma can be explosively ejected into the atmosphere as particles called **pyroclasts**.

- Ash is the smallest particles, less than 2 mm in size, and forms the pyroclastic rock tuff when consolidated.
- Lapilli are particles between 2 mm and 64 mm in size, and form a lapilli tuff when consolidated.
- Bombs and blocks are produced by the coarsest particles over 64 mm in size, and form the pyroclastic rock agglomerate close to the crater of the volcano.

Nuée ardentes are produced by the gaseous magma that forms the fast-moving **pyroclastic flows** that can flow several km, and cool to form the rock *ignimbrite*.

Pumice has already been described as a silicic igneous rock. It is so light that it can travel a long distance from the volcano. When Krakotoa erupted, pumice fell on ships several kilometres from the volcano and floated on the sea looking like snow, as it has low density.

🔺 **Figure 3** Types of lava and pyroclastic material **a** 'blocky' aa lava **b** 'ropy' pahoehoe lava **c** pillow lava

MATHS BOX: CALCULATING VISCOSITY

Viscosity is a measure of how easily, or otherwise, something flows. For example, water flows more easily than wallpaper paste.

Another way to look at viscosity of a liquid is to see how quickly an object, for example a steel ball-bearing, falls through the liquid.

By dropping a ball-bearing into a tall, wide glass tube full of the liquid under test and noting the time it takes to fall a measured distance its average velocity can be found.

We can then calculate the viscosity of a liquid using a variation of Stoke's Law using the formula below:

$$\eta = \frac{2\Delta\rho.\, g.r^2}{9v}$$

In this formula the viscosity is measured in Pa.s [Pascal second].

$\Delta\rho$ is the difference in density between the ball-bearing and the liquid ($kg\,m^{-3}$)

g is the acceleration due to gravity in $m\,s^{-2}$

r is the radius of the ball-bearing in m

v is the average velocity of the ball-bearing in $m\,s^{-1}$

Trying this investigation using different liquids, for example washing up liquid and treacle, allows viscosity to be explored as does using treacle or oil at different temperatures.

QUESTIONS

1 Describe the rock types that form from ash, lapilli and blocks.

2 Prior to eruption, a magma might simultaneously cool, crystallise, and evolve toward a more silica-rich composition. What effect do you think this would have on its viscosity?

2.12 Types of volcano

The viscosity, or resistance to flow, of lava is important because it determines how the lava will behave. It controls the length of lava flows, the velocity of the flow and the shape of the **volcano**. Basaltic lava has a **low viscosity** and flows easily, producing volcanoes with shallow sides. Andesitic and rhyolitic lavas have a much higher viscosity and flow slowly, producing steep-sided volcanoes.

Shield volcanoes

A **shield volcano** is characterised by gentle slopes less than 10 degrees. They are composed almost entirely of thin basalt lava flows built up from a central vent, with **fissure eruptions** on the flanks of the volcano. The low viscosity of the magma allows the lava to travel down the sides of the volcano on a gentle slope and builds up on the lower flanks. Most shield volcanoes have a roughly circular or oval shape in map view. Very little pyroclastic material is found within a shield volcano.

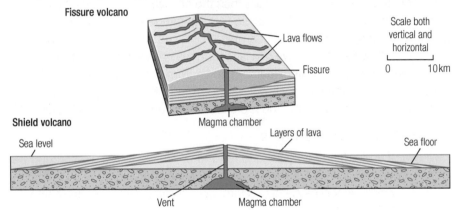

Figure 1 Cross-sections of fissure eruption and a shield volcano

Fissure eruptions

Iceland has a number of fissures, parallel to the axial rift of the mid-ocean ridge, and these can erupt huge quantities of fluid lava that spreads out over a large area. Lava erupted from the 30 km long Laki fissure, which erupted from 1783 to 84, and covered an area of 560 km². The lavas from these fissure eruptions build up to form plateaus. Vast quantities of volcanic gases, particularly SO_2, are released, which are believed to affect the climate.

Columnar jointing

When lava flows are more than 3 m thick, the inside of the flow will cool steadily and slowly, perhaps over weeks or months, instead of the hours or days for the outside of the lava flow. Hexagonal columns form in the centre of these thick lava flows as it contracts during cooling. The cooling originates at equally spaced centres and proceeds in all directions from these centres. Contraction causes tension cracks to start to form half-way between each cooling centre.

Submarine eruptions

When lava is erupted underwater the outer skin cools very rapidly while the inside is still molten. The pressure of more lava from beneath causes the front to break through to form a new pillow shape. Pillow lavas are recognised by their characteristic rounded shape. Each pillow has a rounded top and a sagging bottom, so pillow lava can be used as a way-up structure to identify the youngest rocks. Vesicles may be present towards the outer, upper edge of the pillows and the spaces between pillows may be infilled with fine glassy materials.

KEY FACTS: VISCOSITY

The low viscosity of basaltic lava means:
- It flows easily, with long, thin flows.
- It produces volcanoes with shallow sides.
- It erupts frequently.
- Dissolved gases can be released easily.
- Eruptions are usually quiet.

The high viscosity of rhyolitic and andesitic lava means:
- It flows slowly, with short, thick flows.
- It produces volcanoes with steep sides.
- It erupts infrequently.
- Trapped gases may escape explosively.
- Eruptions are explosive, producing tuffs and pyroclastic flows.

KEY FACT: SHIELD VOLCANOES

75% of annual volume of erupted material is at mid-ocean ridges and the largest volcanoes on Earth are shield volcanoes.

Figure 2 Lava erupting as a fissure eruption

Figure 3 Columnar jointing in a lava flow at the Giant's Causeway in Northern Ireland

0   50
cm

Composite volcanoes (strato-volcanoes)

These form over 60% of the Earth's individual volcanoes. The name comes from the alternating layers (strata) of lava and pyroclastic deposits. These volcanoes typically have a conical shape but in many cases the outline is irregular due to slope failure, **caldera** collapse or explosive damage. New cones build up near the site of an older cone and create complex volcanoes. Each eruption can last hours, days or years and though the viscous lava rarely flows much further than the volcano base, the ash will cover vast areas.

The lava has a wide range of compositions (50 to 71% SiO_2) and when silicic, it may be so viscous that it hardly flows at all, piling up in the vent to form volcanic domes. Lava domes are bulbous and very steep sided and form slowly with new lava pushing up from below. Silicic magmas have a high gas content and this can result in pyroclastic flows.

Each volcano has an eruptive cycle:

- A period of no activity when the volcano may appear dormant. During this time the magma chamber is filling with magma, differentiation may be taking place and pressure is building up inside the chamber. These viscous magmas allow gas pressures to build up to high levels.

- Once the pressure exceeds the mass of overlying rock, including the old lava plugging the vent, there will be an explosion. This blasts away part of the top of the volcano where the vent was blocked and allows the gas-rich pyroclastic material to escape. This creates the layers of ash (tuff), blocks (agglomerate) and pyroclastic flows (ignimbrites).

- The final stage is when lava reaches the surface and forms a layer on top of the pyroclasts. As gas pressure decreases, the lava supply reduces until it cools in the vent, plugging it ready for the next cycle.

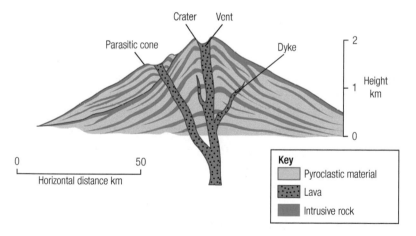

▲ **Figure 4** Cross-section of a composite volcano showing layers of lava and pyroclasts

Calderas

A caldera is a large (1–20 km) circular volcanic depression caused by violent explosions followed by collapse of the top of the volcanic cone. Figure 5 shows three stages in the formation of a caldera:

1 A series of violent explosive (Plinian) eruptions of pyroclastic flows, ash and pumice take place, removing large volumes of magma.
2 The magma chamber starts to empty as the magma is erupted and the top of the volcano starts to collapse down into the weakened area below. This compresses the remaining magma so that eruptions are even more violent.
3 Finally, the entire cone of the volcano collapses and this may cause tsunamis if the volcano is coastal or is an island, such as Krakatoa.

KEY DEFINITIONS

Volcanoes vents at the surface of the Earth through which magma and other volcanic materials are ejected.

Low viscosity where magma or lava are fluid and flow freely.

Shield volcanoes have gentle slopes of less than 10 degrees and a roughly circular shape around a central vent.

Fissure eruptions where magma reaches the surface along long, linear cracks or fissures.

Submarine eruptions where magma comes from a vent or fissure on the sea floor.

Composite volcanoes tall, conical shaped and are composed of alternate layers of lava and ash.

Caldera a large volcanic crater that has undergone collapse, following an eruption.

Effusive the term used to describe the fluid, non-explosive, basalt lava.

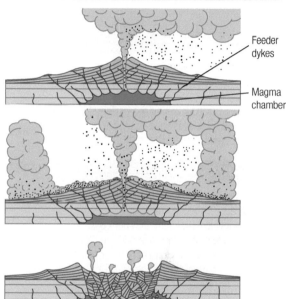

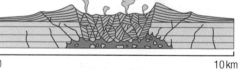

▲ **Figure 5** The formation of a caldera

STUDY TIP

When drawing a shield volcano think about the angle of slope of the sides of the volcano. If you draw a steep angle it is impossible for it to be a shield volcano as the basaltic fluid lava cannot form a steep-sided volcano. It should be less than 10 degrees, and even the composite volcanoes are usually only about 30 degrees.

QUESTIONS

1 Explain how lava from a composite volcano can have a wide range of compositions (SiO_2%).

2 Explain why shield volcanoes form at divergent plate boundaries and composite volcanoes at convergent plate boundaries.

Viscosity and silica content

Mafic magma has a low viscosity because the silica content is low, consequently the basalt lavas are fluid and flow easily. The dissolved gas content of the magma can be high, but the low viscosity allows the gas to escape so that eruptions are quiet and non-explosive (0 or 1 on the **VEI** scale). The volcanic activity is described as effusive and classified as **Hawaiian**. Mafic volcanoes occur at divergent plate margins or hot spots.

Intermediate and silicic magma have a high viscosity because the silica content is high, producing thick and sticky lavas that move slowly. The high viscosity does not allow dissolved gas to escape, so that eruptions are explosive when the gas finally does break through (2 to 8 on the VEI scale). At least 50% of the material erupted is pyroclastic, which can include the incandescent gaseous cloud containing droplets of magma and ash called nuée ardente. The volcanic activity is described as explosive and classified as **Strombolian**, **Vulcanian** or **Plinian** – all named after Italian volcanoes. Intermediate and silicic volcanoes occur at convergent plate margins where subduction is taking place.

Volcanic gases

All atmospheric gases, except photosynthetic oxygen, were released by volcanic eruptions. The gases exsolve during an eruption by pressure release. The main gas is water vapour in the form of steam, which accounts for about 70% of all the gas released. The other main gases are CO_2 (12%), N (7%) and SO_2 (7%), with trace amounts of H, CO, S, Ar, Cl, and F. These gases can combine with hydrogen and water to produce numerous toxic compounds, such as HCl, HF, H_2SO_4 and H_2S. Rain can be very acid close to volcanoes. Fumaroles are often a sign of gas escaping and activity below the surface.

Risk analysis

Studies of the geological history of a volcano are essential to make an assessment of the types of hazards and the frequency at which these hazards have occurred in the past. Geologists examine sequences of pyroclastic deposits and lava flows and their age and characteristics to determine the past behaviour of a volcano.

Hazard maps show zones of danger expected from specific hazards for each volcano and take the following into account:

- Lava flows generally move slowly so rarely kill, but they destroy property and agricultural land.
- Blast damage can be very destructive for areas close to volcanoes, with people in the danger zone killed, trees flattened and buildings destroyed.
- Ash falls can affect areas far from volcanoes, causing damage to property as the mass of ash causes roofs to collapse.
- Pyroclastic flows are very hot and travel rapidly but are mostly confined to valleys. They are responsible for many deaths as they are unpredictable.
- Lahars (mudflows) can be generated by pyroclastic flows or debris avalanches that become 'wet' from rivers or rain, by hot volcanic products that land on snow, by eruptions out of crater lakes or when eruptions occur under ice. They can move rapidly, and may be very destructive and though limited to valleys may affect areas far from the volcano.

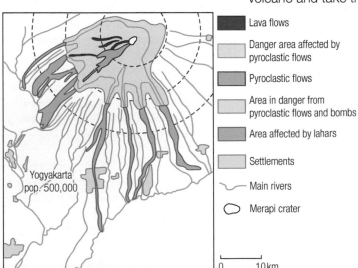

Lava flows

Danger area affected by pyroclastic flows

Pyroclastic flows

Area in danger from pyroclastic flows and bombs

Area affected by lahars

Settlements

Main rivers

Merapi crater

Yogyakarta pop. 500,000

0   10km

🔺 **Figure 1** Hazard map for Merapi, the most active volcano in Indonesia, which produces more pyroclastic flows than any other volcano on Earth.

The Volcanic Explosivity Index (VEI)

The VEI was proposed in 1982 as a way to describe the relative size or magnitude of explosive volcanic eruptions. It is a 0 to 8 index of increasing explosivity. Each increase in number represents an increase of around a factor of ten and so is exponential.

The more explosive the volcano, the less frequently it will erupt, but the greater the volume of pyroclastic material ejected when it does erupt.

VEI	Ejecta volume	Classification	Description	Plume	Frequency
0	$< 10\,000\,m^3$	Hawaiian	non-explosive	$< 100\,m$	constant
1	$> 10\,000\,m^3$	Hawaiian/Strombolian	gentle	$100{-}1000\,m$	daily
2	$> 1\,000\,000\,m^3$	Strombolian/Vulcanian	explosive	$1{-}5\,km$	weekly
3	$> 10\,000\,000\,m^3$	Vulcanian/Peléan	severe	$3{-}15\,km$	yearly
4	$> 0.1\,km^3$	Peléan/Plinian	cataclysmic	$10{-}25\,km$	≥ 10 years
5	$> 1\,km^3$	Plinian	paroxysmal	$> 25\,km$	≥ 50 years
6	$> 10\,km^3$	Plinian/Ultra-Plinian	colossal	$> 25\,km$	≥ 100 years
7	$> 100\,km^3$	Plinian/Ultra-Plinian	super-colossal	$> 25\,km$	≥ 1000 years
8	$> 1000\,km^3$	Ultra-Plinian	mega-colossal	$> 25\,km$	$\geq 10\,000$ years

🔺 **Figure 2** Volcanic Explosivity Index (VEI) chart

Pyroclasts

When an eruption occurs, you would expect that all the area around the volcano would be affected equally. In reality, the pattern is nearly always very uneven so that you could be 3 km in one direction from the volcano and be unaffected or 10 km in the other and have substantial damage. These differences are caused by variables including: energy of the volcanic blast; clast size of the pyroclasts; velocity and direction of winds; gradient of the sides of the volcano and magma viscosity. **Isopachyte** maps are compiled to show the thickness and extent of deposits.

Distribution by grain size

Coarse bombs and blocks are dropped close to the vent and therefore found in a circular pattern around the vent. The finer pyroclastic material will be carried further away from the volcano and could cover hundreds or even thousands of kilometres. The ash gets finer and thinner with distance away from the vent.

Distribution by wind

Where the wind blows in a prevailing direction during an eruption, the ash will be deposited mainly on the leeward side of the volcano. During an eruption, rising columns of ash allow its distribution high into the stratosphere. Jet streams in the upper troposphere can transport the ash around the world. The higher the wind velocity, the further the ash particles can be carried.

Volcanic Explosivity Index (VEI) a measure of the explosiveness of volcanic eruptions, allowing them to be compared. It is a 0 to 8 index of increasing explosivity.

Hawaiian eruptions have large amounts of very fluid basaltic magma from which gases escape, but few pyroclasts.

Strombolian eruptions are more explosive with less fluid basalt and andesite lava. They have regular explosions of gas and pyroclastic material.

Vulcanian eruptions violent with viscous andesitic lava and large quantities of pyroclastic material from large explosions.

Plinian eruptions extremely explosive with viscous gas-filled andesitic and rhyolitic lava and tremendous volumes of pyroclastic material blasted out.

Isopachyte a line joining points of equal thickness of a deposit such as ash. The maps may be called isopach maps.

1 Research where an effusive and an explosive volcano are active today and describe the different hazards.

2 Describe how the geography of an area can affect the distribution of lahars and pyroclastic flows.

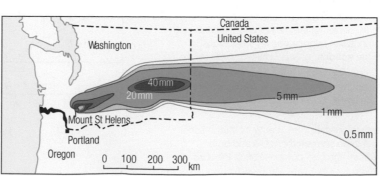

◀ **Figure 3** Isopachyte maps showing thickness of ash deposits around Mt Pinatubo 1991 and Mt St Helens 1980, where wind has affected the pattern of ash deposits. Wind was blowing from west to east.

Reaction rim one mineral surrounding another, as a result of a reaction between the inner mineral with the melt, to form the outer mineral.

Liquidus a phase boundary showing the temperature where the last solid particle melts.

Solidus a phase boundary showing the temperature when a rock first begins to melt when heated.

KEY FACTS

Hornblende belongs to a group of minerals called amphibole.

Augite and **diopside** belong to a group of minerals called pyroxene.

Forsterite and **fayalite** belong to a group of minerals called olivine.

HOW SCIENCE WORKS: BOWEN

Norman L Bowen carried out a huge number of experiments melting powdered rocks of known composition and looking at the temperatures of crystallisation. In 1928 he published his work, which became known as Bowen's Reaction Series.

Mafic rocks form at high temperature as they contain olivine, pyroxene and Ca-rich plagioclase.

Silicic rocks form at low temperature and contain quartz, muscovite, potassium feldspar and some Na-rich plagioclase and biotite.

The reaction series matches with the stability of minerals when exposed to weathering at the Earth's surface:

- Low temperature minerals are the most stable, because they formed in conditions closest to the low temperature surface environment. As a result, the most common mineral in sedimentary rocks is quartz.
- High temperature minerals are the most unstable and quickest to weather at the surface.

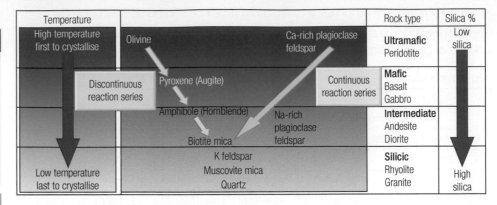

Figure 1 Bowen's Reaction Series

Discontinuous series

The discontinuous series shows the crystallisation of minerals that are rich in iron and magnesium with silica (mafic minerals). In a mafic magma, olivine will be the first mafic mineral to form at a high temperature (>1500°C). As the temperature lowers, *pyroxene*, then *amphibole* and finally biotite will form. If cooling takes place slowly then the early formed, high temperature minerals react with the magma to form the next mineral down the series. For example, olivine reacts with the magma to form pyroxene if there is enough silica present. If the magma is cooled quickly when it is erupted from a volcano, then the reaction will not have time to occur and olivine is preserved. Often the reaction of one mineral with the next will be incomplete and a **reaction rim** will be seen around the edge of a crystal.

Eutectic systems

Crystallisation occurs simultaneously in both the discontinuous and continuous series; however, the minerals that crystallise in each are immiscible and form two solid phases. This is known as a Eutectic system and can be explained by a binary eutectic phase diagram. Ca-rich plagioclase anorthite (**An**) and the pyroxene diopside (**Di**) form such a system.

Figure 2 Binary Eutectic phase diagram for Diopside (DI) and Anorthite (An)

At surface pressure, pure An solidifies at 1553°C and pure Di at 1391°C.

1 A melt containing 70% An (30% Di) cools and the melt reaches the An **liquidus** line. Pure anorthite crystals form, depleting the melt of anorthite and making it more diopside rich so the composition migrates down the liquidus towards the eutectic point. Just above the **solidus**, only anorthite crystals exist.

2 A melt containing 30% An (70% Di) cools and the melt reaches the Di liquidus line. Pure diopside crystals form, depleting the melt of diopside and making it more anorthite rich so the composition migrates down the liquidus towards the eutectic point. Just above the solidus, only diopside crystals exist.

At the eutectic, both diopside and anorthite crystals form in eutectic proportions. The eutectic is the only point at which all three phases, the liquid and both solids, can exist. If all of the melt crystallises, the bulk composition of the solid will match that of the original melt.

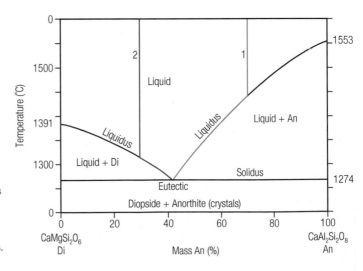

Continuous series

The continuous series explains the crystallisation of plagioclase feldspar (Albite-Anorthite) and is the example we are using here. Anorthite is a calcium plagioclase which forms at high temperatures, and albite is a sodium plagioclase which forms at low temperatures. Intermediate compositions form as the temperature drops. Plagioclase is continuously reacting with the melt to form a more Na-rich crystal as the temperature decreases. Sometimes individual crystals of plagioclase will show zoning. The centre of the crystal is Ca-rich plagioclase and towards the edges the crystal is richer in Na-rich plagioclase and poorer in Ca-rich plagioclase.

Knowing that Ca-rich plagioclase is only found in mafic and ultramafic rocks, while Na-rich plagioclase is found in silicic and intermediate rocks, means that they are used in classification of igneous rocks.

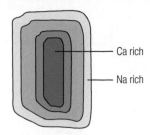

△ **Figure 3** Zoned crystal of plagioclase feldspar. The crystal started growing when the temperature was high, and Ca-rich plagioclase formed. As the magma cooled the crystal continued to grow with new layers forming, gradually becoming richer in Na. Note: Zoning is not always obvious. Rapid cooling limits reactions so crystals may grade with no observable zone. Temperatures in magma chambers can also increase as well as decrease over time.

Solid solution series

The compositional range between the plagioclase end-members albite (**Ab**) and anorthite (**An**) is known as a solid solution series.

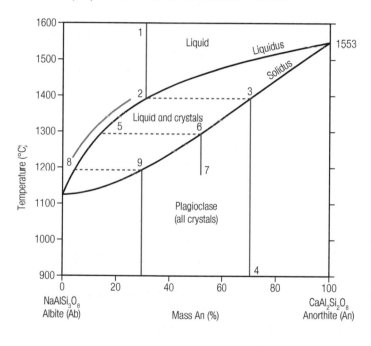

Figure 4 Solid solution phase diagram for Albite (Ab) and Anorthite (An)

1 A melt of composition 30% An (70% Ab) cools.

2 The melt reaches the liquidus line (separating the liquid phase from the liquid+crystal phase) and crystals begin to form.

3 The first formed crystals will be richer in Ca than the original melt composition. This can be seen by reading horizontally across the phase diagram (dashed line) to the solidus line (the line that separates the liquid+crystal phase from the solid phase).

4 These first formed crystals will have a composition of around 70% An (30% Ab).

5 As the crystals remove Ca from the melt, it will become richer in Na and its composition will migrate down the liquidus.

6 The point on the solidus horizontally opposite the liquidus always indicates the crystal composition and the most stable form of plagioclase at that temperature will crystallise (e.g. point 3 opposite 2, 6 opposite 5, etc., on the diagram).

7 The plagioclase crystals continuously react with the melt, allowing free substitution of Ca for Na within the crystal lattice, to form a more Na-rich crystal as the temperature decreases. Thus at any given time, the composition is in equilibrium with the temperature.

8 The composition of the melt and the crystals move down in tandem, directly opposite one another as the melt continues to cool.

9 If the system remains in equilibrium, then the last drop of melt must form a crystal with the same composition as the original melt 30% Ab (70% An).

Olivine

A solid solution series also exists, in the discontinuous series, between the high temperature, Mg-rich olivine (Forsterite) and the low temperature Fe-rich olivine (Fayalite), where magnesium is substituted for iron as the temperature decreases.

Low temperature minerals

When the two reaction series converge at a low temperature, below that shown in either of the phase diagrams above, minerals remain that will not react with the remaining liquid. This final group of minerals to crystallise are the felsic minerals rich in silica. These are potassium feldspar followed by muscovite mica and finally quartz at a temperature of about 700°C.

QUESTIONS

1 In a melt of composition 80% An (20% Di), what will be the composition of the first formed crystals and at what temperature?

2 Research the phase diagram for Forsterite-Fayalite and compare it with the Albite-Anorthite system.

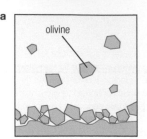

1200°C

Magma composition A

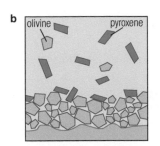

Cooling

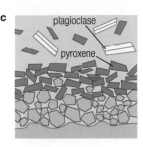

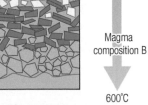

Magma composition B

600°C

Figure 1 Fractional crystallisation and gravity settling. While cooling, different minerals crystallise from the melt as the magma evolves: **a** olivine crystallises; **b** olivine and pyroxene crystallise; **c** pyroxene and plagioclase crystallise; **d** plagioclase crystallises. At the bottom of the magma reservoir, a cumulate forms.

Individual plutonic intrusions and larger igneous provinces (LIP) often display a range of different rock types. This could be a result of:

- Different compositions of the original magma
- Magma mixing
- Contamination of the magma by country rock assimilation
- Evolution of a parent magma to produce one or more daughter magmas through magmatic differentiation.

Magma mixing

Magmas of two different compositions tend to be immiscible, so if they are combined in a magma chamber, various factors such as the density and thermal contrasts act to keep them separated. It generally requires a vigorous stirring mechanism, such as convection within the chamber, to produce a magma that has an intermediate composition between that of the two parent magmas.

Contamination

Stoping is the process where blocks of country rock, from conduits or the walls and roof of the magma chamber, are broken off by rising magma and incorporated into the magma as xenoliths. If these xenoliths melt and become assimilated into the magma, they can contaminate it and change its bulk composition.

Magmatic differentiation

Magmatic differentiation consists of a number of processes that cause a parent magma to evolve into magmas of different compositions. This leads to different igneous rocks being produced from a single parent magma. A mafic magma may crystallise to form ultramafic, basic, intermediate and even silicic rocks.

Fractional crystallisation

As olivine and pyroxene form at high temperatures, they use iron and magnesium from the magma in their crystal lattices. The high temperature plagioclase crystals (anorthite) are rich in calcium. The result is that the magma becomes depleted in iron, magnesium and calcium. The remaining liquid becomes enriched in silica, potassium, sodium and water because the early formed minerals are poor in these elements. Over time, the composition of the magma changes from the original.

Gravity settling

Magma of any given composition is typically 10% less dense than the equivalent composition in solid rock, so in general, crystals will be denser than the liquid and settle out. Early formed minerals such as olivine and pyroxene, have a greater percentage of iron and are significantly denser, so sink to form a layer at the base of the intrusion or magma chamber. This layer of dense, early formed minerals is called a *cumulate layer* and the process is called gravity settling.

Crystals that remain suspended in magma will react with that magma over time, following the continuous or discontinuous paths of Bowen's Reaction Series, as the magma evolves. However, gravity settling removes crystals from the remaining liquid so they can no longer react with the remaining magma changing the composition of the magma.

Filter pressing

During the crystallisation of magma, there is a point where crystals and liquid exist together as a slushy mass. Due to the weight of the overlying crystals, the liquid gets squeezed out, forming a separate layer above. This liquid is depleted in the elements which are incorporated into the early formed crystals and enriched in the elements which form felsic minerals.

Chemical analysis of igneous rock

Mass % oxides	Silicic	Intermediate	Mafic	Ultramafic
SiO_2	70.8	62.5	49.0	41.7
Al_2O_3	14.6	17.6	18.2	0.9
Fe_2O_3	1.6	2.1	3.2	2.9
FeO	1.8	2.7	6.0	5.7
MgO	0.9	0.9	7.6	47.7
CaO	2.0	2.3	11.2	0.7
Na_2O	3.5	5.9	2.6	0.1
K_2O	4.2	5.2	0.9	0.0

Given sufficient time for magmatic differentiation to take place, a single parent magma could produce ultramafic, mafic, intermediate and silicic igneous rocks.

In the classification of igneous rocks, chemical analyses are normally reported as weight percentages of oxides, rather than as separate elements.

The table shows a sample of four oxides, one from each of the four groups of igneous rocks.

Layered intrusions and platinum group elements

Economically mineable deposits of **platinum group elements (PGE)** in the Earth's crust are rare. The largest and most significant geological deposits of PGE, commonly rich in platinum and palladium, are associated with sulfide minerals in layered mafic or ultramafic igneous intrusions (cumulate deposit).

Magmatic sulfide deposits form during the fractional crystallisation process when mafic magma becomes saturated in sulfide (S^{2-}). At the point where the magma can no longer hold sulfur in solution, the sulfur exsolves producing an immiscible sulfide liquid. Droplets from this immiscible sulfide liquid are denser than the magma and so sink through it. Platinum group elements, along with gold (Au) and nickel (Ni) are **siderophile** (iron-loving). They preferentially form metallic bonds with iron and will concentrate in molten iron. The only place a liquid like this exists in significant proportions today is the Earth's outer core. If siderophile elements do not have iron to react with, then they behave like the **chalcophile** (ore-loving) elements silver (Ag) and copper (Cu) and will concentrate in molten sulfide, bonding more readily with sulfur than with oxygen. They are therefore gathered up by the sinking sulfide droplets and removed from the magma, to form layers that range from a few millimetres to a few metres thick.

Geologists have a number of theories on how mafic magma can reach sulfide saturation. Assimilation of country rock that is rich in sulfur has been proposed as the most reasonable mechanism for the largest deposits. Another possibility is silica contamination from assimilation of silica-rich country rock, which is thought to decrease the solubility of sulfur in a mafic magma.

The Skaergaard intrusion in Greenland is one of the most studied in the world and contains magmatic sulfide layers rich in PGE, gold and selenium. It was intruded during the Tertiary, when the North Atlantic was opening.

KEY DEFINITIONS

Platinum group elements (PGE) comprise: platinum (Pt), palladium (Pd), iridium (Ir), osmium (Os), rhodium (Rd) and ruthenium (Ru).

Chalcophile elements combine more readily with sulfur than with oxygen.

Siderophile elements dissolve in iron, either as solid solutions or in the liquid state. They form bonds with carbon or sulfur but not with oxygen.

MATHS BOX: WORKED EXAMPLE

Using density to distinguish between the groups of igneous rocks.

A simple practical activity using a displacement can may be used to measure the specific gravity of rocks:

- Measure the mass of a dry rock for each of the four igneous groups. (Avoid vesicular or amygdaloidal textures.)
- Fill a displacement can so that water is just dripping out of the overflow.
- Carefully lower a rock into the can and catch the overflowing water in a measuring cylinder.
- Repeat for each of the rocks and calculate the densities.
- Density $\rho = \dfrac{\text{mass of dry rock g}}{\text{water displaced cm}^3} \times 100$
- The answers should show a pattern of increasing density from silicic to ultramafic, where the latter contain more Fe and Mg oxides.

QUESTIONS

1 Name the rock with a medium crystal grain size and a composition of:

SiO_2	Al_2O_3	Fe_2O_3	FeO	MgO	CaO	Na_2O	K_2O
49.2%	15.74%	3.79%	7.13%	6.73%	9.47%	2.91%	1.1%

2 Explain why mafic rocks contain more calcium than sodium.

3 Explain why you will rarely get olivine and quartz in the same rock.

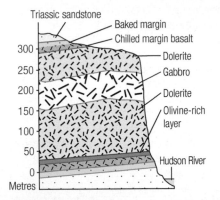

▲ **Figure 1** A cross-section of the Palisades Sill

What happens when molten rock becomes solid? As a magma crystallises, the crystals that form will have a different composition from the melt. When the crystals separate from the rest of the magma as they grow, the chemistry of the magma changes. A single magma could form ultramafic, mafic, intermediate and silicic rocks if there is enough time for differentiation to take place. If the parent magma is mafic then in theory it will produce:

- Peridotite at the base of the intrusion, rich in dense olivine
- Gabbro forming the main part of the intrusion
- Diorite may exist towards the top of the intrusion
- Granite may form veins or a thin layer at the top of the intrusion.

In most cases intrusions are complex, with multiple injections of magma that may vary in composition. It is unusual to have a single volume of magma that crystallises in one period of cooling.

Major layered intrusions

These are large mafic or ultramafic igneous intrusions that cooled slowly below the surface so there was time for them to form distinct layers. Most of the world's platinum and chromium comes from layered intrusions, giving an added incentive to understanding how they formed. Gravity settling is the key process in the formation of layers of magnetite, chromite and platinum.

The Palisades Sill

The Palisades Sill in New Jersey, USA, crops out along the Hudson River and is over 300 m thick. It was intruded into Triassic sediments at a depth of about 3 km, so that it cooled slowly and has now been exposed by erosion. The sill can be divided into three sections, each of which shows a different igneous process at work:

1. The only parts of the sill to cool rapidly are the upper and lower edges, which were in contact with the cold country rock. These chilled margins have fine crystals and most importantly have the same composition as the original magma because they cooled before any differentiation took place. The rock is a basalt.

2. As the main part of the intrusion began to crystallise, the early formed olivine crystals began to sink by gravity settling. Olivine has a density of $3.8\,\mathrm{g\,cm^{-3}}$ compared to $3.0\,\mathrm{g\,cm^{-3}}$ for a mafic magma. The olivine crystals form a layer 10 m thick at the base of the intrusion, just above the lower chilled margin.

3. Crystallisation was taking place from both the top and bottom of the sill as the crystals grew in the cooler areas. The main rock forming the intrusion has medium-sized crystals and is a dolerite. The last part of the magma to crystallise is about 200 m above the base of the intrusion and is a gabbro with coarse crystals, indicating it cooled slowly. As a result of fractionation, the composition is lower in mafic minerals than the original composition because the magma is depleted in iron and magnesium by the time this rock formed. It is richer in plagioclase as the last part of the magma to crystallise is enriched in silica.

Part of Palisade Sill	Composition %			
	olivine	pyroxene	plagioclase feldspar	silica
1 Chilled margin (original composition of magma)	3	50	47	51.8
2 Olivine-rich layer	25	38	34	48.2
3 Last formed part of the sill	0	30	65	52.9

The table shows the chemistry of the sill in each of these sections. The percentages do not add up to 100% because there are small amounts of other minerals present.

The Skaergaard intrusion

The Skaergaard intrusion in Greenland is one of the most studied in the world. It was intruded during the Tertiary, when the North Atlantic was opening, as a magma chamber for basalt volcanoes. Magma was probably intruded in a single injection into a huge conical intrusion.

The Skaergaard is divided into three major units:

1. The Marginal Border Series is a chilled margin with a fine crystal size, indicating that it cooled rapidly. It is no longer the same composition as the original magma, as it has been contaminated by country rock. Crystals grew inwards from the sides of the intrusion.

2. The Layered Series shows rhythmic layering, most easily explained by crystal settling interrupted by periodic large-scale convection. A sequence of denser (olivine or pyroxene) crystals beneath lighter plagioclases is deposited by gravity settling. Then filter pressing causes the expulsion of the differentiated liquid. Then convection mixes the magma, and the process is repeated. Each cycle means that the magma is more evolved due to the removal of the crystals, resulting in **cumulates** forming from the floor.

3. The Upper Border Series is thinner, but mirrors the 2500 m Layered Series with layers that crystallised from the top down.

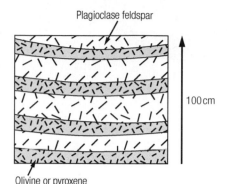

▲ **Figure 3** Cumulate layers of dark olivine or pyroxene crystals with lighter plagioclase feldspar

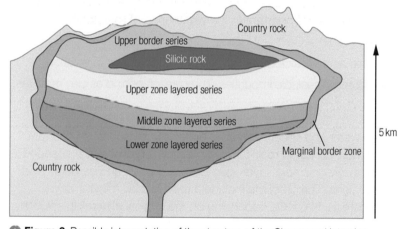

▲ **Figure 2** Possible interpretation of the structure of the Skaergaard intrusion

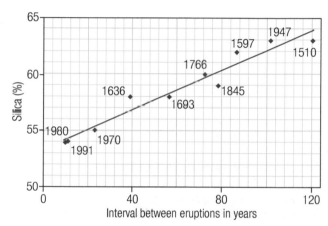

▲ **Figure 4** Time intervals between Hekla eruptions and the silica content

Hekla volcano

Hekla is a large volcano in Iceland near the central rift zone. The magma should be entirely mafic as Iceland is on a divergent plate margin, yet it erupts intermediate lava and has erupted rhyolitic lava in the past. The graph shows the longer the time between eruptions the more acidic the magma that is erupted. If there is a long interval between eruptions, fractionation in the magma chamber causes more mafic minerals, which crystallise at high temperatures, to be found at the bottom and felsic minerals at the top. Each eruption takes magma from the top of the magma chamber and then as the eruption progresses, from lower in the magma chamber.

The 1947 eruption occurred after an interval of 102 years and initially erupted lava with a silica content of 63% but ended like all the eruptions with intermediate andesite (with a silica content of 54%). Medieval, pre-1200CE, lava flows show rhyolitic compositions.

▲ **Figure 5** Layering in the Skaergaard intrusion. Dark layers are mafic and paler layers felsic in composition.

The Bushveld Igneous Complex

The Bushveld Igneous Complex in South Africa is a large layered igneous intrusion, 400 km × 800 km with a volume of at least 1 000 000 km³. It varies in thickness but reaches 9 km in places. Formed around 2 billion years ago, the complex contains some of the richest ore deposits on Earth and the largest reserves of platinum group elements.

QUESTIONS

1. Describe the processes that allow cumulate layers to form.

2. Why are we sure that convection operated in large intrusions such as the Skaergaard to produce layered deposits?

Most rocks are formed below the Earth's surface. When rocks are uplifted near to the surface, and when they are exposed at the surface, temperature, pressure and chemical conditions are different from those where they formed. When this happens, rocks change. **Weathering** is the in situ chemical alteration and mechanical breakdown of rocks by exposure to the atmosphere, water and organic matter.

Chemical weathering

Rocks decompose when the chemical structure of their minerals breaks down. Chemical weathering reactions, all of which involve water, produce ions that are removed in solution leaving an insoluble residue, usually clay minerals. Carbonation and hydrolysis are important reactions because they affect two common rock-forming minerals, calcite and feldspar.

Carbonation

Carbon dioxide gas in the atmosphere reacts with rainwater and pore water in the soil to form carbonic acid. The pore spaces in the soil are rich in carbon dioxide, due to decomposing plant litter. This makes groundwater more acidic than rainwater. **Carbonation** is most important in the weathering of limestone.

$$CaCO_3 + H_2CO_3 \rightarrow Ca^{2+} + 2\ HCO_3^{-}$$

calcite + carbonic acid = calcium + hydrogen carbonate ions in solution

Limestone generally contains insoluble impurities. They are left behind as clay residues.

Hydrolysis

The reaction between water and silicate minerals, especially feldspars, is important because they are the most common rock-forming minerals. **Hydrolysis** is speeded up if the water contains carbonic acid. Hydrogen ions from water and carbonic acid react with the minerals' ions. The products include a residual clay mineral, silica, carbonate or bicarbonate (K, Na or Ca depending on the original mineral) in solution.

Mechanical weathering

Exfoliation

Exfoliation is sometimes known as 'onion-skin weathering', because curved sheets peel off from rocks affected by it. In hot deserts, there is a marked difference between hot daytime and cold night-time temperatures. Different minerals expand and contract by different amounts during heating and cooling, causing the rock to disintegrate.

Frost shattering

Water enters cracks, joints and bedding planes. In climates where daily temperatures fluctuate above and below 0°C, water freezes and expands by 9%. This exerts pressure on rocks, leading to eventual failure, known as **frost shattering**. It produces a residue of angular fragments called scree.

Pressure release

Atmospheric pressure at the Earth's surface is much lower than pressures within the Earth, even at quite shallow depths. When this **pressure** is **released** from rocks due to the erosion of the rocks above them, they expand producing fractures which are more widely spaced the further they are from the surface. Rocks that are well jointed or have many bedding planes are less likely to be affected by this form of weathering.

KEY FACT

Clast a fragment of broken rock produced by mechanical weathering and erosion.

CASE STUDY: STABILITY OF QUARTZ

Ever wondered why most beaches are sand? This is because quartz is not easily broken down and is often the end product of the weathering process, in particular the weathering of granite. This is also a lower temperature mineral from Bowen's Reaction Series.

KEY DEFINITIONS

Weathering the in situ chemical alteration and mechanical and biological breakdown of rocks by exposure to the atmosphere, water and organic matter.

Carbonation the reaction between carbonic acid and minerals.

Hydrolysis ithe reaction between minerals and water, causing the minerals to decompose.

Exfoliation occurs when sheets of rock split off due to differential expansion and contraction of minerals, during diurnal heating and cooling.

Frost shattering caused by the expansion of freezing water in fractures, which forces rocks apart.

Pressure release caused by the expansion and fracturing of rock due to removal of overlying rock.

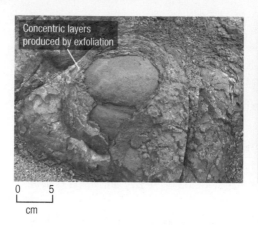

Concentric layers produced by exfoliation

Angular scree fragments

Oxidation of iron minerals

Cracks widened by frost shattering

0 5
cm

0 5
cm

▶ **Figure 1** Weathering processes: exfoliation and frost shattering

Biological weathering

Root action

Tree roots can grow along bedding planes and joints and force them apart mechanically, known as **root action**. They keep surfaces open so that water can penetrate and so make chemical weathering easier. When trees sway in the wind, their roots can prise open fractures in rocks.

◀ **Figure 2** Root action

Tree roots

Fractures in rock forced apart

Burrowing

Burrowing animals include worms, reptiles and mammals. Their activity brings rock material from shallow depths up to the surface, where it can be weathered more easily. The burrows allow atmospheric gases and water to penetrate, making chemical weathering more likely. Fragments that may have been produced by other forms of weathering are usually reduced in size by burrowing organisms.

Climate and weathering		
Climatic zone		Weathering
Arctic	mean annual temp ~2°C mean annual rainfall ~10 cm	Mainly mechanical (frost shattering)
Temperate	mean annual temp ~10°C mean annual rainfall ~50 cm	Mechanical, chemical and biological (frost shattering, carbonation, hydrolysis, root action)
Warm arid	mean annual temp ~20°C mean annual rainfall ~10 cm	Mainly mechanical (exfoliation)
Humid tropical	mean annual temp ~20°C mean annual rainfall ~200 cm	Intense chemical, some biological with greatest amount of residue (carbonation, hydrolysis, root action)

QUESTIONS

1 Describe how carbonation produces soluble products.

2 Explain why the weathering sequence of silicate minerals is similar to Bowen's Reaction Series.

3 Explain how climate affects weathering processes.

What happens to weathered material?

The answer is that some will remain where it is and the rest will be transported away.

Transport can be due to:

- Gravity – moves loose weathered sediment down slopes.
- Wind – lifts and transports finer grained sediment.
- Rivers – lift and transport sediment of all sizes, including heavy and large grains.
- The sea – moves sediment of all sizes.
- Ice – depending on climatic conditions, can carry large amounts of rock debris.

The grains of sediment are involved in the process of **erosion**. They come into contact with rocks over which they are travelling, perhaps in a desert, in the bed of a stream or along the coast. These rocks are subjected to **abrasion**. They may be sandblasted by wind-blown sediments in deserts, ground down by boulders rolling along a riverbed or chipped away by shingle carried by the sea when waves crash into cliffs.

Transported grains are not only in contact with the Earth's surface, but are also in contact with each other. **Attrition** is the wearing down of the transported sediment by rolling, rubbing and crushing together of sedimentary grains during transport. The more contact there is between grains and the longer they have been in transport, the smaller and more rounded they become. Harder minerals, like quartz, survive better than less hard minerals, like mica, which may be completely crushed. Sediments that contain little or no variety of minerals, for example those that contain quartz grains only, are described as **mineralogically mature**.

Methods of transport

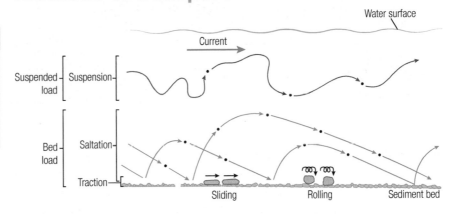

▲ **Figure 1** Methods of sediment transport in water

Figure 1 shows all the methods of sediment transport in water. Most methods also apply to wind transport, though wind does not transport material in **solution**. Ice carries sediment embedded within it, such as in a glacier. Larger grain sizes can be transported when the velocity of the current is greater. Bed loads consist of larger grains that are moved by **traction** and smaller grains that bounce along the bed by **saltation**. Finer grains do not touch the bed and so are carried in **suspension**. Dissolved material transported in solution is invisible and so cannot be shown. Grains become smaller the longer they have been transported.

Clay-sized particles require high velocities for erosion because they are flat and platy in shape and are cohesive so tend to stick together. If you have ever tried washing sticky mud off boots, it's a lot more difficult than washing off sand. The clay particles remain in suspension even at very low current velocities because they are small, have low mass and are buoyant.

KEY DEFINITIONS

Erosion the wearing away of the land surface and removal of sediment by means of transport.

Abrasion the wearing away of the Earth's surface by the action of wind, water or ice dragging sediment over or hurling it at a surface.

Attrition the wearing down of sedimentary grains due to collisions with other grains during transport.

Mineralogical maturity a measure of the extent to which minerals have been destroyed by weathering and attrition.

Solution the transport of ions dissolved in water, particularly K, Ca and Na.

Traction the transport of material by rolling and sliding along a surface.

Saltation the transport of material by bouncing.

Suspension the transport of material in water or air, without it touching the Earth's surface.

STUDY TIP

Sometimes candidates get weathering and erosion confused. They both result in the breakdown of rocks, but in weathering there is no movement of material, it all happens in place (in situ). However, in the case of erosion, there is always a method of transport involved.

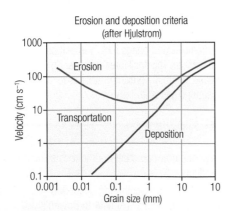

▲ **Figure 2** Relationship between current velocity of water and sediment transport. This is known as a Hjulstrom curve.

Grain shape and roundness

Grains get rounder the longer they have been transported, and the degree of **roundness** can be described using Figure 3.

The shape of grains depends on the type of rock or mineral from which they are made, rather than on transport. There are many ways to describe the **shape**. One way is to compare the grain to a sphere. Figure 3 shows grains that have high and low degrees of sphericity.

Very angular	Angular	Sub-angular	
			High sphericity
			Low sphericity
Sub-rounded	Rounded	Well rounded	
			High sphericity
			Low sphericity

▶ **Figure 3** Roundness descriptions for grains of high and low sphericity

▲ **Figure 4** Grain shapes

Grain size

Measurement and classification of the grain size of sediments is made using the Wentworth–Udden scale. As the range of grain sizes found in nature is large, a logarithmic scale is more practical than a linear scale.

The grain size is obtained from the average diameter of the grains of sediment being studied. The diameter in mm is not always easy to remember, so the **phi (Φ) scale** is also used. This scale also has advantages when making a statistical analysis of sediment grain sizes. Sediments composed of well-rounded grains that are all the same size are described as **texturally mature**.

Diameter (mm)	Phi Φ	Sediment name
>2	−2 to −8	Gravel, pebbles, cobbles, boulders
2	−1	Sand – very coarse
1	0	Sand – coarse
0.5	1	Sand – medium
0.25	2	Sand – fine
0.125	3	Sand – very fine
0.0625	4	Silt
0.0039	8	Clay

🔺 The Wentworth–Udden Scale

Increasing mineralogical maturity ⟶

Wide range of minerals from the source rock: mineralogically immature

Very angular and poorly sorted grains: texturally immature

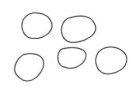

Smaller range of minerals: mineralogically submature

Less angular and more uniform grain size: texturally submature

One chemically stable mineral only: mineralogically mature

Rounded grains of fairly uniform size: texturally mature

Increasing textural maturity ⟶

🔺 **Figure 5** Maturity of sediments

QUESTIONS

1 Describe one difference and one similarity between the processes of weathering and erosion.

2 Explain why quartz grains that have been transported in water for a long time are more rounded than those transported for a short time.

3 Use the graph of current velocity and transport to answer:

   (a) Which sediment grain size requires the least energy for erosion to take place?

   (b) What is the sediment size being deposited, if the velocity is $100\,\text{cm}\,\text{s}^{-1}$?

4 Describe a situation when very high velocities occur. Research the type of material carried by a high-velocity river.

5 Describe the maturity of the sediments deposited when a high-velocity river dries up rapidly.

The analysis of sediments can tell us a great deal.

- The sizes, size range and roundness of sediments are all related to how the sediments were transported, because wind, water and ice vary in terms of energy and viscosity.
- Fine grained, well-rounded sediments have been transported for a longer time than coarse grained angular fragments.
- The shape of grains depends on the type of rock or mineral from which they are made.
- Sedimentary deposits in which all the grains are of similar size have been transported in higher energy conditions than those where there is a wide range of sizes.
- How sediments were transported is related to environmental factors, such as climate.

One method of interpreting sediments is by analysing the grain roundness, the more rounded the sediment, the longer the time of transport. Angular grains suggest rapid deposition with little transport. A second method involves analysing the grain size distribution of a sediment and investigating the degree of sorting of that sediment.

Sorting

A sedimentary deposit of grains that are all the same size would be described as very well sorted. One with a wide range of sizes would be termed very poorly sorted. The range of terms used to describe **sorting** is shown on the photos of sediments.

Sorting provides information about how sediments were transported. Wind does not transport very coarse grains. It deposits well-sorted sands as the energy level and velocity decrease. Finer grains are blown further, forming deposits elsewhere.

Rivers flowing with high energy may transport coarse as well as medium and fine grained sediment. As energy levels fall, larger heavier grains are deposited first, medium-sized grains are transported further and finer grains are deposited last. Poorly sorted sediment has usually not been transported far because the transporting medium lost energy quickly. A desert stream that rapidly dries up due to infiltration or high rates of evaporation will deposit the sediment very quickly. A meltwater stream from a glacier will lose energy as melting stops after the summer thaw has ended and sediment will be deposited over a short period of time.

Grain size analysis of sediments

Sand and gravel-sized sediments are analysed using a sieve bank, a stack of sieves arranged so that the one with the biggest holes (4 mm or 2 φ) is at the top and the one with the smallest (0.0625 mm or 4 φ) is at the bottom. The sieve stack is rotated and shaken. The mass of the sediment trapped on each sieve is measured using a digital balance and recorded. If you use 100 g of sediment it makes calculations very simple.

Results are plotted as histograms. Different methods of transport produce different grain size distributions, shown in the graphs.

The energy of the transporting medium is the key factor that affects the degree of sorting.

Grain size data is plotted as cumulative frequency curves. They show the total percentage of sediment that fails to get through a given sieve size. This kind of graph allows a direct comparison between different samples of sediment, because they can all be plotted on the same graph.

Well-sorted sediments have almost vertical curves. Poorly sorted sediments have curves that stretch across a wide range of sizes. The degree of sorting can be quantified. The coefficient of sorting (ρ) can be measured from cumulative frequency graphs using the following equation:

$$\text{coefficient of sorting } (\rho) = \frac{\phi_{84} - \phi_{16}}{2}$$

where ϕ_{84} is the ϕ value of the cumulative mass of 84% of the sample and ϕ_{16} is the ϕ value of the cumulative mass of 16% of the sample.

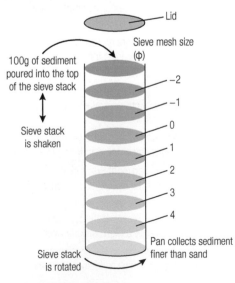

Figure 1 Sediment sorted into well sorted, moderately sorted and poorly sorted. Coin for scale.

Lid

Sieve mesh size (φ)

100 g of sediment poured into the top of the sieve stack

-2
-1
0
1
2
3
4

Sieve stack is shaken

Pan collects sediment finer than sand

Sieve stack is rotated

Figure 2 Grain sizes sorted using sieves

MATHS BOX: CALCULATING THE COEFFICIENT OF SORTING

The dashed red lines on the cumulative frequency graph have been drawn:
- at 84% and the value of phi is 3 for the glacial sediment
- at 16% and the value of phi is −1.7 for the glacial sediment
- using the equation above

$$\frac{3 - (-1.7)}{2} = \frac{4.7}{2} = 2.35$$

The table shows how the coefficient allows sorting to be expressed numerically.

	Value	Degree of sorting
Coefficient of sorting	<0.50	Well sorted
	0.5–1.00	Moderately sorted
	>1.00	Poorly sorted

- The value of 2.35 tells us that the till is very poorly sorted.

Characteristics of sediments related to transport

This table summarises the characteristics of sediments transported by wind, ice, rivers and the sea. Gravity has also been included, but this material is too varied and often too coarse to be analysed using sieving.

Transport	Grain size	Composition	Roundness	Sorting
Wind blown dune sand in a high energy environment	Fine to medium sand	All quartz, sometimes red due to iron oxide	Well rounded	Very well sorted so that most of the sediment is in one sieve, as fine sand is easiest for the wind to carry
Ice: deposited as glacial till in a low energy environment	Varies from very coarse (boulders) to very fine (clay)	Varied: any rock fragments and clay	Angular to sub-angular	Very poorly sorted, so material in every sieve
River deposit of sand in channel Usually high energy with fast currents	Coarse to fine sand, may be coarser nearer to source	Quartz and mica with rock fragments	Angular near to source but sub-angular to sub-rounded downstream	Poorly sorted near to source of river Moderately sorted downstream, gives wide range of sediment size in sieves
Beach or offshore bar in the sea High energy	Medium sand Sometimes coarse (pebbles and gravel) close to shore and on beaches	Nearly all quartz with some shell or rock fragments	Sub-rounded to rounded	Moderately sorted with sediment mainly in a few sieves
Gravity Low energy	Varies from very coarse (boulders) in rock falls to very fine in soil creep	Varied: any rock fragments	Angular to very angular	Very poorly sorted Larger fragments sometimes found at the base of a slope

QUESTIONS

1 Draw diagrams to show well-sorted, moderately sorted and poorly sorted sediments.

2 Compare and contrast the grain characteristics of sediments transported by ice and by wind.

3 Calculate the coefficient of sorting for the dune, beach and river sediments on the cumulative frequency graph.

4 Explain the advantages and disadvantages of qualitative and quantitative methods of grain size analysis.

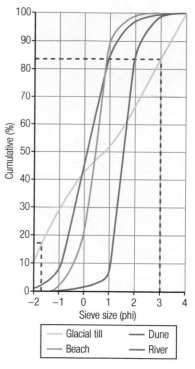

Figure 3 Cumulative frequency curves of the four sediments

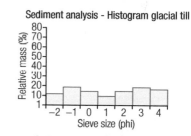

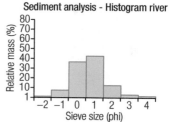

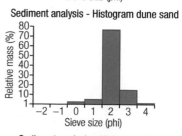

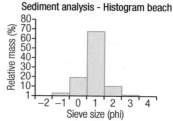

Figure 4 Histograms showing grain size distributions of four sediments

Using observations to identify rocks

You have seen that sedimentary rocks can be classified by observing grain size, grain shape and mineral composition. Now these observations are used to identify some **mechanically formed** sedimentary rocks. The flow diagram shows how the observation of characteristic features can be used to identify **siliciclastic rocks**.

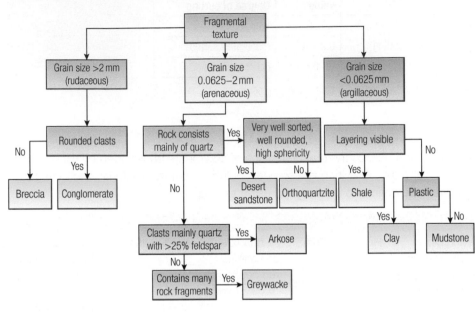

Figure 1 Identifying siliciclastic sedimentary rocks

Describing and identifying coarse grained siliciclastic sedimentary rocks

Breccia

Breccia is a coarse grained (average grain size >2 mm diameter) siliciclastic rock. The clasts are angular, which means that they were not transported for very long. The coarse clasts are often mixed with finer grains and set in a **matrix**. This means that they are poorly sorted, so we can infer that the transporting medium lost energy quickly. Breccias commonly form as scree, alluvial fans and wadi deposits. Volcanic breccias are pyroclastic rocks from a volcanic vent.

Conglomerate

Conglomerate is a coarse grained siliciclastic rock. The clasts are rounded, which means that they have been transported for a longer period of time than the angular clasts in a breccia. The large clasts may be surrounded by a finer grained matrix and be poorly sorted. In some conglomerates, the clasts are held in place by mineral cement. Conglomerates commonly form as beach and river channel deposits.

Sandstones

Sandstones (average grain size 0.0625 to 2 mm diameter) are a very important group because they are one of the main oil reservoir rocks and aquifers for water supply, as well as being used as building stones. They may be well cemented, in which case they will have low porosity, or be poorly cemented, in which case they will have high porosity. The main component is quartz but other minerals low on Bowen's Reaction Series, such as muscovite mica and K feldspar, are also common. Sandstones are found in virtually all sedimentary environments.

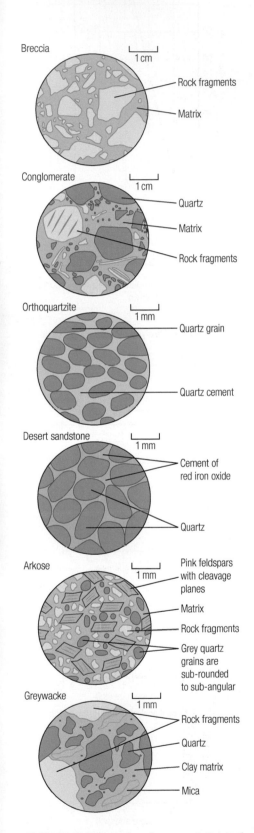

Figure 2 Thin section diagrams of siliciclastic sedimentary rocks

Orthoquartzite

Orthoquartzite consists only of quartz grains held together by quartz cement, so it is white or grey in colour. Orthoquartzite is >90% quartz. The grains are well sorted and well rounded, indicating transport over a long period. This indicates extensive weathering so that less stable minerals are not present. Orthoquartzites commonly form in beach and shallow marine deposits.

Desert sandstone

Desert sandstone is red in colour due to iron oxide coating the quartz grains. It is very well sorted, with very well-rounded, coarse sand-sized (around 1 mm) grains having high sphericity, sometimes referred to as 'millet seed sand'. The grains have a frosted appearance when examined using a hand lens. Unlike orthoquartzite, desert sandstone is not composed entirely of quartz due to the iron oxide around the grains. The clasts are cemented together by silica or iron minerals. These rocks commonly form in arid environments, as wind-blown sands.

Arkose

Arkose is a medium to coarse grained sandstone that contains at least 25% feldspar (K feldspar and plagioclase). The other main mineral is quartz, though rock fragments and mica are also present. Arkose is often pink in colour due to the feldspars and is usually moderately sorted. The grains will be sub-angular to sub-rounded, showing that the sediment has not been transported for very long. It is commonly formed in alluvial fan environments in arid areas.

Greywacke

Greywacke sandstones have fine to coarse grain sizes. They are dark coloured and poorly sorted, with angular to sub-angular clasts that mainly consist of rock fragments (lithic clasts) with some quartz and K feldspar (orthoclase). More than 15% of the rock is clay matrix. They commonly form as turbidite deposits (deep water, deposited by turbidity currents), so often show graded bedding.

Fine grained siliciclastic rocks

All these argillaceous siliciclastic rocks and sediments are deposited in low-energy environments, commonly marine but also on the flood plains of rivers or in lakes. They are the most abundant, accounting for about 75% of all sedimentary rocks and may contain fossils. None of them has significant amounts of cement between the grains.

Clay

The grains are clay size (average grain size up to 0.0039 mm) and mainly consist of clay minerals. Clay may have a variety of colours, including dark brown, red and green, depending on the carbon or iron content. The higher the organic content, the darker the clay. Clay is plastic and can be moulded. It forms layers, with distinct bedding planes.

Mudstone

Mudstones are dark grey, very fine grained siliciclastic rocks. They contain **clay minerals**, mica and quartz, but you will not be able to identify individual minerals even with a hand lens. The minerals do not have any preferred alignment and so mudstones are not layered. Also, mudstone does not have **plasticity**.

Shale

Shale is a dark-coloured, fine grained siliciclastic rock that has distinctive layers, due to the alignment of minerals. Clay minerals are flat and platy and align parallel to the beds at 90 degrees to the pressure from the mass of overlying rocks. The rock easily splits along the layers or laminations and for this reason it is described as **fissile**. It is composed of clay minerals, mica and quartz. Shale is not plastic but hard, brittle and impermeable.

▲ **Figure 3** Breccia. Hammer for scale

▲ **Figure 4** Desert sandstone

KEY DEFINITIONS

Mechanically formed sedimentary rocks result from the processes of erosion, transport and deposition of clasts.

Siliciclastic rocks form from sediments composed of silicate minerals and rock fragments.

Matrix the background material of small grains in which larger grains occur.

Clay minerals a group of sub-microscopic platy aluminium silicates related to mica.

Plasticity the ability of a material to permanently change shape without fracturing.

Fissile refers to the tendency of a rock to split into thin layers.

KEY FACTS

Rudaceous sedimentary rocks are those in which the grain size of clasts is greater than 2 mm.

Arenaceous sedimentary rocks are those in which the grain size of clasts is 0.0625 to 2 mm.

Argillaceous sedimentary rocks are those in which the grain size of clasts is less than 0.0625 mm.

Texture refers to the interrelationship of grains in a rock. It includes grain size, sorting, roundness, shape and packing of sedimentary grains.

Fossil the remains of an organism that lived more than 10 000 years ago, including skeletons, tracks, impressions, trails, borings and casts.

QUESTIONS

1 Describe the texture of the breccia shown in Figure 2.

2 Describe one similarity and one difference between arkose and greywacke.

3 Explain the characteristic features of desert sandstone.

Sedimentary rocks containing a high percentage of the minerals calcite ($CaCO_3$) or dolomite ($CaMg(CO_3)_2$) are classed as carbonate rocks. They are also known as limestone, or dolomitic limestone depending on their mineral content.

Limestones

Limestones fall into two groups:
- Chemical limestone formed from the precipitation of $CaCO_3$ from sea water.
- Biological limestone formed from organic remains such as shell fragments.

A rock containing more than 50% calcium carbonate is limestone. Most limestone contains more than 90% $CaCO_3$, usually in the form of calcite, so observing its mineral composition is a good method of identification. Modern limestone is composed of aragonite, a form of $CaCO_3$ that is unstable in rocks over time and changes to calcite. Calcite reacts vigorously with dilute HCl and has two cleavage directions, not at right angles, making it easy to identify.

Oolitic limestone

These are chemically formed limestones containing sub-spherical sand-sized grains about 1 mm in diameter, called **ooliths**. In cross-section, ooliths show concentric layers of calcium carbonate surrounding a nucleus, which may be a grain of sand, shell fragment or pellet. The ooliths are surrounded either by a fine grained calcite mud matrix (**micrite**) or a crystalline calcite cement (**sparite**). They form in tropical or sub-tropical seas, in shallow water agitated by high-energy waves.

Fossiliferous limestone

These biologically formed limestones are composed of fossils or fragments of fossils. They can be described as bioclastic if they are made of fossil fragments. They may have a micrite matrix or sparite **cement**. One common type is a crinoidal limestone that is made up of stem sections or single ossicles with rare plates formed from the calyx or arms of a crinoid. Reef limestones contain fragments of corals as well as many brachiopod or bivalve shells. Many other limestones can be described as shelly, made almost entirely of bivalve or gastropod shells. Fossil-rich limestones can form in a range of environments from low energy freshwater lakes or lagoons, such as the gastropod *Viviparus* limestone of the Jurassic, to marine beds, such as the *Ostrea* bed full of bivalves, also of the Jurassic. If the fossils are whole then the rock is likely to have been formed in a low energy environment, while broken fossils suggest high energy.

Chalk

This is a biologically formed limestone composed of coccolithophores (coccoliths), the calcareous disc or oval shaped platelets that form part of the skeletons of single-celled algae. The chalk is white as it is pure calcium carbonate. Chalk is commonly formed in low energy, deep water shelf environments.

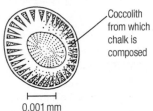

Figure 1 Diagrams of carbonate rocks

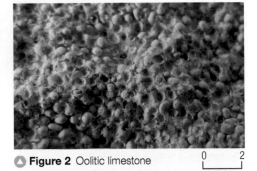

▲ **Figure 2** Oolitic limestone

0 — 2
mm

▲ **Figure 3** Fossiliferous limestone from the Jurassic that is over 90% whole and broken shells of the bivalve *Ostrea*

0 — 5
cm

> **STUDY TIP**
>
> If asked to draw a labelled diagram to show features of sedimentary rocks, make sure that your labels have lines that touch the feature you are describing. Diagrams of rocks should always have a scale, because grain size is an important identification feature.

> **KEY FACT: TESTING FOR LIMESTONE**
>
> Limestone fizzes (reacts to evolve carbon dioxide) when you add hydrochloric acid. Either 0.5 M or 1 M is best.

Classification

Folk classification

Robert Folk devised a scheme for the classification of limestones based on the type of grains and the amount of sparite or micrite they contain. The grains are described in prefixes and the matrix or cement as a suffix. A microscope is usually needed to identify grain types.

Folk classification	Examples
Prefixes	
oo – for ooliths	
intra – for intraclasts	A limestone composed of ooliths in a coarse grained crystalline cement is an *oosparite*
bio – for fossils	
pel – for **pellets**	A limestone composed of fossil fragments in a mud matrix is a *biomicrite*
Suffixes	
micrite	
sparite	

Dunham classification

Robert Dunham classified limestones based on their texture and mud content. The classification can be used in the field because there is no need to identify grain types. Mud is largely calcareous (fine grains of microcrystalline calcite), but may also contain silt and clay minerals.

Dunham classification			
Original components not bound together during deposition			
Contains mud (particles of clay and fine silt)			Lacks mud
Mud supported		Grain supported	
Less than 10% grains	More than 10% grains		
Mudstone	Wackestone	Packstone	Grainstone

Key ▢ Mud ● Grains

KEY DEFINITIONS

Ooliths spherical grains showing concentric banding of carbonate material. These are less than 2 mm in diameter, if larger they are called *pisoliths*.

Micrite a microcrystalline calcite, a depositional matrix of lime mud.

Sparite a coarse grained (>0.01 mm) crystalline calcite cement, formed after deposition.

Cement the minerals precipitated between grains in sedimentary rocks binding them together.

Pellets are carbonate material excreted by animals. They lack concentric structures and are normally 0.04 to 0.08 mm diameter.

KEY FACT

Intraclasts are grains formed by erosion of material within the basin of deposition.

QUESTIONS

1 Choose appropriate terms to describe the rocks shown in Figures 2 and 4 using:
   • Folk's classification.
   • Dunham's classification.
   Explain why you chose these terms.

2 Explain why fossiliferous limestones are not described as clastic sedimentary rocks.

3 If the limestone in Figure 3 is 24% whole fossils and 76% broken fossils, describe the possible conditions of deposition of this rock.

▲ **Figure 4** Photograph of limestone for use with question 1

Diagenesis is the group of physical and chemical processes that change sediments into sedimentary rocks. It takes place at low temperatures and pressures, at or near the Earth's surface. The main processes involved are compaction and cementation. Sediments eventually become lithified, meaning they become sedimentary rocks. Diagenesis may continue after the sediment has been converted into a rock, so **lithification** is part of diagenesis.

Compaction

As layers of sediment accumulate, one on top of another, their mass produces load pressure. This acts vertically and affects the sediments below causing compaction. Chemical processes involving **pressure dissolution** and crystallisation also occur. Grains become more closely packed and this reduces the **porosity** of the sediment.

Mud and sand

Mud and clay are much more affected by compaction than sandstones. The original thickness of sediment can be reduced by as much as 80% when a mudstone is formed from a mud. If the minerals show preferred alignment, it becomes shale. The diagram shows how porosity varies with depth of burial and therefore compaction in sand and mud.

KEY DEFINITIONS

Diagenesis all the changes that take place in sediments at low temperature and pressure, at or near to the Earth's surface.

Lithification the process of changing unconsolidated sediment into rock.

Pressure dissolution (sometimes called *pressure solution*) occurs where minerals dissolve as the result of applied pressure. Because pressure dissolution leads to a reduction of volume of the rock in which it occurs, it is also called chemical compaction.

Porosity the volume occupied by spaces in between sedimentary grains. A reduction in porosity squeezes fluids from pore spaces.

Peat partly decomposed plant remains with high water content.

Coal a carbon-rich rock formed from fossil plant remains.

Permeability is the ability of a rock to allow fluids such as water to pass through it.

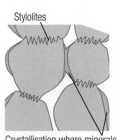

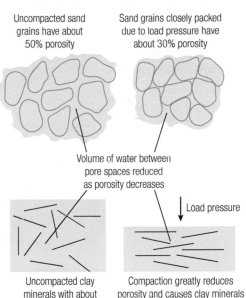

Uncompacted sand grains have about 50% porosity

Sand grains closely packed due to load pressure have about 30% porosity

Volume of water between pore spaces reduced as porosity decreases

Load pressure

Uncompacted clay minerals with about 80% porosity

Compaction greatly reduces porosity and causes clay minerals to be aligned perpendicular to the direction of load pressure

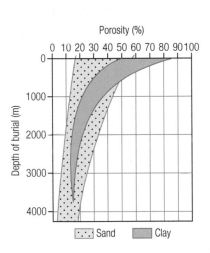

Porosity (%)

0 10 20 30 40 50 60 70 80 90 100

Depth of burial (m)

∴∴∴ Sand   ▨ Clay

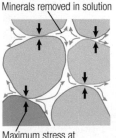

Minerals removed in solution

Stylolites

Maximum stress at grain contacts

Crystallisation where minerals come out of solution

Marks caused by pressure dissolution of quartzite

🔵 **Figure 1** Burial, compaction and porosity. Hammer for scale in photograph.

Pressure dissolution

Although most commonly seen in coarse grained rocks, sediments of all grain sizes are affected by the chemical process of pressure dissolution. Load pressure is concentrated at the contact points between grains. Minerals dissolve more easily when they are affected by stress. The dissolved minerals may recrystallise and be deposited nearby, or they may be removed in solution by groundwater flow. Pressure dissolution can produce irregular grain contacts called stylolites, especially in carbonates (limestones).

Plants and coals

Where plant remains fall into swamps, the process of decay uses up the available oxygen. Anaerobic bacteria change the plant material into **peat**. Woody material (lignin), resins and waxes are preserved. If peat is buried beneath other sediments it is subjected to increased pressure and higher temperature, which expels water and volatiles such as CH_4 and CO_2. This reduces the volume and increases the proportion of carbon it contains. As diagenesis continues, the peat gradually takes on the properties of **coal**. These gradual changes result in different types of coal, which are ranked by the proportion of carbon they contain.

Cementation

Sandstones

Sands and many biologically formed types of sediments have greater **permeability** than muds. Groundwater containing minerals in solution flows through the pore spaces and where conditions are right the minerals are precipitated forming a cement, which binds grains together to form sandstones and limestones. The most common cementing minerals are:

- Quartz, from pressure dissolution, e.g. in orthoquartzites
- Calcite, from solution of carbonate shells, e.g. in fossiliferous limestone
- Iron minerals, often hematite (Fe_2O_3) or limonite (FeO(OH))
- Clay cement from the breakdown of feldspars.

When a rock becomes compacted the grains are pushed together reducing the pore spaces where water or oil can be stored. If the sediment is then cemented by minerals, then both porosity and permeability will be reduced.

Unconsolidated sand	Sandstone compacted but not cemented	Well-cemented sandstone
57% porosity	35% porosity	3% porosity

This means that most aquifers where water is stored and reservoir rocks for oil and gas are in younger rocks that are most likely to have a high porosity. However, losing porosity can be very useful. A dam built on uncemented sandstone would leak water and the foundations of the dam may be liable to subsidence. A tunnel through a porous rock is likely to let in water, which could be very dangerous.

Limestones

Modern limestones are composed of aragonite ($CaCO_3$) which is unstable. Over time, it changes to become the more stable form of calcite. Some ancient corals had hard parts composed of calcite. The details in these fossils are better preserved than in fossils of creatures that had hard parts composed of aragonite that converted to calcite during diagenesis. Pore spaces between shell fragments may be filled by coarse grained sparite. Spaces left when the soft parts of trapped organisms decay are also filled by calcite crystals that grow from the internal surfaces of the fossil fragments. Parts of crinoid fossils, made from large single crystals of calcite, develop large overgrowths of calcite.

Increasing rank		Composition	Characteristics
Peat Thickness of peat before compaction		Water/gas	Spongy plant debris
Lignite Relative thickness of lignite from compaction of peat		Carbon Water/gas	Contains recognisable woody material in brown, crumbly coal
Bituminous coal Relative thickness of bituminous coal from compaction of lignite		Water/gas Carbon	Black in colour and has bright and dull layers
Anthracite Relative thickness of anthracite from compaction of bituminous coal		Carbon	Hard with bright metallic lustre

(Increasing depth of burial)

▲ **Figure 2** Diagenesis and coal

Fossil plant

▲ **Figure 3** Plant remains in coal

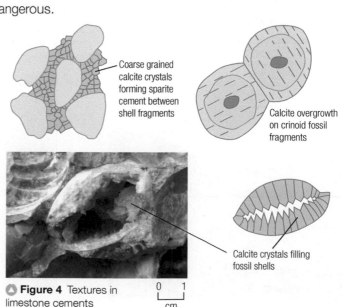

Coarse grained calcite crystals forming sparite cement between shell fragments

Calcite overgrowth on crinoid fossil fragments

Calcite crystals filling fossil shells

▲ **Figure 4** Textures in limestone cements

QUESTIONS

1 Describe and explain the changes shown by the graph in Figure 1.
2 Describe the processes that change sand into orthoquartzite.
3 Describe the conditions in which plant remains are preserved.

Interpreting past environments

Facies

Facies comprises all the characteristics of a rock that are produced by its environment of deposition, and that allow it to be distinguished from rocks deposited in an adjacent environment. This includes mineral content, grain size, sorting, structures, fossil content and any other observable property. Physical, chemical and biological factors influence environments of deposition and this is recognised in facies descriptions. The physical characteristics of a rock, for example mineral composition and grain size, are used to describe **lithofacies**. Fossil content is used to define **biofacies**. Both are related to the environment of deposition.

Facies associations

A **facies association** is group of sedimentary facies that occur together and typically represent one depositional environment. For example, all the facies found in a shallow sea environment may be grouped together to define a shallow marine facies. It is usual to use terms such as reef facies, fluvial facies or other environmental descriptions in this way. Many facies associations imply the process that produced the rock in its environment, for example turbidite facies or evaporite facies. Facies are the basic unit of sedimentary geology because one aim of sedimentary geology is to interpret the environments in which sedimentary rocks accumulated.

CASE STUDY: GLACIAL PROCESSES AND PAST ENVIRONMENTS

Glacial environments exist in arctic climates. At the present day they are restricted to polar regions and high mountains where the temperature averages are close to zero. They produce characteristic sediments and structures.

Fragments of rock at the base of glaciers scratch striations on underlying rock surfaces.

Melting ice deposits boulder clay (till), a poorly sorted sediment ranging from boulders to clay size.

Sediments produced by meltwater streams flowing from a glacier (fluvio-glacial sediments) are better sorted and show evidence of being transported by water.

When glacier ice melts, lakes are formed. During the spring thaw, streams transport fine sand, silt and clay. The sand and silt size grains settle to the floor of the lake first. The very fine grains remain in suspension, settling when the lake freezes in winter.

Temperate regions do not experience glaciation at the present day. However, using uniformitarianism when examining sedimentary rocks produced in the past has enabled geologists to discover that glaciation has been more widespread.

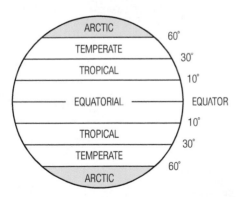

Figure 1 Location of arctic climates and photograph showing a glacier

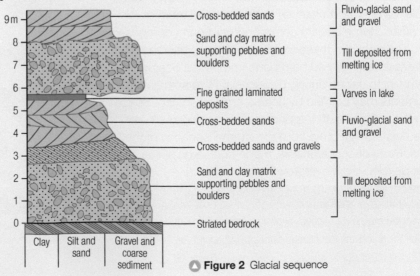

Figure 2 Glacial sequence

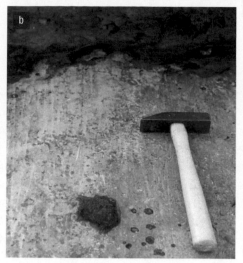

🔺 **Figure 3** Photos of glacial products: **a** varves (coin for scale); **b** striations (hammer for scale) and **c** boulder clay (coin for scale).

CASE STUDY: MARINE EVAPORITES AND COASTAL DESERTS

Sabkhas are low-lying coastal sand flats in hot arid regions. Very gentle slopes mean that during extreme high tides the whole coastline is under water.

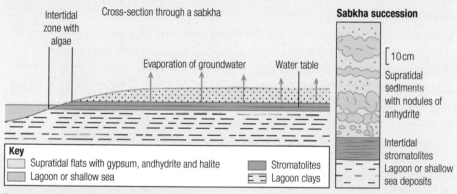

🔺 **Figure 4** Evaporite deposition in a sabkha

- Very rapid evaporation occurs so that as the sea water approaches, the coast salinity increases.
- Much of the calcium carbonate in the sea water is removed by shelled organisms living offshore.
- Evaporation of groundwater from the supratidal area draws sea water into the sediment along the shore.
- Evaporites develop above the high water line as groundwater is evaporated. Algae grow on the shore between high and low tide levels and are preserved as stromatolites. Evaporite minerals are found above the stromatolites in sabkha sequences.
- The first mineral to crystallise out within the sediments is gypsum ($CaSO_4.2H_2O$).
- Anhydrite ($CaSO_4$) crystallises out next, forming nodules in the sediment.

Figure 5 shows rocks exposed in a coastal cliff in a temperate environment. The succession is the same as in Figure 4 and the rocks are what would be produced from the deposits in a modern sabkha, so they can be said to belong to a sabkha or evaporite facies.

STUDY TIP

Ancient glacial deposits can be recognised by:
- **Striations** scratches formed by rocks carried in moving ice. They can be used to tell the direction of movement of ancient ice sheets.
- **Tillite** rock formed from very poorly sorted and varied material dropped by the ice.
- **Varves** annual lake clays and silts in distinctive thin layers.
- **Fluvio-glacial sands and gravels**.

🔺 **Figure 5** Photo of part of a sabkha succession

QUESTIONS

1 Explain how uniformitarianism and facies are linked.

2 Describe the sequence of events in Figure 2.

3 Research the events and sediments formed when the Mediterranean Sea dried up 25 Ma.

Figure 1 Polymictic conglomerate containing clasts from a mixture of different rock types (coin for scale).

Transport and facies

We might expect the lithology of sediments to be the same as the rocks in their source area. It's easy to see the different rock types in the clasts of coarse grained sediments. Conglomerates containing clasts from a mixture of lithologies from their source area are called **polymictic**.

Not all conglomerates are like this. The clasts in an **oligomictic** conglomerate are made up of a few different lithologies; whilst those of a **monomictic** conglomerate are made of only one lithology. Two processes operate during transport to affect the characteristics of sediments found in any environment; deposition and weathering. Some grains are deposited before or after others due to their size, shape or density and some grains are destroyed by weathering due to their mineral composition. The characteristics of the transported load and the deposited sediments change gradually during transport.

We have already seen how sorting of sediments is related to transport. This section looks at the relative importance of deposition and weathering during transport and how it affects the characteristics of facies.

Transport by rivers

In fluvial transport, sorting by size is the more important process. Coarser grains are often deposited near the source of a river, sand size grains are usually deposited in the middle course and finer grained materials, such as clay and silt are deposited nearer to its mouth. Some sediments are transported out into shallow seas. Although affected by river transport, they will be deposited in a marine environment. River transported sediments are usually sub-mature.

Transport by wind

Sediments transported by wind have a small size range and are mineralogically mature. Wind transports sand and silt, and only the minerals found in these grain sizes are present in wind transported deposits. Quartz is the most common, but calcite can also survive because solution is rare in arid environments where wind transport is often important. Softer minerals such as mica are destroyed by attrition and transported away as fine dust spread over very wide areas and so do not form significant deposits. The limited size range and resistance to mechanical weathering of the grains result in texturally and mineralogically mature sediments.

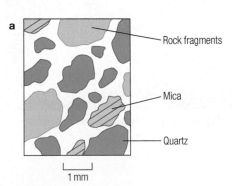

a

Rock fragments

Mica

Quartz

1 mm

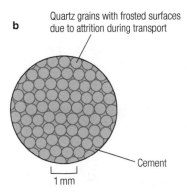

b

Quartz grains with frosted surfaces due to attrition during transport

Cement

1 mm

c

Figure 2
a River sands are texturally and mineralogically immature.
b Wind-blown sands are texturally and mineralogically mature.
c Large dunes are produced by wind blown sand deposits. Dunes approximately 20m high.

Glacial transport

Weathering is more important in glacial transport. Many sediments transported by glaciers are very poorly sorted. They often include a fine matrix of non-resistant rock types and the coarser pebbles and cobbles of more resistant rocks. Glacial sediments are often transported over relatively short distances and the characteristics of the transported load change with the rock type over which the ice is moving. Sediments deposited by glaciers are called till and when lithified they are tillites. **Till fabric analysis** shows that boulders within till tend to lie with their long axes parallel to the direction of movement of the ice. When a glacier melts, the transported sediments are liberated and are sorted by water. They are called **fluvio-glacial deposits.**

▼ **Figure 3**
a Till deposited from a glacier. Large boulders are approximately 30 cm.
b A meltwater stream transporting mainly sand and gravel size grains but also some pebbles. Person for scale.
c Fluvio-glacial deposits often show cross-bedding. Large clasts about 8 cm.

Shallow marine environments

Deposition of sediments in shallow marine environments occurs in the **littoral zone** and on the continental shelf. In littoral deposits sorting by size is opposite to that found in rivers because coarser grains are sometimes transported for greater distances, for example storm beaches composed of pebbles are on the landward side of a beach. Weathering is important in shallow marine transport because grains must be able to withstand high-energy waves. Because of this, deposits in the littoral zone are likely to contain more resistant grains. Less resistant grains are broken down into finer sizes and deposited further offshore as silty clays, often containing marine fossils.

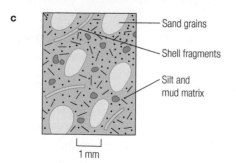

c
- Sand grains
- Shell fragments
- Silt and mud matrix

|— 1 mm —|

◄ **Figure 4**
a Modern storm beach
b Ancient storm beach with variety of rock types in the clasts
c Thin section diagram of shallow marine deposit
d Fine grained rock with marine fossils, part of a shallow marine facies

QUESTIONS

1 Explain the importance of transport, weathering and deposition in influencing the characteristics of sedimentary facies.

2 Evaluate the importance of the surface processes that produced the rock shown in Figure 1.

3 Suggest a data collection methodology to test the hypothesis that lithology is the main influence on grain shape in storm beach environments.

4 Explain why the glacier in Figure 3b is black.

What are sedimentary structures?

When sediments are deposited, they form a series of horizontal layers or beds. Each bed is separated from the next by surfaces called bedding planes. Sedimentary structures are features found on bedding planes or within the beds themselves. They are produced by a variety of sedimentary processes in a range of environments. They are **palaeo-environmental indicators** as some structures only form in specific conditions and some of them can be used to indicate whether beds are the right way up.

Cross-bedding

Sand grains will be moved by wind, river or marine currents in high-energy environments. The sand particles are deposited to form dunes or sand bars. The flows or currents are unidirectional, as there is movement in one main direction over time. The result is a gentle slope on the windward/upstream side (side from which currents flow) and a steep slope on the lee side where the sand grains avalanche down the slope to settle at a maximum angle of 37 degrees from the horizontal. This is the maximum angle of rest.

Dunes or bars constantly migrate in the direction of the current or flow, but only the lee side or slip face is preserved. A new dune migrating on top of an original dune cuts off the first set and produces a new cross-bedding set.

Wind blown sand in deserts is found in large dune structures so the cross-bedding is measured in metres. Beach and shallow sea sand bars produce cross-bedding that is about 10 to 15 cm high, while the smallest scale is in rivers at about 3 to 10 cm high. Cross-bedding can be used as a **palaeocurrent indicator**, as the cross-bedding planes always dip down in the direction of current flow.

Cross-bedding is the right way up when it is concave upwards. The cross-bedding pattern flattens out at the base but is cut off at the top. This is due to erosion, which can only occur at the top.

Ripple marks

Ripple marks form as sand grains are transported by saltation in high-energy conditions:

- Symmetrical ripple marks are formed by oscillating currents, so they are most common on beaches where tidal action moves sediment up and down the beach, known as *bi-directional flow*.
- Asymmetrical ripple marks are formed by currents that travel in one direction, known as *uni-directional flow*. They are common in rivers, shallow seas and desert environments where the wind acts in the same way as water, the medium to move the grains. These ripple marks can be used as palaeocurrent indicators.

Graded bedding

The largest and heaviest particles are on the bottom of the bed and finer particles are at the top. In each layer, the grains become progressively finer towards the top but there is an abrupt change in grain size at each bedding plane. This makes graded bedding useful as a **way-up structure**.

Graded bedding forms when the energy levels of the water drop and sediment settles out of suspension. Turbidity currents that flow onto the abyssal plain from the continental slope, or rivers that flow into lakes, bring sediment into the calm water where it settles out to form a distinct bed. Graded bedding is found in sandstones and conglomerates.

Cross-bedding

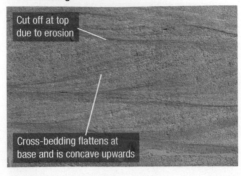

Cut off at top due to erosion

Cross-bedding flattens at base and is concave upwards

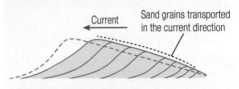

Current

Sand grains transported in the current direction

Ripple marks

Symmetrical ripple marks formed by the tides on a beach

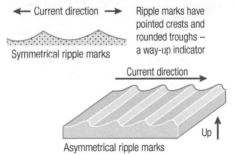

← Current direction →

Symmetrical ripple marks

Ripple marks have pointed crests and rounded troughs – a way-up indicator

Current direction

Up ↑

Asymmetrical ripple marks

Graded bedding

Size of grains varies from coarse at the bottom to finer at the top

🔺 **Figure 1** Cross-bedding, ripple marks and graded bedding

Desiccation cracks

These cracks only form in clay-rich sediments. Loss of water due to evaporation by solar heating causes mud to contract, forming polygonal-shaped blocks separated

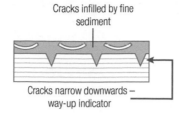

Cracks infilled by fine sediment

Cracks narrow downwards – way-up indicator

by cracks. Each crack has a V-shaped cross-section with a wide top and a narrow base, which makes them good way-up indicators. The cracks are wider at the top because evaporation of water from the mud is greatest near the surface, which receives most solar heat energy.

Cracks are only preserved if they are infilled by sediment, usually silt or mud of a different colour. Desiccation cracks are good palaeo-environmental indicators of arid areas with high evaporation rates, such as playa lakes in arid environments.

Salt pseudomorphs

Cubic halite crystals grow at the surface of a bed due to evaporation of salty water in arid environments. The crystals may become partly embedded in sediment deposited by evaporating saline water. The sediment dries out in the heat. When the dried-up lake fills with water again, the halite crystals are dissolved, leaving cubic-shaped moulds. These are then infilled by sediment, which takes on the shape of the halite crystals. Hence the name pseudomorph, meaning imitation form.

Imbricate structure

Pebbles are rolled along a stream bed and then pile up against each other. Flat (disc and blade shape) pebbles stack against each other so that their long axes are roughly parallel. This orientation provides maximum resistance to movement. The pebbles are inclined (dip) in an upstream direction, the tops of the pebbles pointing downstream.

However, in a density flow, such as a debris flow, the long axes will point in the flow direction (downstream).

KEY DEFINITIONS

Palaeo-environmental indicator a sedimentary structure formed in specific environmental conditions in the ancient past.

Palaeocurrent indicator a sedimentary structure that allows the direction of an ancient current to be deduced.

Way-up structures allow geologists to work out whether rocks are in their original orientation or whether they have been inverted by folding.

QUESTIONS

1 Draw a diagram of a recumbent fold showing one bed with graded bedding. Label the inverted limb.

2 Describe the mode of transport of the grains shown in the cross-bedding diagram.

3 Identify the sedimentary structures shown in Figure 3 and describe the palaeo-environmental conditions that produced them.

Desiccation cracks
Plan view show polygonal blocks separated by cracks. Sizes are variable.

0   5
cm

Salt pseudomorphs
Halite forms cubic crystals which may be infilled by sediment if the halite has dissolved. Cubes approximately 1 cm³.

0   1
cm

Current direction

Imbricate structure
In river sediments, imbricated pebbles 'lean' in the direction of the current.

🔺 **Figure 2** Desiccation cracks, salt pseudomorphs and an imbricate structure

🔺 **Figure 3** Sedimentary structures in the High Atlas Mountains, for use with question 3.

The usefulness of sedimentary structures

As we have seen, sedimentary structures can provide evidence for the environments in which they formed. This includes the direction of currents in water, or wind directions in air and some of them can be used to indicate whether beds are the right way up.

Where sedimentary structures are found	Used as way-up indicator	Used as palaeo-current indicator	Used as palaeo-environmental indicator
Within beds			
Cross-bedding	Yes	Yes	Sometimes – large-scale desert dunes
Graded bedding	Yes	No	Sometimes – in turbidite deposits
Imbricate structure	No	Yes	Yes – river channels
Salt pseudomorphs	No	No	Yes – arid evaporation
On bedding planes			
Ripple marks	Yes	Yes – if asymmetrical	Yes – beach, shallow sea or dune
Desiccation cracks	Yes	No	Yes – arid evaporation
Flute casts	Yes	Yes	Yes – turbidite deposits on deep ocean floor

Structures that are useful to determine the palaeo-environment, palaeocurrent and way-up indicators include flute casts.

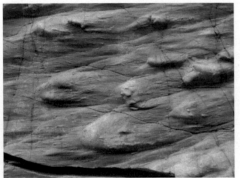

Close up of flute casts

0   10
cm

Flute casts

A flute cast is a structure found at the base of a bed. They form as a result of erosion caused by turbulent flow, and are often associated with powerful turbidity currents, although they can form in any environment where water flows with high energy over soft mud. They form parallel to the current and are deeper and pointed at the upstream end. This makes them good palaeocurrent indicators. They are preserved as they are infilled by overlying sediments.

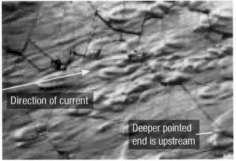

Direction of current

Deeper pointed end is upstream

Base of a bed showing flute casts

0   5
cm

Palaeocurrents

Measurements made on sedimentary structures can be used to show information about a palaeocurrent, plotted in the form of a rose diagram:

- Measuring the orientations of flute casts using a compass and plotting these data on a rose diagram is very good at telling us the direction of flow of the turbidity current that formed these structures.
- Cross-bedding dip directions can also be measured as the sediment is only preserved on the leeward slope of a dune or sand bar. Measuring sets of cross-bedding over a large area gives data that can then be plotted to show the current direction. The wind direction that formed the sand dunes of the Triassic can be identified using this method.

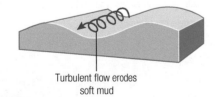

Turbulent flow erodes soft mud

▲ **Figure 1** Flute casts

MATHS BOX: INTERPRETING A ROSE DIAGRAM

Sample measurements of cross-bedding dip directions were taken in a series of desert sandstones to help determine the wind direction that formed them.

Orientation (degrees from North)											
1–30°	31–60°	61–90°	91–120°	121–150°	151–180°	181–210°	211–240°	241–270°	271–300°	301–330°	331–360°
Number of dune bedding readings											
4	0	0	0	0	2	10	15	3	7	6	3

These data are plotted on the rose diagram by shading in the appropriate area.

The modal value plots between 210° and 240°. Cross-bedding dips in a downwind direction, this means that the prevailing wind was from the north-east (30° to 60°). In desert environments, wind directions may change due to circulation around areas of atmospheric high pressure. This explains the spread of palaeowind directions.

If the strike direction of cross-bedding is plotted, the 180° to 360° part of the rose diagram is a repeat of the 0° to 180° part. This is because compass bearings along the line of strike can be taken in both directions, for example 30° from North is along the same line as 210° from North.

Strike directions are not the same as palaeocurrent directions.

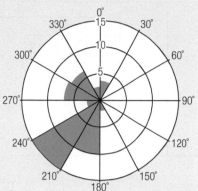

▲ **Figure 2** Rose diagram

▲ **Figure 4** Cross-bedding in a desert sandstone

STUDY TIP

Practise drawing sedimentary structures the right way up and upside down so that you can easily recognise them. Cross-bedding is one that can prove difficult. Make sure it is concave upwards. When you draw sedimentary structures, watch for differences in scale and try to get realistic angles of dip. You need to be able to explain how sedimentary structures form as well as being able to recognise them. Just give the answer you are asked for. Describing the characteristic features of sedimentary structures does not require you to describe the environment in which they form.

HOW SCIENCE WORKS

Uniformitarianism allows us to use modern sedimentary sequences and processes to study ancient environments. Investigating palaeowind directions involves using a sampling strategy and deciding on a sample size for collecting field data. Analysis is carried out, in this case by using a circular graph, the rose diagram. Using theories, selecting appropriate methodology to carry out an investigation and then using analytical methods to interpret the data is how science works in this and many other instances.

Way-up indicators

Way-up indicators are needed where rocks have been inverted by folding, making it difficult to identify the bottom and top of a bed.

	Correct way up	Inverted	Identification
Graded bedding	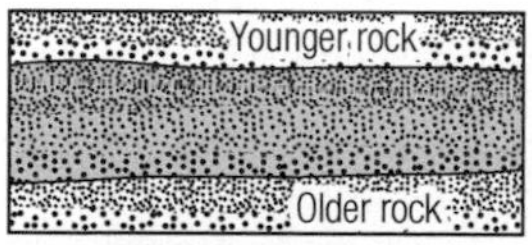 Younger rock / Older rock	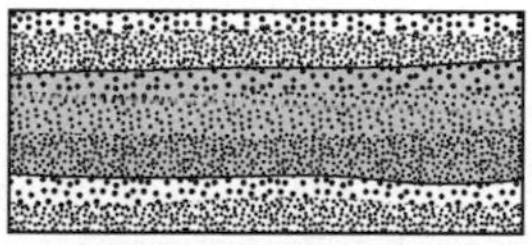	In graded bedding the heaviest / largest particles settle out first, so the sequence is coarse to fine upwards
Cross-bedding	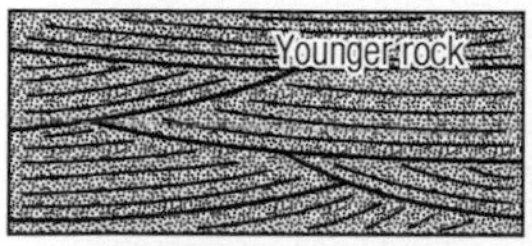 Younger rock	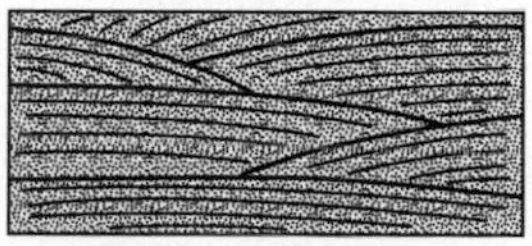	Each cross-bedding plane is concave upwards. Truncated cut off planes are at the top.
Ripple marks	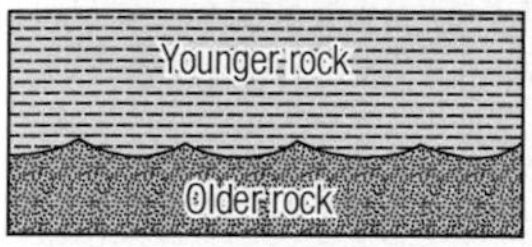 Younger rock / Older rock	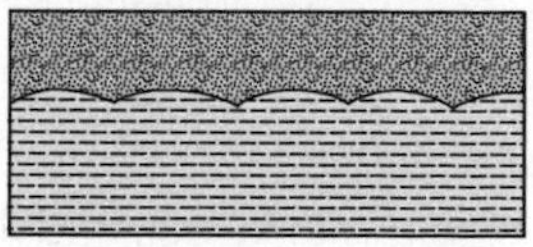	Each ripple has a crest that points up.
Desiccation cracks	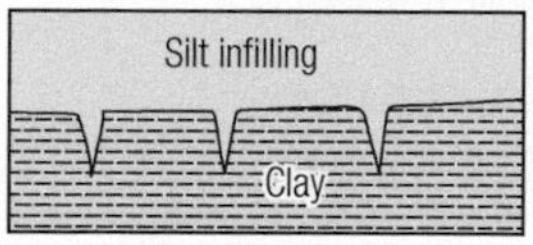 Silt infilling / Clay	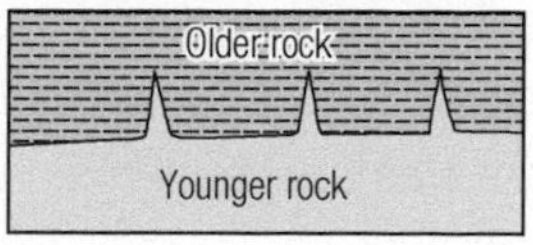 Older rock / Younger rock	Cracks are wide at the top and narrow downwards. They are filled by younger sediment

▲ **Figure 3** Way-up structures

QUESTIONS

1 Explain why desiccation cracks are wider at the top than the base.
2 Name the sedimentary structures that are useful in determining palaeocurrent directions.
3 Explain why sedimentary structures can help to put the beds in age order in areas where the rocks are deformed.
4 Draw and label the sedimentary structures present in Figure 5. Explain what information can be deduced from Figure 5.

▲ **Figure 5** For use with question 4

3.11 Graphic logs

A graphic log is a graphical method of recording observations and measurements of a sequence of modern sediments or sequences of sedimentary rocks, often exposed in cliff or quarry faces.

The first step is to use a sheet with headings like the one shown in Figure 1, which also shows some of the symbols that can be used.

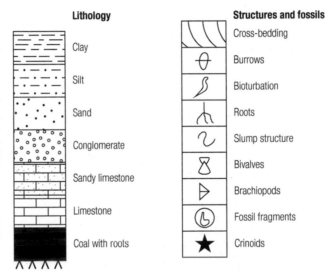

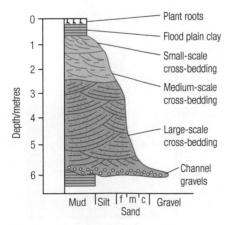

Figure 1 Graphic log plotting sheet and symbols

Record the characteristics of the sediments or rocks in the field. Start at the base and work upwards. Where possible, record the following information for each bed:

- Composition of both grains and cement or matrix
- Grain size, sorting and shape with colour, and the degree of cementation or weathering
- Identification of the sedimentary rock
- Thickness of beds in cm and describe any lateral variation
- Nature of bedding planes as uneven, erosional or gradational where one bed gradually changes to another
- Record and measure any sedimentary structures present
- Any fossils should be identified and variations in size, abundance, distribution and orientation recorded.

The grain size controls the graph. The top scale is used to draw a bed to extend out to the correct grain size. So, a conglomerate will extend to the right (or to the left, depending on which way round you draw your graph) further than a fine sandstone.

Note that the graphic log in Figure 2 includes the lithology symbols within the grain size bars. This is an alternative presentation to Figure 1 which has a separate column for lithology. You will see both types.

Figure 2 Graphic log of a fluvial sequence

Drawing and evaluating graphic logs

Practise drawing and evaluating the information in your own graphic log by following the steps below:

1 Complete a graphic log for the beds A (oldest) to G (youngest) described below. Take the complete height of the section and choose an appropriate scale to represent this cliff section.
2 Plot each bed separately on a blank graphic log, starting with the oldest at the bottom of the page.
3 Show each bed at the correct grain size and with the correct symbol. The cross-bedding symbol of curved lines is drawn on top of the sandstone symbol.
4 Where the base of a bed is eroded, show it as an uneven line on your log.
5 Add descriptive details of the beds in the appropriate column – colour, composition, sedimentary structures, bedding details and fossil content.
6 A bed may change from bottom to top, so use a combination of symbols from the edge of the bed up to show a change from one grain size to another.
7 Interpret all the evidence to describe the changing environments in which beds A to G formed.

Bed A is a purple siltstone with thin beds (laminations), contains muscovite mica and is 12 cm thick.

Bed B is 52 cm thick and is a red, medium sandstone, well sorted with large scale cross-bedding.

Bed C is cross cutting and uneven at the base and is a conglomerate with pebbles up to 27 mm across. The matrix is a red silt and the bed is 18 cm thick.

Bed D is a pebbly sandstone with all the pebbles at the base and grading up into a brown, coarse sandstone. Total thickness 32 cm.

Bed E is a fine sandstone, yellow–white in colour, 28 cm thick.

Bed F is a mud, light grey in colour with black, carbonaceous roots 42 cm thick.

Bed G is a medium sandstone nearly white in colour with small brachiopods. It is 15 cm thick.

Evaluating fossil data

In some areas, large numbers of fossils can be found and either collected or measured in the field. Data can then be used to analyse the nature of the assemblages by observing the orientation of fossils, whether they are whole, broken or disarticulated, and the sizes of the fossils. Where there are many fragmented fossils it is likely to be a death assemblage where fossils have been transported and then deposited. The graphs in Figure 3 show two populations of fossils that can be analysed.

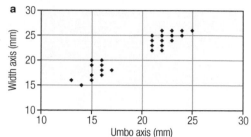

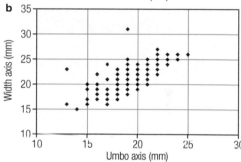

Figure 3 Graph a shows two different species of brachiopod.
Graph b shows a single species with individuals of all growth stages from small young brachiopods to larger, older ones. Two of the fossils measured are anomalous.

QUESTION

1 Figure 4 shows two fossil assemblages. Analyse the photographs and give reasons for classifying each one as either a life or a death assemblage.

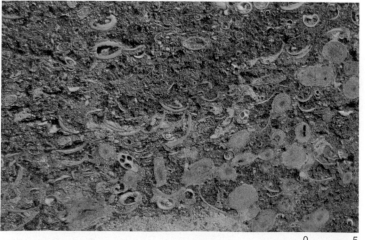

Figure 4 Fossil assemblages

a

Flood plain
Gentler gradient deeper channel no boulders meandering channel velocity and discharge increased

Source
Steep gradient shallow wide channel contains boulders velocity and discharge low

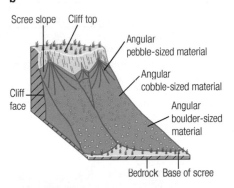

Mouth
Low gradient wide, deep channel minimises friction maximum velocity and discharge

	Sea
	High land
	River channels

b

Scree slope   Cliff top

Angular pebble-sized material

Cliff face

Angular cobble-sized material

Angular boulder-sized material

Bedrock Base of scree

🔺 **Figure 1 a** Fluvial environments **b** Deposits in source areas

KEY FACT: DEBRIS FLOWS

Debris flows are hyper-concentrated flows containing soil, unsorted rocks and sediments which are mixed with water to produce a thick slurry. They consist of approximately 40 to 50% solids and the rest is water.

KEY DEFINITIONS

Point bar deposits of sand or coarser grained sediments on the inside of a meander bend.

Channel lag coarse grained sediment left in a channel after finer grained particles have been transported away.

Flood plain flat land adjacent to a river over which it spreads when in flood.

Upward fining describes a series of layers in which average grain size decreases upwards as energy decreases.

Fluvial environments include all parts of rivers. They are generally high-energy environments, with only the flood plain lacking a strong current. When heavy rainfall causes flooding, energy levels rise rapidly so that even small streams will carry large boulders. In dry conditions, very little water will flow and energy levels will be low. As particles are transported, they get smaller and rounder due to attrition and abrasion.

Velocity and deposition

The gradient down which a stream flows is steeper near its source in mountainous areas, and shallower across the flood plain near the mouth of the stream. Velocity of the water is slower near a stream bed or bank than in the centre of the channel. This is due to friction with the channel floor and sides. The current tends to flow in a straight line, so the bank is eroded where the current meets the outside of a river bend.

Deposition takes place where velocity is reduced:

- By a sudden change in slope
- On the inside of meander bends
- Where a tributary joins a river
- Where rivers enter lakes or seas that do not have strong currents.

Products of sedimentation

Alluvial fan breccias and conglomerates

Where mountain streams flow onto a flat valley floor or plain, the marked decrease in gradient and reduction in stream velocity and energy cause large volumes of sediment to be deposited as alluvial fans. Breccias (poorly sorted, angular and coarse clasts) are formed from scree or debris flows. Conglomerates (moderately sorted, more rounded clasts) are deposited in the streams themselves (channel lag) or over large areas as sheet flooding.

Channel conglomerates and sandstones in meandering and braided rivers

Meandering streams flow in channels with looping curves. The outside bank of a meander bend is eroded, and deposition occurs on the inside of the meander on a **point bar**. As a result, the channel migrates laterally.

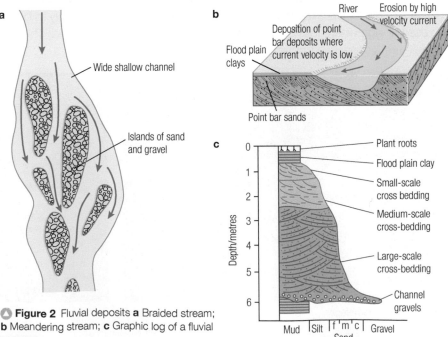

🔺 **Figure 2** Fluvial deposits **a** Braided stream; **b** Meandering stream; **c** Graphic log of a fluvial sequence

MATHS BOX: STATISTICAL TESTING

A sample of grain size measurements was taken in two different quarries (x and y) containing conglomerates formed in alluvial fans.

Field measurements	
x (mm)	y (mm)
25	61
28	35
30	10
14	12
16	22
32	46
36	38
12	11
21	40
43	22
26	
35	
63	
45	
19	
$n_x = 15$	$n_y = 10$

Ranked measurements	
r_x	r_y
11	24
13	16.5
14	1
5	3.5
6	9.5
15	23
18	19
3.5	2
8	20
21	9.5
12	
16.5	
25	
22	
7	
$\Sigma r_x = 197$	$\Sigma r_y = 128$

The null hypothesis (H_0) tested was that the two samples are both part of the same deposit.

For this to be true there should be no significant difference between the two sets of values.

H_0 can be evaluated using the Mann-Whitney U test using the formulae:

$$U_x = n_x n_y + \frac{n_x (n_x + 1)}{2} - \Sigma r_x \quad \text{and} \quad U_y = n_x n_y + \frac{n_y (n_y + 1)}{2} - \Sigma r_y$$

For quarry x the value of $U_x =$

$$15 \times 10 + \frac{15 (15 + 1)}{2} - 197 = 270 - 197 = 73$$

For quarry y the value of $U_y =$

$$15 \times 10 + \frac{10 (10 + 1)}{2} - 128 = 205 - 128 = 77$$

Because H_0 doesn't specify which of x or y will have the higher mean rank, it is evaluated using tables of U for a two-tailed test, and the value needed is the smaller of the two, in this case U_x.

The critical value at the 0.1 confidence level, where $n_x = 15$ and $n_y = 10$ is 44. Because the value of U_x is > 44 H_0 cannot be rejected and we can say, with a high degree of confidence, that the two samples are both part of the same deposit.

If H_0 does not specify which of x or y will have the higher mean rank it is evaluated using tables of U for a two-tailed test. The value needed is the smaller of U_x and U_y.

If H_0 states that one of the mean values (either x or y) is greater than the other it is evaluated using tables of U for a one-tailed test. If H_0 states that x > y the value needed is U_x. If H_0 states that x < y the value needed is U_Y.

Conglomerates (moderately sorted, more rounded clasts) are deposited in the streams themselves (**channel lag**) or over larger areas from sheet flooding. They can show imbricate structures if the clasts are disc or blade shaped. Sandstones form from point bar deposits on the inside of meander bends in the channels. They are often cross-bedded, moderately well sorted and with sub-rounded grains. They may contain muscovite mica as well as quartz, so may be described as micaceous sandstones. The sandstones have rare masses of lignite, which are fossil tree trunks and branches washed into the rivers. Clays and mudstones formed in the low energy environment of the **flood plain**.

Braided streams and rivers are found in semi-arid areas where discharge may be low due to lack of rainfall, and as glacial meltwater streams, where melting of the glacier may be slow. They transport large amounts of sand and gravel by traction or saltation and so have wide shallow channels that maximise the bed surface area. Banks of sand and gravel are deposited in the channel when energy is reduced during low discharge. The channel becomes braided as it divides up, reunites and then divides again. The deposits are coarse grained.

Meandering rivers and braided rivers result in **upward fining** cycles of sediments, as shown in the graphic log in Figure 2.

Flood plain clays and silts

At times of high flow, rivers flood the surrounding flat area adjacent to the channel, forming a flood plain. Clays rich in organic material are deposited from the suspended load of the river and may contain fossil plants. Siltstone is also common on the flood plain and may show very small-scale cross-bedding. The mudstones may show evidence of sub-aerial exposure to the elements, with desiccation cracks and the development of soils. Any fossils will be terrestrial in origin, for example vertebrate footprints, insects or plant fossils.

STUDY TIP

Ancient river deposits can be recognised by:
- The limited lateral extent – as the channels move the sandstone in lenses, cross cutting into older beds.
- Lignite plant fossils in sandstones but generally the lack of other fossils.
- The mix of conglomerate, sandstone and shales all together.

QUESTIONS

1 Describe the fluvial environments in which the sedimentary rocks shown in the graphic log were deposited.

2 Describe the texture of sediments deposited in alluvial fan environments.

3 Explain with the aid of diagrams how meandering river channels migrate laterally.

4 Research the sediments that were laid down in Britain under fluvial conditions during the Triassic.

Many, but not all, of the hot deserts on Earth are in tropical latitudes. Outside these latitudes, there are extensive deserts in continental interiors where there is a wide diurnal temperature range due to lack of cloud cover. When rainfall does occur, it is often torrential. The bare, sun-dried ground allows rapid run-off, causing **flash floods** that give streams high energy for short periods. Perhaps surprisingly, water is one of the main agents of erosion and transport in deserts. The other is the gradual effect of the wind, which moves sand and finer grained sediment, leading to the formation of dunes.

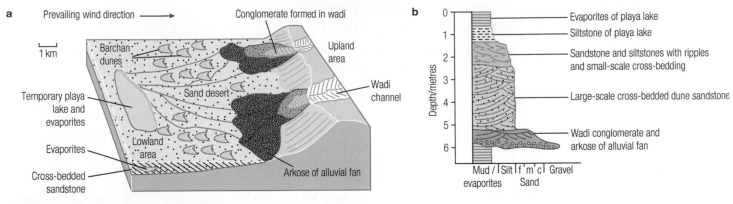

Figure 1 a Geography of a hot desert environment **b** Graphic log of hot desert deposits

Processes and products in hot desert environments

Wadi conglomerates

Desert stream channels are known as **wadis**. Their source is usually in mountains, so they have steep gradients. Together with the intensity of the rainfall, this gives the streams very high energy and so they can transport coarse grained fragments. Energy is quickly lost when the rain stops and because water sinks into the porous rocks of the channel. Deposition is rapid, leaving poorly sorted conglomerates. Grains may be sub-rounded if they have been transported on several occasions. Grains that have only been briefly transported, perhaps during one flash flood, may be angular. Desert conditions often lead to grains having a red coating of oxidised iron minerals.

Alluvial fan arkose

Coarse grained sediments deposited near to the mountains (Figure 1) become breccias but in the middle part, a fan arkose is formed from sand-sized material deposited rapidly in arid conditions. It contains more than 25% feldspar, which shows little sign of weathering. The usual source of the feldspar is granitic rock upstream.

Aeolian sandstones

Sand grains are transported by high energy winds and affected by attrition. This makes them:

- Very well sorted
- Very well rounded
- Have high sphericity
- Frosted due to the frequent collisions between grains.

Figure 2 a Poorly sorted clasts in a wadi conglomerate – grain size varies from cobbles to fine sand. Hematite coating the grains gives the rock a red colour (coin for scale) **b** Wadi at Gorges Todra, Morocco (tree for scale).

They are composed entirely of quartz grains because other minerals present in the source rocks have been removed by mechanical and chemical weathering. Quartz remains because it is unreactive and hard. The frosted surfaces of the grains are due to attrition during transport. Because of their shape and size, they are sometimes described as 'millet-seed sand'.

Desert sandstones are formed from these sands. The grains are coated with oxidised iron minerals so that they have a red colour and desert sandstones often have silica cement. They show large-scale cross-bedding on a scale of metres rather than centimetres because they are often deposited in large dunes. The dunes may be crescent-shaped barchan dunes or straight seif dunes with crests transverse to the prevailing wind. The dunes move in the direction of the prevailing wind and sand is blown up the windward side and slips down the leeward side. It is this face that can be preserved to make the cross-bedding. Some dunes can be up to 200m high.

Playa lake evaporites

Due to the hot climate, infrequent rainfall, high rates of evaporation and infiltration, desert streams tend not to flow all the way to the sea. Hot deserts are often regions of inland drainage where streams flow into temporary **playa lakes**. Stream water flowing into the lakes contains ions of calcium, sodium and potassium in solution. They are products of the weathering of rocks upstream. When the water evaporates in the hot sun, all the dissolved ions become more concentrated. The least soluble are the first to precipitate out to form layers of **evaporite minerals**. The most soluble will be in the centre of the playa lake.

Evaporites			
Solubility	Mineral name	Properties	Order of precipitation
Less soluble	Calcite ($CaCO_3$)	White; hardness = 3; rhombic crystals and good cleavage; vigorous reaction with dilute HCl	Precipitated first
	Gypsum ($CaSO_4\ 2H_2O$)	White; hardness = 2; crystals or may form nodular layers	
	Anhydrite ($CaSO_4$)	Colourless to pale blue; hardness = 3.5; crystals usually in layers	
	Halite (NaCl)	Cubic 'hopper' crystals; cubic cleavage; salty taste	
More soluble	K minerals, such as Sylvite (KCl)	Colourless or white: hardness = 2; salty bitter taste	Precipitated later

Fine grained sediment is also deposited from suspension to form mudstones in playa lake environments. These may contain desiccation cracks, ripple marks and salt pseudomorphs, as well as lenses of evaporites, usually gypsum.

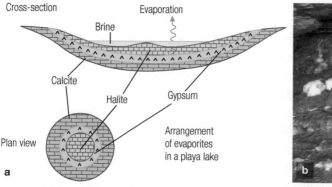

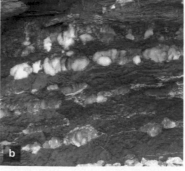

Figure 3 a Diagram showing evaporite deposits in a playa lake (variable in size, up to several hundred km) and **b** lenses of gypsum in mudstones from a playa lake (layers 2cm thick).

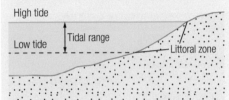

Most sand and finer grained sediment deposited on beaches and in shallow seas has been transported there by rivers. Shallow seas extend from the extreme low water level of the **spring tide** to the edge of the continental shelf, about 200 m below sea level. Their extent varies, but there are large areas of shallow seas around the coast of Britain. Transport in shallow seas may be by **longshore drift**. The currents are uni-directional and take water and sediment along the coast or back out to sea. Storm events can also be very important in transporting sediment.

The littoral zone is the area between extreme high and extreme low water level of the spring tide, so it is covered by the sea for part of the time. It includes beaches and is a high energy area, affected by waves and tides, which are bi-directional. Coarser grained sediment may have been supplied by the erosion of coastal cliffs. Pebbles and cobble-sized sediments are usually found at the back of a beach, where they have been transported by high-energy waves during storms.

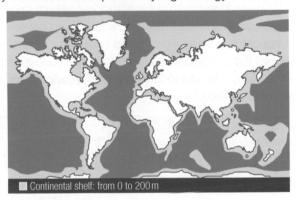

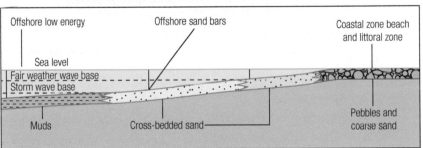

Figure 1 Map showing the extent of the continental shelf and a cross-section showing where the sediment is deposited

Beach environments

Beaches are mainly composed of sand or sand and gravel, though some may contain mud. Sand grains are regularly picked up and moved around by waves and this makes them well sorted. The grains are usually rounded as a result of attrition whilst being transported by waves and tides. The rocks commonly formed are light coloured orthoquartzites, as they are made entirely of quartz.

A few specially adapted burrowing organisms such as bivalves can survive in these extreme environments, which vary between being land and being covered by the sea, according to the tides. Tracks, trails and burrows of beach organisms as well as broken shells may be found scattered through the sand or in shell banks.

The bi-directional movement of water by tides leads to the formation of symmetrical ripple marks with crests that are parallel to the crests of the waves.

In high-energy areas, beach gravel, pebbles or cobbles are deposited. These are less frequently worked than the beach sands because they are only affected by

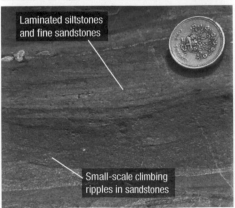

Figure 2 Shallow sea sediments

the highest tides and highest energy storm waves. They are less well sorted than beach sands. Traces of life are rare in this environment, though there may be some shell fragments washed in. The rocks commonly formed are conglomerates. They produce coarsening upward sequences in some beach deposits. The fact that they formed on a beach is usually shown by the presence of shell fragments.

Shallow seas

The material deposited in marine sediments has its origin on the land:

- Rivers carry a suspended load of clay minerals, a dissolved load of salts, and some bed load of sand.
- Wind carries atmospheric dust, which can be deposited anywhere in the sea.

Below the littoral zone is the shallow water of the continental shelf where the average water depth is 130 m. Here sediments are mainly affected by currents. In general, sediment size decreases as depth and distance away from the coast increase.

Marine sand sheets deposited offshore are reworked by uni-directional currents (e.g. tide or wind-driven currents) forming asymmetrical ripple marks. Sandstones produced in these environments often contain **glauconite**, a distinctive green potassium iron silicate mineral.

Mudstones and clays form in areas of lower energy offshore and commonly contain a wide range of fossils.

On continental shelves sediment supply controls the rate of sedimentation. It varies according to how much sediment is being transported to the sea by rivers, and this depends on rates of weathering and erosion on land.

Changes in sea level influence the areas close to shore. A rise in sea level causes the sea to flood the land and the zones of sediment move inland so that mud may be deposited on top of sand. This is called a marine **transgression**. Submerged forests around the coast of Britain are evidence of the rise in sea level since the last glacial period. Fossil forests are where trees were submerged and preserved by replacement of the woody tissue by silica or calcite. A fall in sea level means that the sea retreats, causing a marine **regression** and renewed erosion on land takes place bringing more sediment to the sea. Raised beaches are evidence of marine regression.

Sediment transport can be increased by changes in a current direction or rate of flow. Offshore sand bars can move several hundred metres in a year. This causes a danger to shipping and shipping lanes have to be dredged to keep them clear.

Sea water contains suspended sediment, reducing daylight penetration closer to shore. Abundance of life is closely linked to the availability of sunlight. The area of the continental shelf where water depth is less than 100 m has abundant life.

KEY DEFINITIONS

Glauconite a green-coloured mineral, an iron and potassium silicate of the mica group, formed on continental shelves.

Transgression occurs where the sea spreads over the land.

Regression occurs where the sea retreats from the land.

Beach sediments with shell fragments in a cliff, now above beach level

🔺 **Figure 3** Raised beach

STUDY TIP

Ancient shallow sea deposits can be recognised by:

- Beds of well-sorted and -rounded sandstone with symmetrical ripple marks or small-scale cross-bedding
- Shales with abundant fossils
- Lateral variation, as all the sediments are being laid down at the same time on the continental shelf.

0   10
⌐___⌐
cm

0   2
⌐_⌐
cm

🔺 **Figure 4 a** Fine sandstone with vertical burrows (*Skolithos*) and **b** Cretaceous Gault Clay with fossils

QUESTIONS

1 Explain why you would not expect to find whole fossils in a beach conglomerate.

2 Describe one similarity and one difference between sandstones formed in a river channel and in a shallow sea.

3 Explain why both symmetrical and asymmetrical ripples are found in shallow seas.

4 Research the sediments that were laid down in Britain as the Greensands under shallow sea conditions during the Cretaceous.

KEY FACTS: FLINT AND CHERT (SiO₂)

- Flint and chert are composed of SiO_2.
- They are so fine grained that their texture is described as **cryptocrystalline**.
- They are not carbonates but are found in association with them.
- Flint is common in the south and east of England because it is found in chalk.
- Chert is found in limestones and is common in the Carboniferous Limestone in Derbyshire.
- There is debate about their origins.
- Flint nodules form in chalk during diagenesis. Sites of organic activity, such as burrows or fossil shells act as sites for silica precipitation, forming flints. The source of the silica is thought to be sponge spicules.
- Chert may be of organic or inorganic origin. Some varieties are composed of the microscopic skeletal remains of **radiolaria** deposited in low energy conditions.
- Because chert is cryptocrystalline and resists weathering and metamorphism, it is suitable for the preservation of early life forms, including bacteria.

Outer surface becomes brown when weathered. Irregular shape is typical.

🔺 **Figure 2** Flint nodule weathered out from chalk in Lincolnshire (coin for scale).

Shallow carbonate seas

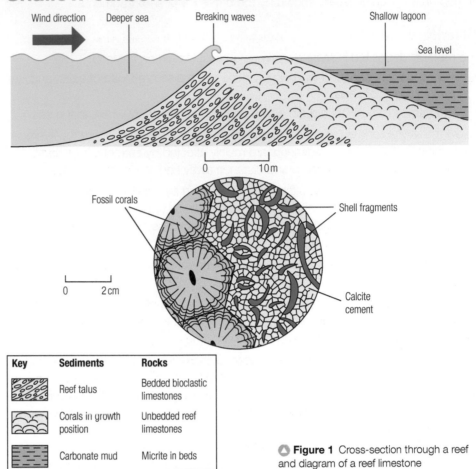

Key	Sediments	Rocks
	Reef talus	Bedded bioclastic limestones
	Corals in growth position	Unbedded reef limestones
	Carbonate mud	Micrite in beds

🔺 **Figure 1** Cross-section through a reef and diagram of a reef limestone

The rocks that form in clear, non-clastic, shallow marine environments are biologically and chemically formed limestones. Most are formed from the remains of organisms living in the sea, and often contain visible **macrofossils**, **microfossils** and fossil fragments. But some are composed of calcium carbonate that has been directly precipitated from sea water by chemical processes and may be oolitic.

Bioclastic or fossiliferous limestone

Up to 75% of these rocks may be composed of invertebrate skeletal remains from crinoids, bivalves, brachiopods, gastropods or a mixture of fossils. The rest of the rock is formed of calcareous mud and ordinary **detrital** mud. The rock is commonly grey in colour, such as the Carboniferous Limestone of Derbyshire, which has a high chemical purity. They are hard, well jointed and form in thick, massive beds. They may contain the broken remains of stem segments (ossicles) of crinoids (and are called crinoidal limestones). Bioclastic limestones are also common from the Middle Jurassic. They are often crumbly and cream in colour. Fossils such as *Ostrea*, a bivalve, can be picked out whole. Skeletal remains are broken into fragments during transport by the action of marine currents or waves.

Reef limestone

Corals provide the main framework of most reefs. Algae and some bivalves may also form reefs, but other organisms also live on and around reefs. Carbonate-secreting algae encrust the reef, cementing it together. Crinoid, brachiopod, echinoid, bivalve, gastropod and trilobite remains can all be found in reef limestones. Reef limestones are typically unbedded because reefs form by growing upwards.

Reef-building corals live in warm, shallow, high-energy, well-oxygenated tropical waters. Water temperatures need to be between 23 and 29°C. Reef-building corals do not grow well at depths greater than 25 m. Although reefs and atolls occur in the middle of oceans, the reef-building corals only live in the shallow water near land masses. In the Pacific, volcanic islands are colonised by corals and as the islands sink the coral may grow up at the same rate so that the coral forms a barrier reef around the island. Eventually the island sinks and a ring of coral called an atoll forms.

Oolitic limestone

These rocks are formed from ooliths, and are typically white but may be yellow or orange if iron stained. Current and wave action means they are often cross-bedded. They form in shallow seas where tiny grains of sand, shell fragments or pellets are rolled in carbonate mud by tidal currents and wave action. Around this nucleus, concentric layers of calcium carbonate, in the form of aragonite, are precipitated from the sea water. The ooliths are therefore formed by chemical processes in warm tropical climates. They form in water usually less than 2 m deep, where wave agitation and tidal movements are active. Ancient ooliths have recrystalised to form calcite, a more stable polymorph.

Fossils such as brachiopods and bivalves, which were attached to the sea floor to withstand the high-energy conditions, are common. Organisms such as irregular echinoids and bivalves, survived by burrowing. Most fossils are broken by the high-energy conditions.

Deep water carbonate seas

Carbonates are preserved in sediments only where the sea floor is above the carbonate compensation depth (CCD). This is the depth at which the rate of solution equals or exceeds the rate of supply of skeletal remains.

> **HOW SCIENCE WORKS: CARBONATE COMPENSATION DEPTH (CCD)**
>
> The carbonate compensation depth (CCD) is the depth in the oceans at which the rate of solution of solid $CaCO_3$ equals the rate of supply. Above this depth calcareous plankton may accumulate as an ooze. Below this depth the $CaCO_3$ is dissolved and remains in solution, as shown in the formula below:
>
> $$CaCO_3 + CO_2 + H_2O \rightleftharpoons Ca^{2+}(aq) + 2HCO_3^-(aq)$$

Chalk

Chalk is white, hard, often massive and well jointed. It is a biologically formed limestone composed of coccolithophores (coccoliths), tiny calcareous discs or oval-shaped platelets that formed part of the tests of single-celled planktonic algae. These are measured in microns, usually 5–8 μm. It may contain the microfossils of calcareous foraminifera such as *Globigerina*. There may also be macrofossils of echinoids, belemnites, brachiopods and bivalves. Chalk is formed in low-energy, deep water shelf environments (<200 m), but only where very little sediment was being supplied from the land. Flint nodules are common in chalk.

Micritic limestones

Micrite is a fine, hard crystalline limestone formed from calcite mud, which has undergone diagenesis. In shallow seas calcite mud is produced by chemical precipitation from sea water or by carbonate secreting algae. In the deep oceans, the **tests** of planktonic foraminifera fall to the sea floor to form a **calcareous ooze**. Most oceanic ridges are above the CCD, allowing carbonate to accumulate. Burial and diagenesis of calcareous ooze produces micritic limestones. They are buried as the oceanic crust moves away from the ridge to greater depths.

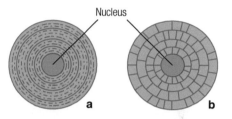

Nucleus

a   b

▲ **Figure 3** Internal structure of **a** modern ooliths and **b** ancient ooliths showing radial structure in oolitic limestone (both about 1 mm in diameter)

▲ **Figure 4** Beds of chalk in a cliff   1 m

Some sedimentary processes are infrequent or may be difficult to observe but they can be understood and explained using scientific principles.

Deposition and settling velocity

We can observe the ways in which grains are transported, by suspension, saltation and traction; but what are the factors that decide whether a grain is transported or deposited? The factors are: the size and density of the particle and the velocity, density and viscosity of the fluid. The way in which viscosity affects the settling velocity (v) of particles can be modelled and analysed using Stokes' law.

MATHS BOX: STOKES' LAW

The Stokes' law equation is $v = \dfrac{gd^2 (\rho_p - \rho_w)}{18\eta}$

Where ρ_p is density of particle; ρ_w is density of fluid (which would be water in most cases)

η is viscosity of fluid; g is acceleration due to gravity; $9.8\,\mathrm{m\,s^{-2}}$; d is particle diameter.

Stokes law shows that the terminal settling velocity of small spheres in a fluid is inversely proportional to its viscosity and directly proportional to the density difference between the fluid and solid, the radius of the spheres, and gravity.

Practical investigation

Measure particle velocity by using a measuring cylinder containing glycerol. Glass spheres (ballotini or marbles) of different sizes are used instead of sedimentary grains. Their diameters can be measured in micrometres. A thermometer is required because the glycerol must be maintained at the same temperature, so as not to affect its viscosity. A metre rule and stopwatch are used to measure distance travelled and time taken. Ensure that the terminal velocity is reached (the particles are not still accelerating). Results of velocity (y axis) against size are plotted on log graph paper, which should show a linear relationship between grain size and velocity over most of the graph.

Most sediments are deposited from flowing water. For both air and water transport at low speeds, small particles obey Stokes' law, but for larger particles the relationship changes.

Bedforms

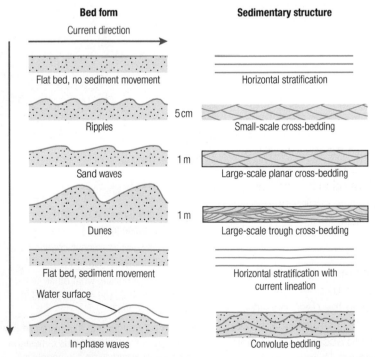

Figure 1 Bed forms and sedimentary structures

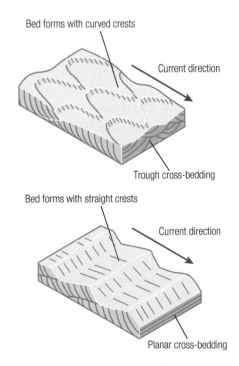

Figure 2 Curved and straight crested ripples

Observations made in flume tanks and natural streams show that different bedforms are produced as water velocity increases. If the bed forms have linear crests, tabular cross-bedding is formed. Bed forms with curved crests produce trough cross-bedding. Bed forms are related to grain size in addition to current velocity. Small-scale cross-bedding is seen only in sediments no coarser than medium sand, because ripple bed forms are not produced in coarser sands.

Flocculation of clay particles

Clay-sized particles have very slow settling velocities (10^{-6} to $10^{-5}\,\mathrm{m\,s^{-1}}$). Flocculation is a process in which clay particles adhere to one another to form larger groupings or aggregates which have higher settling velocities. Clay particles have a negative charge. This attracts cations which are adsorbed onto the surface of the clay. When two particles come close together they are repelled by the similar charge that they carry. In estuaries and especially in the open sea, the water is a strong electrolyte. It reduces the size and effectiveness of the layer of adsorbed cations and clay particles come close enough for flocculation to occur. This can only happen if they are brought together by turbulence, Brownian motion or by attraction to larger grains with higher settling velocities.

Practical investigations to model sedimentation processes

The formation of graded bedding can be modelled by filling a coffee jar about one third full with sediment that contains a range of grain sizes. Fill the jar up with water and put the lid on, then shake it so that the sediment is suspended in the water. The different settling velocities of the various grain sizes produce a graded bed in the jar.

To model the formation of symmetrical ripples put a layer of fine sand evenly spread 5mm thick in the bottom of a fish tank and cover with 50mm depth of water. Place the tank on a pencil with a circular cross-section or a drumstick. Make waves by rocking the tank. Symmetrical ripples will form.

Use a circular trough to model the formation of asymmetrical ripples. Place a beaker of sand in the middle of the trough. Half fill the rest with water. Add clean sand to form a layer about 5mm thick. Stir the water around the trough. Asymmetrical ripples will begin to appear in the layer of sand. This is because the flow is uni-directional.

When a distributary channel on a delta becomes blocked by deposited sediment the flow of water switches into a new channel. The process of delta mouth switching can be modelled using a tray with a 10mm bed of beach sand. Tilt the tray gently so that the end overhangs a sink and create a sediment trap in the sink. Pour water on at the other end of the tray. A river system will develop in a few minutes and several processes, including delta mouth switching, can be seen.

The phi scale

You met the phi scale when analysing the grain size distribution of modern sediments. It is a logarithmic scale calculated from the following equation:

$$\phi = -\log_2\left(\frac{D}{D_0}\right)$$

Where D is diameter of particle; D_0 is a reference diameter equal to 1 mm.

You should be familiar with it because, in addition to being used to define grain sizes, it has the advantages of allowing the use of statistical methods directly to construct cumulative frequency curves and making the description, plotting and analysis of grain size data more straightforward.

▲ **Figure 3** Curved crested ripples (hammer for scale).

▲ **Figure 4** Beach sediments with flat bed and ripples

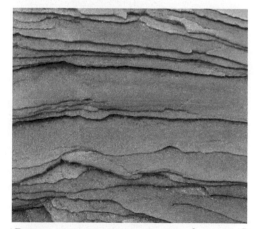

▲ **Figure 5** Horizontal stratification in sandstone.

0   5
cm

QUESTIONS

1 Suggest reasons why Stokes' law may not apply to large grains.

2 How could the phi equation be rearranged to find the diameter of a particle?

3 Explain the occurrence of both a flat bed and ripples in Figure 4.

4 Describe how you could model the formation of desiccation cracks.

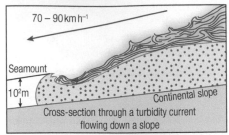

Cross-section through a turbidity current flowing down a slope

70 – 90 km h⁻¹

Seamount

10²m

Continental slope

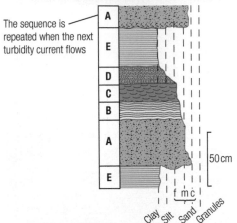

The sequence is repeated when the next turbidity current flows

50 cm

Clay Silt Sand Granules

Figure 1 Deposition from turbidity currents

KEY FACTS

- **Continental shelf** is next to the land and slopes gently out to the top of the continental slope at a depth of less than 200 m.
- **Continental slope** has a steeper gradient and in many places is cut by submarine canyons eroded by turbidity currents.
- **Continental rise** has a gentle gradient and it is where turbidity currents deposit much of their load.
- **Abyssal plain** is an almost level area of the deep ocean floor covered by deep sea sediments.

KEY FACT: ARNOLD BOUMA

In the 1960s Arnold Bouma, a Dutch geology student, was working on greywackes in the Alps and realised that they were deep water sediments. He proposed his turbidite model to account for the ideal upward fining sequence.

Sediment brought to the sea by rivers accumulates on the continental shelf. If the sediment at the edge of the slope becomes unstable and loses shear strength, it will begin to move down slope under the force of gravity. Rapid sedimentation, stress from storm waves and earthquakes can cause material to become unstable. Submarine gravity flows, known as **turbidity currents**, are turbulent and incorporate more water into the flow. Its density and low friction means that it flows with high velocity down the gentle gradient transporting huge volumes of clastic material. The sediment includes coarse and medium grained particles, as well as large quantities of mud. Much of the sediment is deposited in submarine fans on the continental rise or in ocean trenches. However, they can also spread thousands of kilometres across abyssal plains. As the gravity flows becomes more dilute its capacity for work reduces; heavier coarse grains are deposited first and then the finer sand and silt settles out. The background sedimentation in the ocean is called pelagic fallout, a gentle rain of fine mud is deposited from suspension to form shale.

The Bouma turbidite model		
Bouma layers	**Structures**	**Interpretation**
E Shale, which may contain pelagic marine fossils such as graptolites	Parallel laminations in mud 0.5 cm	No current. Suspension settling. Interturbidite.
D Fine sandstone and siltstone	Parallel laminations in sand and silt 0.5 cm	Low-energy current. Suspension settling. Changes in current energy cause alternating laminations of coarser and finer grains.
C Sandstone (greywacke)	Cross-bedding; climbing ripples 1 cm	Sufficient energy to carry sand by saltation. **Climbing ripples** form when deposition exceeds rate of migration of ripples. Energy insufficient to cause complete erosion of stoss side of ripples.
B Coarse, then medium sandstone (greywacke)	Flat bed (high velocity); graded bedding; sole marks including flute casts and **tool marks** Sole structures 2 cm	Sufficient energy to carry sand grains by traction. Sole structures occur on the base of a bed. They may take the form of pits or grooves that act as moulds for sediment deposited from above. When filled, they produce casts.
A Coarse bed of pebbles / granules (conglomerate) in a sandy matrix	Graded bedding, rip-up clasts, erosional base 1 cm Shale rip-up clasts	High energy. Coarsest grains in suspension. Erosion at the base of the flow tears up shale from underlying beds. Shale **rip-up clasts** are included at the base.

MATHS BOX: SPEARMAN'S RANK

A survey of submarine debris flows collected data on thickness and maximum lithic clast size for each bed. Spearman's Rank correlation coefficient was used to test the null hypothesis (H_0) that there is no correlation between bed thickness and maximum lithic clast size.

Bed thickness (m)	Rank x (highest = 1)	Lithic clast size (max) (mm)	Rank y (highest = 1)	d (x − y)	d²
2.00	7	12	11.5	− 4.5	20.25
0.54	16	21	9	7	49
1.89	9	29	6	3	9
1.87	10	43	2.5	7.5	56.25
4.90	4	36	5	− 1	1
18.25	1	41	4	− 3	9
1.90	8	43	2.5	5.5	30.25
0.55	15	10	14	1	1
0.65	13	6	17.5	− 4.5	20.25
0.83	12	12	11.5	0.5	0.25
0.05	18	6	17.5	0.5	0.25
0.60	14	17	10	4	16
0.06	17	7	15.5	1.5	2.25
4.31	5	25	7	−2	4
9.15	2	57	1	1	1
1.15	11	7	15.5	− 4.5	20.25
3.12	6	11	13	−7	49
7.81	3	22	8	−5	25

$$\Sigma d^2 = 314$$

Results for r_S vary between +1 = a perfect positive correlation, and −1 = a perfect negative correlation. A value of 0 indicates no correlation. The equation is as follows:

$$r_s = 1 - \frac{6\Sigma d^2}{n^3 - n}$$ where n is the number of pairs of variables, in this case n = 18, so

$$= 1 - \frac{6 \times 314}{5832 - 18} = 1 - \frac{1884}{5814} = 1 - 0.32 = + 0.68$$

Degrees of freedom for r_S = n − 2, which in this case is 16. Using a table of statistical significance, the critical value at the 0.005 confidence level, where degrees of freedom (df) =16 is 0.666. Because the value of =0.68 is > 0.666 the null hypothesis can be rejected at the 0.005 confidence level. This means that in the debris flows sampled, there is a strong positive correlation between bed thickness and maximum lithic clast size. As bed thickness increases so does the size of the largest lithic clast it contains.

Biogenic deposits

These are calcareous and **siliceous oozes**. When **planktonic** organisms die, their tests slowly sink to the sea floor to be preserved as microfossils. The composition of the ooze is dependent on the composition of the planktonic organisms.

We have already met calcareous oozes in the deepwater carbonate section. Below the CCD, any carbonate material dissolves before it reaches the sea floor, due to the higher CO_2 content and lower temperature of the water.

Siliceous oozes form from the skeletons of **diatoms** in deposits nearer the poles and radiolaria in deposits nearer the equator. Silica dissolves at a slow rate in sea water. Oozes accumulate where the rate of deposition is greater than the rate of solution.

Sedimentation rates on abyssal plains are estimated at between 1 mm and 3 cm per 1000 years.

KEY DEFINITIONS

Turbidity current high-velocity current that flows down gentle gradients because the sediment dispersed within it makes it denser than sea water. These currents are triggered by earthquakes or slope instability.

Turbidites an upward fining deposit of greywacke deposited from a turbidity current.

Bouma sequence an idealised sequence of sediments and sedimentary structures seen in a turbidite deposit.

Climbing ripples a series of cross-laminae formed by superimposing migrating ripples. They form when deposition exceeds the rate of migration of ripples.

Tool marks impressions made on the surface of soft sediment by the dragging or bouncing of an object (tool) in a current.

Rip-up clasts pieces of shale or mudstone eroded by a current containing suspended sediment. They are preserved when the current deposits its sediment.

Siliceous ooze pelagic clay containing >30% biogenic skeletal material made of silica.

Plankton minute organisms living in the surface layers of the ocean, that are transported by currents.

Diatoms planktonic algae that secrete siliceous material.

🔺 **Figure 2** Turbidite sequence

QUESTIONS

1 Name the sedimentary features eroded into the sea floor by turbidity currents.

2 Explain the relationship between changes in energy level and the rocks shown in the graphic log sequence.

3 Make annotated diagrams of tool marks and explain their formation.

4 Research the rocks that formed in shelf and deep sea conditions in the Welsh Basin during the Silurian.

Deltas occur where a river flows into a sea or lake and deposits its load of sediment. There must be little wave or tidal action to allow the sediment to build up at the mouth of the river and not be washed away. It is the decrease in energy that causes deposition of sediments, with the coarsest grains settling out first.

Deposition of sediment causes the river channels to become blocked so that the river switches its course and splits into **distributary** channels.

The upper surface of a delta is just above sea level and in between the distributary channels there are usually swampy islands. In equatorial and humid tropical climates vegetation grows very rapidly and is abundant. Preservation of plant remains requires anaerobic conditions, and swamps provide these conditions. Ancient plant remains are preserved as coal in equatorial deltaic sequences.

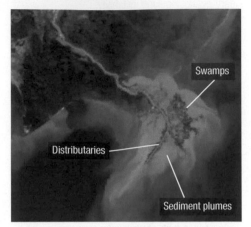

Figure 1 Satellite photo of the Mississippi Delta showing sediment flowing in the Gulf of Mexico through distributary channels

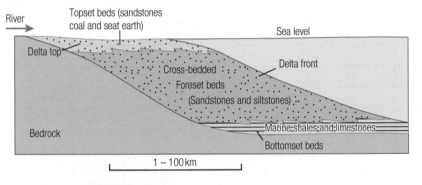

Figure 2 Photo of a river channel containing course grained sands and gravel cutting through older beds

Figure 4 Fossil plant remains (bark from a tree) (coin for scale).

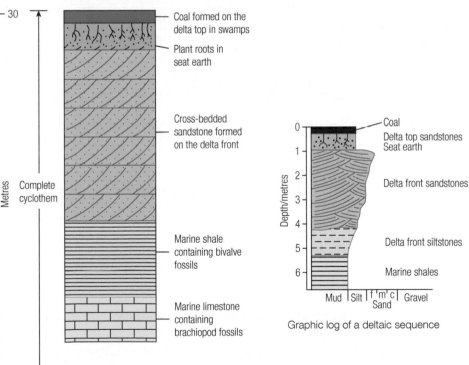

Figure 3 Cross-section through a delta

HOW SCIENCE WORKS: ARCHITECTURE OF DELTAS

The earliest deltas to be studied were in playa lakes. They have a simple architecture: alluvial fans topset beds, foresets of steeply dipping graded beds, and laminated clay bottom set beds. Most deltas however are more complex as river avulsion leads to delta mouth switching, producing sequences called **cyclothems**.

The delta top

The delta top is dominated by distributary channels with areas of swamps, bays or flood plains in between. The channels change course frequently, a process known as delta switching that leads to lateral changes in the pattern of sedimentation.

Sediments deposited on the delta top are called the **topsets**. Coarse grained sands and gravels are deposited in the channels. They make up the bulk of sediments in the delta top and form channel sandstones, which may contain cross-bedding.

Clays are deposited in areas between the channels. In the swamps, anaerobic conditions allow peat formation. Diagenetic processes lead to the formation of coal. Soils in which the trees grew may be preserved as **seat earth**.

The delta front

As the river meets the sea, the coarsest sand grains are deposited first and the finest silt last. Deposition occurs on the delta front. As the front of the delta advances into the sea, a vertical succession forms.

Sediments deposited on the delta front are called **foresets**. Coarse sand, often cross-bedded, is deposited at the top in crescent shaped **bars** that form at the **mouths** of the main distributaries. Here, energy is lost as the channels meet the sea. The crests of sand bars contain very clean sand because they are under the influence of waves and currents that remove fine material. Lower down there are likely to be finer sands and silts. Marine fossils may be found.

The prodelta

Deposition takes place in low-energy deeper water at the bottom of the delta front. These sediments are called **bottomsets**. They consist mainly of clays and silts, which are thinly bedded and lack sedimentary structures. They lithify to form shales, which may contain marine fossils.

Deltaic sequences

The sequence produced in a delta is a coarsening upwards sequence. It is repeated many times and each cycle is known as a cyclothem. It starts with marine shales and silts then fine to coarse sandstones, and finally coals, sandstones, seat earth and clay. Limestones are sometimes found at the base of the succession, representing normal marine conditions before the deltaic conditions begin.

Bioturbation and trace fossils are common, with marine fossils in the bottomsets and non-marine fossils in the higher parts of the deltaic sequence. Plant fossils are often found and sometimes, if the surface of the delta is exposed, soils and coal may form. There is often very rapid deposition of mud covering the plant material before it decays. Many of the swamps are anaerobic, which increases the chances of the plant material being preserved to form coal. The cyclothem, or part of it, may be repeated many times in succession, due to subsidence or emergence, leading to marine transgressions and regressions. These are caused by:

- Changes in sea level
- Local isostatic changes, due to the mass of sediment causing the sea floor to sink
- Changes in position of the delta lobes, due to migration of the channels
- Changes in the rate of sedimentation, allowing the delta to build out or be inundated by the sea.

KEY DEFINITIONS

Distributary a stream channel that takes water away from the main stream channel.

Cyclothems represent layers repeated due to cyclic sedimentation.

Topsets the uppermost horizontal layers of a delta commonly consisting of channel sandstones, coal and seat earth.

Seat earth a sandy or clay-rich fossil soil found beneath a coal seam. It represents the soil in which coal-forming plants grew and frequently contains carbonised traces of plant roots.

Foresets the inclined layers formed on the delta front, commonly consisting of cross-bedded sandstones.

Mouth bars crescent-shaped deposits of sand and silt forming below sea level, where distributaries enter the sea.

Bottomsets the lowest horizontal layers of a delta, commonly consisting of shales.

Bioturbation the disturbance of sediment by the activities of organisms (e.g. burrowing).

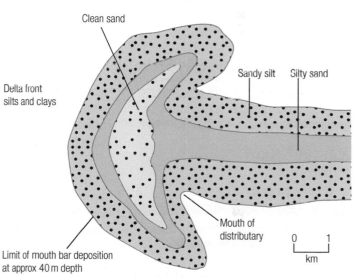

Clean sand
Sandy silt   Silty sand
Delta front silts and clays
Mouth of distributary
Limit of mouth bar deposition at approx 40 m depth
0   1
km

🔺 **Figure 5** Mouth bar sediments

QUESTIONS

1 Name the climatic zone where coal is likely to form.

2 Explain why there are many layers of coal in deltaic sequences.

3 Describe the succession of sediments that may form when deltas prograde.

4 Research the sediments that were laid down in Britain under deltaic conditions during the Upper-Carboniferous Coal Measures.

Banded iron formations (or BIFs) are units of sedimentary rock of **Archean** and **Palaeoproterozoic** age. They consist of repeated, thin layers (a few mm thick) of dark iron oxides, mainly hematite (Fe_2O_3) with some magnetite (Fe_3O_4), both of which contain ferric (Fe^{3+}) iron, alternating with lighter coloured bands of chert.

BIFs are an important commercial source of iron ore. There are large-scale mining operations at Pilbara in Western Australia, Minnesota USA and elsewhere.

HOW SCIENCE WORKS

BIFs have been studied for decades. The iron found in BIFs is Fe^{3+}, which is relatively insoluble. Because of this, early research was concerned with the problem of transport of Fe^{3+} from the land in solution in rivers for later deposition in seas. Theories involving the acidity of river waters were put forward. Later, following exploration of the ocean floor, new research indicated that the source of the iron was more likely to be hydrothermal. It came from vents and moved up into the oceans rather than down from rivers. Both the atmosphere and oceans lacked free O_2. Recently researchers looked for explanations in the **Great Oxygenation Event** at 2.3 billion years ago. At that time, cyanobacteria began to produce oxygen as a by-product of photosynthesis. This could have oxidised Fe^{2+}, which is soluble in water, to Fe^{3+} which would then be precipitated. However, there are BIFs that are older than the GOE and cyanobacteria do not live at the depths at which iron from a hydrothermal source would be found. More recently, some researchers have examined the role of **photoferrotrophs**. The search for explanations using theories, models and ideas is how science works.

BIFs and bacteria

Photosynthetic bacteria, photoferrotrophs, oxidise ferrous iron, Fe^{2+} into ferric iron Fe^{3+} using sunlight. They live in iron-rich environments, typical of Precambrian oceans before the GOE. They probably lived below the surface layer that was inhabited by cyanobacteria. The water in which the two organisms lived did not mix due to density differences at the **pycnocline**. The Fe^{2+} in Precambrian oceans is now thought to be from submarine hydrothermal vents. Research seems to show that photoferrotrophs alone could have precipitated enough Fe^{3+} to account for all the iron minerals present in BIFs. The chert bands may have been deposited by silica-forming organisms. The rhythmic nature of deposition suggests a cyclic variation in conditions. By producing oxygen as a by-product of photosynthesis, cyanobacteria are thought to have converted the early oxygen-poor, reducing atmosphere, into an oxidising one, causing the GOE which led to the near-extinction of anaerobic organisms, the remains of which could account for some of the chert.

KEY DEFINITIONS

Banded iron formations units of sedimentary rock of Precambrian age consisting of alternating layers of iron oxides and chert.

Archean an era within the Precambrian, covering a period from 4.0 billion to 2.5 billion years ago.

Palaeoproterozoic an era within the Precambrian, covering the period from approximately 2.5 billion to 1.6 billion years ago.

Great Oxygenation Event (GOE) the biologically induced appearance of dioxygen (O_2) in Earth's atmosphere. It occurred around 2.3 billion years ago.

Photoferrotrophs photosynthetic bacteria that oxidise ferrous iron Fe^{2+} into ferric iron Fe^{3+} using sunlight to create iron oxides and hydroxides instead of molecular oxygen.

Pycnocline the boundary separating two liquid layers of different densities in the oceans. It forms a lower limit to turbulence caused by mixing processes at the surface.

KEY FACT

Soluble ferrous iron, supplied from mid-ocean ridges and hydrothermal vents, was oxidised by **photoferrotrophs** to ferric iron in the following way:

$$4Fe^{2+} + CO_2 + 11H_2O + light \rightarrow [CH_2O] + 4Fe^{3+}(OH)_3 + 8H^+$$

CH_2O = Formaldehyde

STUDY TIP

Oxidation involves a loss of electrons, which gives ions an increased positive charge. Reduction is the reverse.

To remember which reactions are oxidising and which are reducing think of OILRIG.

(**O**xidation **I**s **L**oss, **R**eduction **I**s **G**ain)

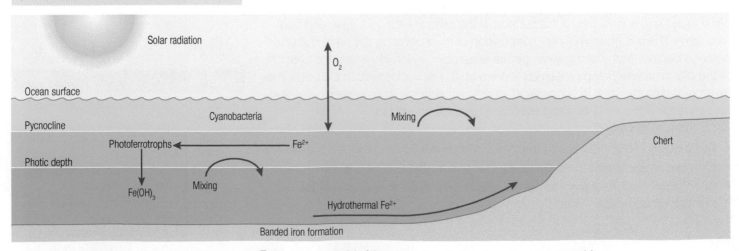

△ **Figure 1** A model of the origin of banded iron formations (not to scale)

Walther's law of facies

In a sedimentary sequence without unconformities, a vertical succession of facies represents sedimentary environments that once existed side by side and have migrated over one another through time. This law was first stated by Johannes Walther, a German geologist, in 1894. Looked at in another way, it implies that to be conformable, vertically adjacent facies must reflect those environments which occur side by side.

Carboniferous delta sequences

The previous spread on deposition in deltaic environments is an example of the way in which Walther's law can be applied. It's clear that the facies arranged vertically in a cyclothem are the same as the ones that exist side by side in present-day deltas. Cyclothems are characteristic of Carboniferous delta sequences in areas of Britain once mined for coal. The best known examples are the Yoredale Cycles of the north Pennines.

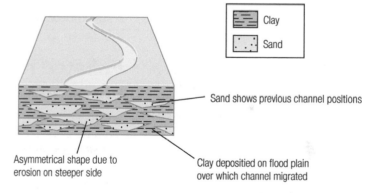

Vertical sequence in borehole is same as lateral seqence on the surface

▲ **Figure 2** Vertical and lateral facies variation in a delta (not to scale)

Fluvial environments and facies

In fluvial environments, meandering streams migrate across their floodplains due to bank undercutting on one side of the meander and deposition of point bar deposits on the other side. In between are the channel deposits. As the channel migrates, later sediments are deposited on top of earlier ones.

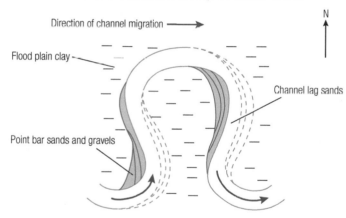

Clays covered by channel lag and then point bar deposits as stream channel migrates east.

◀ ▲ **Figure 3** Lateral and vertical facies variation in fluvial environments (not to scale)

Coastal environments and facies

Shallow siliciclastic seas and beaches along the coast migrate laterally during marine transgressions and regressions. The former positions of the coastal environments can be seen in the vertical sequence of facies. Here, the time lines cut through facies that were deposited at the same time. The facies boundaries cross the time lines, so in this case they are **diachronous**.

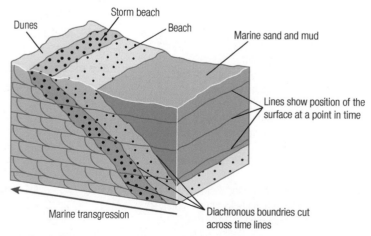

▲ **Figure 4** Migration of coastal sediments in a marine transgression (not to scale)

QUESTIONS

1 Draw annotated diagrams to explain how Walther's law of facies could be applied to hot desert environments.

2 Explain why the GOE is sometimes called the Oxygen Catastrophe.

4.1 Metamorphic rocks

Metamorphism is the **isochemical** process by which rocks are changed by either heat or pressure, *or* both heat and pressure. The chemical composition of the parent rock will be the same as the metamorphic rock produced.

Formation of metaquartzite and marble by contact metamorphism

The parent rock of metaquartzite is orthoquartzite, a type of sandstone composed of quartz grains held together by quartz cement. Quartz grains in the sandstone recrystallise, forming interlocking quartz crystals with irregular boundaries. The quartz crystals are equidimensional so there is no foliation. This texture is described as **granoblastic**. Any sedimentary structures or fossils in the parent sandstone are destroyed. The colour of metaquartzite is white or grey, unless there were other minerals in the original rock. If any iron oxide was in the parent rock, there will often be a pink colour.

The parent rock of marble is limestone. It is composed of calcite, which is stable over a wide range of temperatures and pressures. Metamorphism of limestone causes the original calcite crystals to grow larger. Calcite grains and fossil fragments in the limestone parent rock recrystallise to form an interlocking mosaic of equidimensional calcite crystals. There can be no foliation. Marble has granoblastic texture, but the crystals of calcite make it look sugary. Fossils are destroyed during metamorphism. Marble from pure limestone is white. Impurities in the parent limestone give some marble a range of coloured streaks. If there are clay minerals in the limestones, green or red minerals such as garnet may form. If there are sand grains present, a chemical reaction between calcite and quartz produces wollastonite ($CaSiO_3$), which can be light green, pinkish, brown, red or yellow.

Foliated rocks produced by regional metamorphism

These rocks have all been affected by some degree of pressure during regional metamorphism. Any platy minerals they contain take on a preferred alignment known as **foliation**. All the rocks below are foliated. The most common platy mineral is clay, so they all have shale as the parent rock. Slate, phyllite, schist and gneiss are all formed by the progressive metamorphism of a shale.

Slate

Shale is composed of clay minerals and fine quartz particles and is the parent rock of slate. Clay minerals are rich in aluminium and so are the metamorphic minerals in slate. Slate is mainly composed of clay minerals and mica (although chlorite and quartz may also be present). It is fine grained (grains <1 mm diameter) and shows **slaty cleavage**. Traces of original bedding may still be preserved as *relict bedding*.

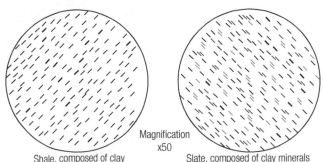

Alignment of minerals has changed due to pressure and recrystallisation during metamorphism

Magnification x50

Shale, composed of clay minerals aligned by compaction during diagenesis

Slate, composed of clay minerals and platy metamorphic minerals mica and chlorite

▲ **Figure 1** Shale and slate

Phyllite

Phyllite is produced by higher temperatures and pressures than slate. It is fine grained and crystalline. It often has a silky sheen due to the parallel orientation of platy minerals (mainly recrystallised mica and chlorite) and contains quartz. Phyllite is fissile (tends to split into sheets) and the foliation is commonly crinkled or wavy in appearance. Phyllites are usually grey or pale greenish grey in colour.

Schist

Schist is produced by higher temperatures and pressures than those producing phyllite. It is medium grained (1 to 5 mm) and crystalline. It can occur in a variety of colours, but always has a shiny appearance where the flat surfaces of muscovite and biotite mica crystals are visible. Schist typically contains mica and garnet. The garnets often form large crystals called **porphyroblasts**. The mica crystals are all aligned at right angles to maximum pressure, forming the texture **schistosity**.

Gneiss

Gneiss is formed by the highest temperatures and pressures during regional metamorphism. It is a coarse grained (>5 mm), crystalline rock with **gneissose banding**. Gneiss is typically composed of quartz and feldspar in the light bands and biotite mica (and other mafic minerals) in the dark bands. There is replacement of chlorite and existing micas by crystalline mafic minerals and biotite mica.

KEY DEFINITIONS

Porphyroblasts large crystals that have grown during recrystallisation in a metamorphic rock and surrounded by a finer grained groundmass of other crystals.

Schistosity the texture in medium and coarse grained metamorphic rocks formed by the preferred alignment of flat/tabular minerals. The alignment is perpendicular to the direction of stress applied during metamorphism. No traces of original bedding remain.

Gneissose banding the segregation of light- and dark-coloured minerals into layers or bands at the scale of mm to cm in thickness. The light band is normally granoblastic (granular) and the dark band normally shows schistosity.

KEY FACT: MYLONITE

This forms at depth, within the crust, where ductile deformation leads to recrystallisation producing a fine grained and highly deformed rock.

Summary table of metamorphic rocks

Parent rock	Metamorphic rock	Colour	Texture	Mineral composition	Type of metamorphism
Unfoliated rocks					
Limestone composed of calcite ($CaCO_3$)	Marble	White	Granoblastic Medium grain size (1–5 mm) grain size increases with metamorphic grade	Calcite (reacts with dilute HCl)	Contact or regional
Sandstone composed of quartz (SiO_2)	Metaquartzite	White or grey		Quartz	
Shale composed of clay minerals, mica and quartz	Spotted rock	Grey, purple, green or black – with darker spots	Poorly developed slaty cleavage Fine grain size (<1 mm)	Clay minerals and mica. Poorly formed minerals (mica, andalusite, graphite) in spots	Contact
Foliated rocks					
Shale, composed of clay minerals	Slate	Grey, purple, green or black	Slaty cleavage, fine grain size (<1 mm)	Clay minerals and muscovite mica, with some chlorite and quartz	Low grade regional
	Phyllite	Green / grey sheen	Wavy foliation, fine grain size (<1 mm)	Muscovite mica, chlorite and some quartz	
	Schist	Silvery sheen	Schistosity, medium grain size (1–5 mm)	Muscovite and biotite mica, quartz, garnet, and kyanite	Medium grade regional
	Gneiss	Dark and light bands	Gneissose banding, coarse grain size (>5 mm)	Biotite mica, mafic minerals, quartz, K feldspar and sillimanite	High grade regional

QUESTIONS

1 With the aid of labelled diagrams, describe the differences between schist and slate.

2 Describe one similarity and one difference between schist and gneiss.

3 Write a balanced equation for the formation of wollastonite marble. Wollastonite's formula $CaSiO_3$.

Grade and contact metamorphism

Metamorphic grade describes the temperature and pressure conditions under which metamorphic rocks form. As a body of rock is affected by a temperature and/or pressure increase, the grade of metamorphism increases.

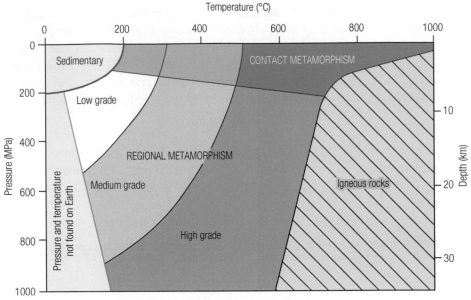

△ **Figure 1** Relationship between temperature, pressure and metamorphic grade

Contact metamorphism occurs when the **country rock** is affected by heat from a large igneous intrusion. Because temperature differences between the country rock and the intruded magma are greater at shallow levels in the Earth's crust, where pressure is low, contact metamorphism is described as high temperature, low pressure metamorphism. High temperature, not pressure, leads to the formation of altered, recrystallised, unfoliated rocks in a zone surrounding the intrusion. This zone is the **metamorphic aureole**. Around a large igneous intrusion, such as a batholith, the metamorphic aureole may be up to 10 km wide. Temperature decreases with distance from the contact with the intrusion and for this reason the effects of contact metamorphism are greatest near to the contact and decrease with distance. The outer part of the metamorphic aureole is only weakly metamorphosed and forms **spotted rock**. Some recrystallisation occurs, causing clusters of dark minerals to grow in separate spots. Metamorphic grade increases in all directions towards the intrusion.

a

b

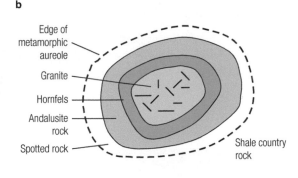

c

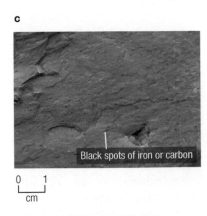

△ **Figure 2** **a** Andalusite porphyoblasts, **b** a metamorphic aureole showing contact metamorphism and **c** shale with spots of partial recrystallisation

The thermal gradient and index minerals in a metamorphic aureole

When a batholith is intruded into beds of shale, increases in metamorphic grade are marked by the appearance of an **index mineral**:

• Index minerals are metamorphic minerals, which are stable under specific temperature and pressure conditions. They indicate the metamorphic grade.

• Because contact metamorphism is caused by temperature only, an increase in grade represents a thermal gradient.

The dip of the sides of an intrusion has a major effect on the width of the metamorphic aureole. A shallow angle of dip gives a wide aureole and a steep angle of dip gives a narrow aureole. If the sides of the intrusion dip at different angles, then the metamorphic aureole will be asymmetric.

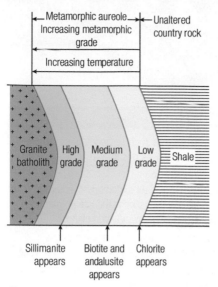

Figure 3 Sketch map showing index minerals and metamorphic grade

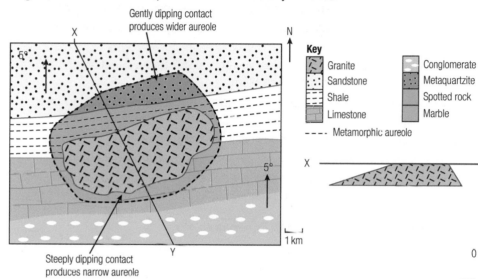

Figure 4 Map of an intrusion with dipping sides (contact)

Grades of regional metamorphic rocks

Regional metamorphism of shale produces the following rocks (all described earlier):

• Low grade: slate
• Medium grade: schist
• High grade: gneiss.

Regional metamorphism of orthoquartzite and limestones produces the same products as contact metamorphism – metaquartzite and marble. Each of these rocks is monomineralic as they are composed of one mineral only, quartz and calcite, respectively. The minerals are equidimensional, so they cannot align under pressure.

Grade and metamorphic minerals

Low-grade metamorphic rocks contain a higher proportion of hydrous minerals (clay minerals and chlorite contain H_2O in their crystal structure). High-grade metamorphism takes place at higher temperatures. H_2O is lost and anhydrous minerals (garnet for example) become more common. However, biotite, a hydrous mineral, is stable at very high grades of metamorphism.

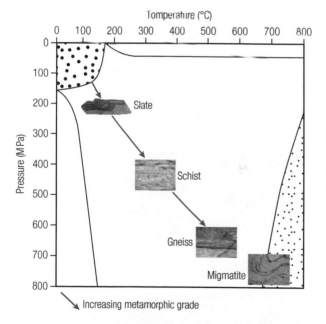

Figure 5 Regional metamorphic rocks and their relationship to pressure and temperature

QUESTIONS

1 Explain why metamorphic grade is not the same in contact and regional metamorphism.

2 Why does the metamorphism of shales provide so much information on metamorphic grade?

3 What units would you expect to use in measuring a thermal gradient?

4 Research the work of George Barrow in the Scottish Highlands and write a summary.

Metamorphic rocks are classified mainly based on their fabric (texture). This is because grain size and orientation tell us a lot about the conditions of metamorphism.

Over time, if rocks are subjected to directed pressure, a preferred orientation of the minerals develops at 90° to the pressure. If the minerals are flat or platy, foliation is produced. Foliation results from pressure and is a characteristic of rocks formed by regional metamorphism.

Slaty cleavage

Rocks with slaty cleavage will split into thin sheets along the cleavage planes. It occurs in fine grained rocks formed by low-grade regional metamorphism:

- It can only form in rocks consisting of platy minerals such as clay minerals, chlorite and micas.
- At the microscopic scale, these minerals align at 90° to the direction of maximum pressure during metamorphism.
- Slaty cleavage may be at any angle to bedding, but is usually parallel to axial planes of folds.
- It cannot occur in rocks with rounded grains, such as quartz in sandstones.

Bedding and fossils may not be completely destroyed by low-grade metamorphism, leaving traces or **relict structures**. Fossils may be deformed due to the high levels of compressive stress. Slates are common in North Wales and the Lake District.

a

Foliation produced by the alignment of flat minerals e.g. mica

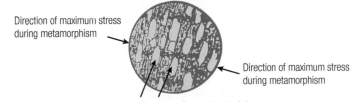

Direction of maximum stress during metamorphism

Direction of maximum stress during metamorphism

Flat minerals like mica align so that their long axis is at 90° to the direction of pressure

b

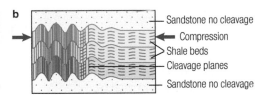

- Sandstone no cleavage
- Compression
- Shale beds
- Cleavage planes
- Sandstone no cleavage

Slaty cleavage

c Relict bedding is at a different angle from the cleavage

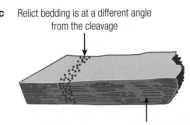

Slaty cleavage developed at 90° to maximum stress

▶ **Figure 1 a** Foliation produced by the alignment of flat minerals e.g. mica. **b** and **c** formation of slaty cleavage

Schistosity

Schistose texture is found in phyllites (fine grained) and schists (medium grained) and formed by regional metamorphism. Schistosity results from the alignment of flat, platy minerals, commonly muscovite mica, at 90° to the direction of maximum pressure during metamorphism. Light-coloured muscovite mica is concentrated into thin parallel bands, giving the rock a characteristic shiny appearance (micaceous sheen) where flat surfaces of mica are visible.

Garnet porphyroblasts are often present and they disrupt the alignment of mica minerals. Schists are found in the Highlands of Scotland and phyllites in the north-west of Arran, both in **Dalradian** rocks.

Garnet porphyroblast

Crenulation cleavage gives a wavy appearance

Schistosity caused by alignment of micas

▲ **Figure 2** Porphyroblastic fabric and crenulation cleavage in a schist (coin for scale)

Gneissose banding

Found in gneisses (coarse grained rocks formed by regional metamorphism), gneissose banding is formed when light- (usually quartz and feldspar) and dark-coloured minerals (usually biotite mica and mafic minerals) are separated into bands. The mica-rich layer is foliated and the pale layer is granoblastic. The bands may be contorted or folded but are roughly at 90° to the maximum pressure direction.

Porphyroblastic fabric

This fabric occurs in both regional and contact metamorphic rocks. Porphyroblasts are large crystals that grow during metamorphism and are surrounded by a finer grained groundmass. Metamorphic rocks that contain these large crystals are described as porphyroblastic. Pyrite porphyroblasts can develop in slate, often forming cubic crystals. Garnet porphyroblasts found in schists may contain **inclusions**, as shown in Figure 4. Examples like this provide clues about the timing of events during metamorphism. The early formed inclusions are not in a straight line. This suggests that the garnet must have been rotating as it grew, in response to shear stress. The foliation shown by the brown mica crystals splits and goes around the garnet rather than through it, so the garnet must have formed before the stress that aligned the mica.

Granoblastic fabric

This is an **unfoliated** fabric and is formed by thermal metamorphism. Pressure is not a factor in the formation of a granoblastic fabric. The main characteristics are randomly orientated, equidimensional crystals usually in rocks with few, and sometimes only one, mineral. **Hornfels** is an example of a fine grained rock with granoblastic fabric. Marble and quartzite are also granoblastic. Because of their medium grain size and white colour, their texture is sometimes described as sugary as the grain boundaries are indistinct.

Crenulation cleavage

Regional metamorphism continues over long time periods, during which the amount and direction of stress applied to rocks is likely to vary. So, several foliations may be present in a rock, not all in the same orientation, giving it a crenulation cleavage. It is a fabric formed in metamorphic rocks such as phyllite, schist and some gneisses where two or more stress directions cause the formation of one or more foliations superimposed on earlier ones, giving a wavy appearance to the fabric. Information such as this allows us to determine the number of phases of metamorphism that have affected a rock.

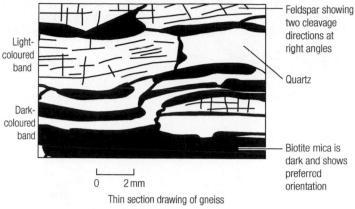

Light-coloured band

Dark-coloured band

Feldspar showing two cleavage directions at right angles

Quartz

Biotite mica is dark and shows preferred orientation

0   2 mm

Thin section drawing of gneiss

▲ **Figure 3** Gneissose banding

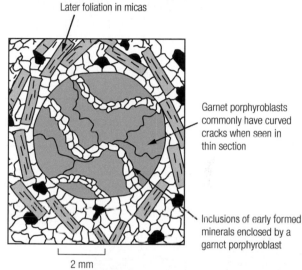

Later foliation in micas

Garnet porphyroblasts commonly have curved cracks when seen in thin section

Inclusions of early formed minerals enclosed by a garnet porphyroblast

2 mm

▲ **Figure 4** Thin section drawing of a garnet porphyroblast

▲ **Figure 5** Pyrite porphyroblast in slate, approximately 6 cm long

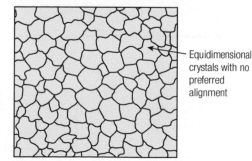

Equidimensional crystals with no preferred alignment

▲ **Figure 6** Granoblastic fabric (not to scale)

QUESTIONS

1 With the aid of labelled diagrams explain how crenulation cleavage is formed.
2 Explain why slaty cleavage commonly has a different orientation from relict bedding.
3 Explain how garnet porphyroblasts can be used to place metamorphic events in a time sequence.

The aspects of metamorphic fabric we have considered so far can be used as diagnostic properties. Figure 1 uses the presence or absence of foliation, grain size, mineral content, banding and spots and allows metamorphic rocks, and the type of metamorphism that produced them to be identified.

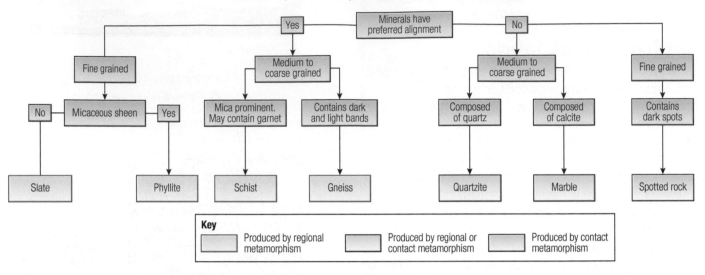

▲ **Figure 1** Identifying metamorphic rocks using their fabrics using a flow chart.

▲ **Figure 2** Gneissose banding

0 — 2
cm

More than this, fabrics can help us to reconstruct conditions of metamorphism during mountain building.

Fabrics, stress and time during mountain building

As temperature increases during mountain building, new metamorphic minerals grow. Higher temperatures increase the rate at which ions diffuse between minerals, but it is still a slow process because ions are moving through solid rock during metamorphism. The process is speeded up by water, which allows the ions to diffuse more rapidly. These reactions take place over millions of years, so the pressure and temperature conditions that produce metamorphism must exist over long periods of time. It also means that conditions will change over time. There are likely to be several periods of deformation. Long time scales mean that there will be changes during the process. Minerals that form early in metamorphism sometimes grow over existing sedimentary features and this can help to identify their parent rock. This evidence can be destroyed by deformation and later metamorphic recrystallisation as pressure and temperature change and increase. Using a microscope allows more detailed information about the relationship between the growth of new minerals and the effects of deformation to be studied.

Time and deformation

If stress is applied to rocks very rapidly, for instance in fault zones, mineral grains suffer brittle fracture and are crushed and ground mechanically as the fault moves. This process is cataclasis. It produces a variety of fabrics ranging from coarse grained angular clasts in a fine grained matrix, to rock consisting almost entirely of finely crushed particles. These fabrics are characteristic of dynamic metamorphism, but will occur in fault zones as part of mountain building.

If stress is applied to rocks slowly, over a long period at low strain rates, the rock does not fracture but deforms gradually. Although temperature is important, foliation develops largely in response to stresses. Later phases of deformation may produce foliation with a different orientation and by recognising these differences in metamorphic fabric, a history of changing pressure and temperature during regional metamorphism can be determined.

Using fabrics to reconstruct conditions of metamorphism

Interpreting the history of metamorphism during mountain building uses the same principles you will become familiar with later on in the course, when you work on the geological history of an area shown on a map or cross-section diagram. If you want to have a preview, look at the section on relative dating now. Metamorphic minerals can be related to a time scale that is based on successive phases of deformation. Some of this can be done in the field. The most useful minerals are porphyroblasts. If they cut across a cleavage, or just grow randomly on it, this shows that they must have grown later than the cleavage. Veins of metamorphic minerals that cut across a fold must have been formed later than the fold. Foliated minerals in a cleavage that has been folded must be earlier than the folding, providing the minerals have not recrystallised.

Fabrics in thin section

Using a microscope allows more detailed information about the relationship between the growth of new minerals and the effects of deformation to be studied. Whether minerals were present before, during or after each phase of deformation can be worked out in many cases by using thin sections.

- Pre-kinematic metamorphic minerals appear to have been affected by later deformation.
- Syn-kinematic metamorphic minerals show that an episode of deformation took place at the same time as mineral growth.
- Post-kinematic minerals grew across an earlier foliation. They grew after the deformation took place.

Working out the history of an orogenic belt may involve studying many hundreds of samples.

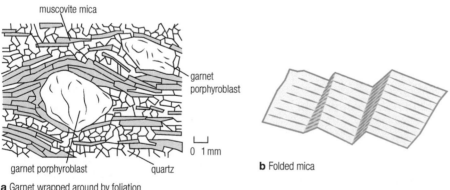

b Folded mica

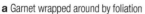

muscovite mica

garnet porphyroblast

0 1 mm

garnet porphyroblast   quartz

a Garnet wrapped around by foliation

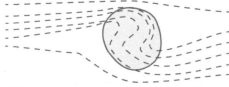

d Mineral growth during shear along the cleavage

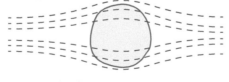

e Mineral growth during compression at right angles to cleavage

Fabrics and chemical reactions

One group of minerals can be related to other groups by looking for evidence of chemical reactions. Metamorphism rearranges the chemistry of rocks, but the bulk chemistry of the parent rock is retained. By using experiments and theories about the thermodynamics of mineral reactions it is possible to suggest pressures and temperatures involved in the growth of metamorphic minerals. The evidence for these chemical reactions comes from metamorphic fabrics.

🔺 **Figure 3** Field exposure of metamorphic rock. Porphyroblasts shown are 5mm in diameter.

Garnet porphyroblasts

🔻 **Figure 4** Metamorphic fabrics in thin section diagrams
a and **b** pre-kinematic minerals; **c**, **d** and **e** syn-kinematic minerals; **f** post-kinematic mineral

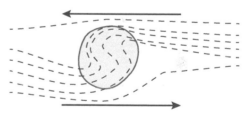

c Mineral growth during shear along the cleavage

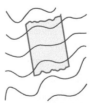

f Mineral enclosing earlier folds

QUESTIONS

1 Describe the metamorphic history of the rock shown in Figure 2.

2 Draw a sketch of Figure 4d and add arrows to show the direction of shear stress.

3 In Figure 3 the diameter of the porphyroblast is 5 mm. Calculate the magnification used for this photograph.

Index minerals and contact metamorphism

Metamorphism can be described as low, medium and high grade and can be determined by the use of index minerals.

In contact metamorphism, biotite is the low-grade index mineral found in spotted rocks.

Andalusite is the low to medium temperature, low pressure Al_2SiO_5 **polymorph** found in andalusite slate. It indicates medium grade of metamorphism.

With increasing metamorphic grade, contact metamorphism follows a path from andalusite to sillimanite on the Al_2SiO_5 polymorph phase diagram. Sillimanite indicates high grade and is found in hornfels.

Kyanite, the high pressure, low temperature Al_2SiO_5 polymorph, is not found in contact metamorphic rocks due to the lack of pressure.

Mapping the Dalradian Supergroup

- In 1893, George Barrow mapped a sequence of highly deformed regionally metamorphosed rocks in the south-eastern part of the Scottish Highlands. The metamorphism and deformation occurred during closure of the Iapetus Ocean and during the Caledonian orogeny, about 400 Ma. These Precambrian rocks are known as the Dalradian Supergroup.

- Clay-rich sedimentary rocks such as shale produce a variety of metamorphic minerals, as temperature and pressure conditions change. When Barrow mapped rocks like these, he noticed that there was a pattern to the occurrence of metamorphic minerals. He used the first appearance of these minerals, which he termed index minerals, to plot **isograds** and map metamorphic zones. Some of the minerals that crystallise at low grades are stable at higher grades so more than one index mineral may be found in one rock.

- He was able to map **metamorphic zones** using index minerals and isograds, which define the boundaries of the zones. Although he did not do all the mapping personally, the system he devised was named after him and the zones are called **Barrovian zones**.

Index minerals and metamorphic zones

	Metamorphic grade		
	low	medium	high
Rock type	slate	schist	gneiss
Index minerals and metamorphic zones	Chlorite Biotite	Garnet Kyanite	Sillimanite

The chlorite zone represents low-grade (low pressure and low temperature) regional metamorphism. The rock is slate where most of the rock has recrystallised but some clay minerals may still exist.

Schists develop as a result of increasing temperatures and pressures and can be found in both the biotite and garnet zones. The crystal sizes increase with metamorphic grade. Schists formed at lower temperatures and pressures are composed of quartz, muscovite mica and biotite mica. Medium-grade metamorphism results from higher temperatures and pressures and many schists formed at this grade contain garnet, and less commonly, kyanite porphyroblasts.

Andalusite slate

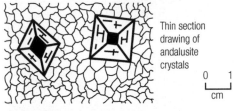

Dark-grey fine grained rock

Relict bedding

Slaty cleavage

Crystals of andalusite

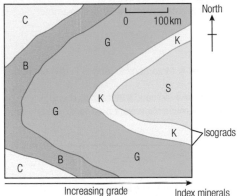

Thin section drawing of andalusite crystals

0   1
cm

▲ **Figure 1** Andalusite and sillimanite

▲ **Figure 2** Index minerals, isograds and metamorphic zones

North

0   100 km

Increasing grade

Index minerals
S = Sillimanite
K = Kyanite
G = Garnet
B = Biotite
C = Chlorite

Isograds

KEY DEFINITIONS

Polymorphs minerals that have the same composition but occur in different crystal forms.

Isograd the name given to a line on a map joining points of equal metamorphic grade. They join places where the first appearance of an index mineral occurs.

Metamorphic zone the name given to an area between two isograds. The zone is named after the lower grade isograd. All locations within a metamorphic zone experienced the same metamorphic grade.

Barrovian zones are metamorphic zones mapped using index minerals identified by George Barrow.

The Al_2SiO_5 polymorphs kyanite and sillimanite are found in regional metamorphic rocks. A rock formed at high pressure and low temperature may contain kyanite.

Kyanite is typically found in gneisses and the kyanite zone represents high-grade regional metamorphism. With increasing metamorphic grade, regional metamorphism follows a path from kyanite to sillimanite, shown on the Al_2SiO_5 polymorph phase diagram.

The sillimanite zone represents high-grade regional metamorphism with very high temperatures and pressures. The rocks are gneisses. An estimate, based on the sillimanite zone, indicates a maximum temperature of about 700°C and maximum pressure of about 700 MPa. This pressure exists at a depth of about 25 km below the surface of the continental crust. It gives us a geothermal gradient of about 28°C km^{-1}.

Quartz and plagioclase feldspar are stable throughout the whole range of grades. This makes them no use as index minerals.

Metamorphic facies

In metamorphic rocks, changes in mineral assemblages are due to changes in temperature and pressure conditions. These mineral assemblages are an indication of the temperature and pressure environment during their formation. A metamorphic facies is a set of metamorphic minerals that were formed under similar pressures and temperatures.

So, rocks containing a particular assemblage of minerals can be linked to certain tectonic settings and times, and can be used to reconstruct the conditions of metamorphism. The fields for each facies in Figure 5 have spaces between them because the change from one facies to another is gradual.

The sequence of metamorphic facies seen in an area, depends on the geothermal gradient during metamorphism. Gradient C in Figure 5 represents an area around an igneous intrusion, and rocks there would belong to the hornfels facies. Gradient R represents an area of regional metamorphism. Rocks would show zeolite, greenschist, amphibolite, and eclogite facies as grade of metamorphism and depth of burial increased.

Retrograde metamorphism

Metamorphism we have looked at so far is **prograde**, during which fluids such as H_2O and CO_2 are driven off. These fluids are necessary to form the hydrous minerals that are stable at the Earth's surface. Retrograde metamorphism involves mineral hydration, carbonation and oxidation as metamorphic rocks are returned to the surface. Chemical reactions take place more slowly as the temperature is decreased.

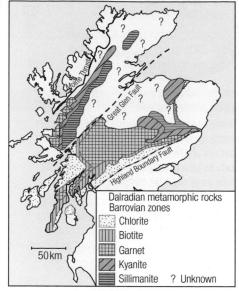

▲ **Figure 3** Regional metamorphic zones

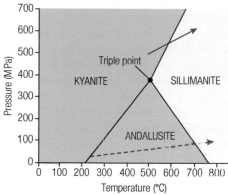

▲ **Figure 4** Temperature and pressure fields for the Al_2SiO_5 polymorphs

KEY DEFINITION

Prograde metamorphism is the recrystallisation of a rock in response to an increase in the intensity of metamorphism.

HOW SCIENCE WORKS: AL_2SIO_5 POLYMORPHS

These are Kyanite (high pressure, low temperature), Sillimanite (high pressure, high temperature) and Andalusite (low pressure, high temperature). The presence of one or more of these minerals tells us a lot about the pressure and temperature regimes during the rock's formation.

QUESTIONS

1 Explain why andalusite is not formed by the contact metamorphism of pure limestone.

2 Explain why clay-rich parent rocks are the most useful in mapping metamorphic zones.

3 Explain the probable tectonic setting for rocks metamorphosed along thermal gradient S, in Figure 5.

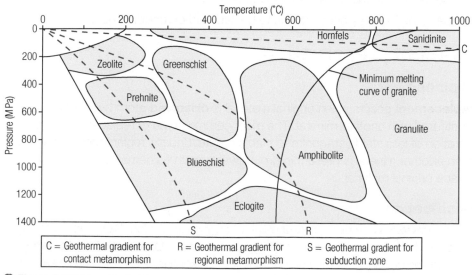

▲ **Figure 5** Metamorphic facies

5.1 Preservation of fossils

KEY DEFINITIONS

Fossil used to describe any trace of past life. Fossils may be parts of organisms, such as teeth or shells or whole organisms such as body fossils. Fossils may also be traces of organisms, such as footprints and burrows.

Organism an individual life form, such as an animal, plant, fungus or bacteria. Organisms may be a single cell or multicellular (many cells). Organisms may be preserved as fossils.

Body fossil the hard parts of an organism, such as the skeleton or shell.

Replacement atom by atom substitution of one mineral for another.

Dissolution the process whereby minerals that make up the fossils are dissolved away and removed in solution by groundwater.

Taphonomy the study of the process of fossilisation from the death of the organism to discovery of the fossil. It is an overlap area between biology and sedimentary geology.

Mould the impression of the outside or inside of a fossil.

Cast an in-filled fossil void, usually with another mineral.

What are fossils?

Fossils are the preserved remains of living **organisms** or the traces of those organisms when they were alive. Most fossils are the hard parts of whole or fragmented organisms, such as the shells of a bivalve or the skeleton of a dinosaur. These are called **body fossils** as they represent the skeletal remains, or 'hard parts' of the organism. The other main types of fossils are trace fossils including tracks, trails and burrows.

Factors that affect fossilisation

The type of fossilisation varies with the conditions during the life, death, transport and burial of an organism and there are many changes in the sediments after burial. These changes are due to taphonomic processes, which are the biological, chemical and physical processes that change organisms after their death, resulting in the preservation of the organism as a fossil in a rock.

The most important factors are listed below:

Original composition. Many fossils are made of calcite or aragonite, which can be easily altered. Hard parts made of silica, such as radiolarian microfossils and some sponges, may be preserved unaltered.

Energy levels. High energy produces lots of fragments due to collisions causing breakage; low energy produces more complete fossils due to lack of water movement and lack of collisions.

Transport distance. Fossils are fragmented during transport. This is due to abrasion or collisions causing breakage.

Rapidity of burial. Faster burial means more chance of whole body fossils being preserved and less chance of scavengers eating them.

Amount of oxygen. The presence of oxygen accelerates the breakdown of organisms due to bacterial decay and encourages scavengers.

Size of sediment. Fine sediment preserves organisms better than coarse sediment. Only poor-quality fossils can be found in coarse sandstone or gravel, but fossils are found in a wide variety of sediments.

Diagenesis. These are the changes within the sediments after burial. The composition and acidity of the percolating groundwater is important, as it may dissolve or replace the fossil with another mineral.

Compaction. This may cause fossils to be flattened on bedding planes.

Types of preservation

Replacement

Replacement occurs when original material is **dissolved** atom by atom and substituted with another mineral. The most common type of replacement is the alteration of less stable aragonite to stable calcite. Equally, replacement can occur when another mineral is present in groundwater, such as hematite, which can replace original material.

Silicification

This occurs when percolating groundwater rich in silica dioxide (SiO_2) moves through the rock. The minerals dissolved in the groundwater crystallise out of solution and fill any pores or voids that may be present. If the fossil has been dissolved away by

CASE STUDY: TAPHONOMY

This is the study of organisms and how they have become fossilised. It often includes details of bacterial decay, which destroys soft tissue. Taphonomic processes can be biological, chemical or physical.

Biological processes (after death) include alteration of the organisms by micro-organisms such as bacteria in the process of decay and by macro-organisms such as the actions of predators or scavengers.

Chemical processes include diagenetic changes in the sediment that eventually form the rock.

Physical processes include the damage to an organism's skeleton after death, for example by movement on the sea floor and collisions with other debris.

earlier movement of groundwater, then the SiO_2 simply fills the voids in the rock. However, if wood, shell or bone is still present, the pores may be filled by a mineral, increasing the density of the rock. Petrified wood is a common example.

Carbonisation

The process occurs during burial as the overlying mass of rocks increases the pressure and temperature in the rock. This allows volatiles within the organic material to be driven off. This reduces the amount of oxygen, methane, carbon dioxide and water and increases the carbon content overall, preserving organisms such as leaves as a thin film of carbon within shale or mudstone.

Pyritisation

This is the replacement of original material by iron pyrites. It takes place when the environment is devoid of oxygen (anaerobic) and the only live organisms are sulfur bacteria. The bacteria use sulfur in the environment to respire, which reduces the sulfur to bisulfate. The bisulfate then reacts with any iron in the environment to form iron pyrites. This then replaces the fossil material. Pyritisation commonly occurs in deep sea environments with an anoxic sea bed, or in shallow swamps.

Mould and cast formation

Moulds are formed when fossils are dissolved out of the rock they are in. This leaves a void or hole in the rock. Breaking this rock open will reveal an external mould of the fossil. If the fossil was filled with sediment before complete burial, the sediment may reveal an internal mould showing internal structures along with any impressions of soft tissues within.

A **cast** is formed when the void is in filled with another mineral, such as iron pyrites or silica. These can be seen as counterparts to external moulds when breaking the rock open. Equally, casts can be made in the laboratory by in filling moulds with latex or modelling clay.

▲ **Figure 1** Pyritisation of an ammonite

▲ **Figure 2** Carbonisation of a fern

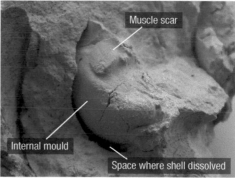

▲ **Figure 4** Internal mould in Jurassic Portland limestone. The space is where the original shell has been dissolved away.

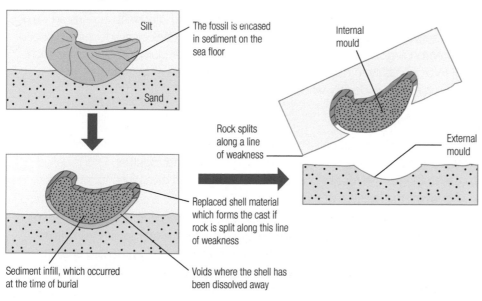

▲ **Figure 3** Diagram to show mould and cast formation

HOW SCIENCE WORKS: ARAGONITE VERSUS CALCITE

Aragonite and calcite are polymorphs of one another, with the formula $CaCO_3$. This means that the bonds are arranged in a slightly different way but the minerals have the same composition. Some shells such as bivalves and corals are made from aragonite, which is less stable than calcite. Aragonite rearranges its chemical structure to form calcite over time. Unaltered aragonite is only found in recent deposits and can usually be recognised by a milky appearance (translucent) compared to calcite, which is usually transparent.

QUESTIONS

1 Explain how silicification, pyritisation and carbonisation differ.
2 What is the difference between moulds and casts?
3 Explain how transport distance, size of sediment and diagenesis can affect the quality of fossil preservation.

Organisms have evolved to exploit all environments. They live in various parts of the water column, on or in the sediment. They may hunt, scavenge or filter feed. They may swim, float, crawl or remain fixed in one position. The flow diagram shows the technical terms used to describe different modes of life of **extant** and **extinct** organisms.

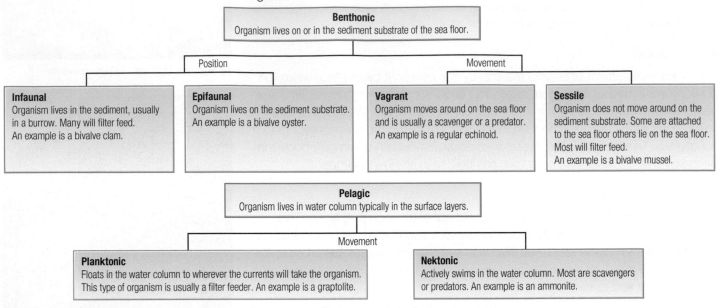

Figure 1 Flow diagram showing terminology of different modes of life

What is a species?

You will be familiar with the definition of a species from GCSE Biology; defined as a group of organisms that can interbreed to produce fertile offspring. We have no idea if organisms that looked similar from the fossil record could interbreed, so instead we use an alternative definition, a morphological species. A morphological species is grouped using morphological similarities (shape) and ignores other biological differences such as DNA or the ability to reproduce. The morphology is the most important factor in classifying fossils, especially when there are no similar living organisms.

Assemblages

It is rather unusual if you only find one type of fossil in a rock. If conditions for life were good, then many different types may be found together. This is called an assemblage. The fossils found in assemblages can give clues to the palaeo-environment that existed at the time.

When organisms die, their shells are usually transported and deposited elsewhere. This is called a **death assemblage** because the fossils are not in their original 'life' positions. Death assemblages can be recognised by fragmented shells that are sorted by size or aligned by a current. Death assemblages are made up of **disarticulated** shells.

Fossils found fossilised in their living positions form a **life assemblage**. For example, a reef community or burrowing organisms preserved in their burrows. They are rarer than death assemblages. These are also called **geopetal structures** as these can be used to determine the way up of the rock in which they are found.

Derived fossils were originally preserved in an 'older' sedimentary layer, then eroded, transported and deposited in a 'younger' sedimentary layer. For example, a Jurassic ammonite found in Quaternary sediments must be derived. They are often poorly preserved.

**HOW SCIENCE WORKS:
EXTINCT ORGANISMS**

Problems occur when we do not really know the mode of life of a fossil group. This is certainly true for the extinct trilobites, ammonites and dinosaurs. For some groups, we can only infer the mode of life, by comparing them with modern-day equivalents. For example, belemnites are similar to modern-day squid, and so we infer the same mode of life. However, there are unusual fossils that do not resemble any modern-day organisms which are more problematic.

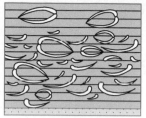

Mudstone

Sandstone

0 cm 4

Life assemblage   Death assemblage

▲ **Figure 2** Life and death assemblages

Adaptations and palaeo-environments

As a rule, *thick*-shelled organisms, with a lot of **ornament** or very **robust** forms were strong enough to withstand *higher energy* conditions. *Thin*-shelled organisms which lacked ornament and were less robust usually inhabited *lower energy* conditions as they would be destroyed by higher energy conditions. This could then be backed up by other evidence, such as pockets of broken shells concentrated in layers and the type of sediment itself. Times of low energy may result in finer grained sediments, such as micritic mud, whilst times of higher energy may result in layers of silt and sand. The table below shows some possible assemblages and an interpretation for each assemblage.

Other evidence, such as the nature of the sensory organs in an organism, may give us extra evidence about the environment. For example, trilobites lacking eyes may have lived in the dark at depth or in burrows, suggesting an oxygenated substrate which consisted of soft sediment. Those with standard compound eyes probably lived in the photic zone, needing eyes for hunting or scavenging.

Palaeo-environment and assemblages

Environment	Possible assemblage	Description
High energy continental shelf	Thick-shelled brachiopods and bivalves. Fragments of trilobites, or other broken fossils such as corals. Microfossils.	Fragmentation suggests high energy. This is supported by the thick-shelled fauna.
High energy shallow marine	Thin-shelled burrowing bivalves. Thick-shelled bivalves or those which show methods to attach to the substrate. Broken fossil remains. Microfossils.	Fragmentation suggests high energy. Burrowers do not need thick shells, but this suggests soft sediment. May be in the **littoral zone** or shallow marine.
Low energy shallow marine	Brachiopods or epifaunal bivalves with wide shells which are mostly **articulated**. These do not show any method of attachment. Irregular echinoids. Burrowing bivalves. Microfossils.	Wide shells suggest the need to spread the weight on a soft substrate. Burrowers also suggest a soft substrate.
Low energy deep marine	Complete specimens of graptolites or cephalopods. May have many microfossils.	No bottom dwellers. This could mean that the conditions on the substrate were **anoxic**. Complete specimens suggest low energy as they have not been broken up.
Low energy deltaic / terrestrial	Plant stems, leaves and spores. Insects and gastropods.	Presence of plant material may mean terrestrial or deltaic conditions. It certainly signifies a close proximity to land.

KEY DEFINITIONS

Extant organisms are still alive today.

Extinct organisms are no longer alive today.

Death assemblage a collection of organisms found in a different place and position than they occupied in life, such as a collection of disarticulated shells.

Disarticulated organisms found as fragments, such as separate shells or parts of trilobites.

Life assemblage a collection of organisms found within sediments in the same position as they would have been when they were alive, such as a bivalve in a burrow.

Geopetal structures allow us to see the way up of a rock, for example a coral or bivalve in life position.

Derived fossil weathered out of one rock and re-deposited in another. Different fossils may give conflicting dates.

Ornament expressed on the surface of the fossil, such as ribs, tubercles, spines and growth lines.

Robustness the ability of the fossil to resist abrasion. Robust forms are more likely to be preserved whole or with slight damage only.

Articulated organisms found whole or connected, as in life.

Littoral zone the high-energy area between high and low tide.

Anoxic or anaerobic conditions lack oxygen and are unsuitable for life.

HOW SCIENCE WORKS: THIN SHELLS AND THE INTERPRETATION OF PALAEO-ENVIRONMENTS

Thin-shelled organisms are usually interpreted as living in a low-energy environment, as the organism did not need to be robust to survive. However, if they lived within the sediment itself, then they are protected from the high-energy conditions and so may also be thin shelled. In such cases the sediment type can help us decide the energy levels, as sand would indicate high-energy conditions and mud low-energy conditions.

QUESTIONS

1 Explain why it is useful to know the rock type when looking at fossil assemblages.

2 Describe how shell thickness and ornament can help us interpret the palaeo-environment. Give reasons for your answer.

KEY DEFINITIONS

Trace fossils formed by benthonic infaunal and benthonic epifaunal organisms. These can be aquatic or terrestrial, although the preservation potential for terrestrial traces is poor.

Tracks footprints of an organism made when it moved along the sediment.

Trails impressions of animals which were travelling. This could have formed due to part or all of the animal dragging along the surface of the substrate, for example a trilobite tail.

Resting traces a type of trail as the whole body of the animal had stopped moving (rests).

Terrestrial refers to anything formed on land.

Bioturbation refers to burrowing or working the sediment in a way that disrupts the bedding. This is caused by the activity of living organisms.

Substrate the name given to the sediment or rock on the sea floor.

What is a trace fossil?

Trace fossils provide glimpses of the nature and behaviour of ancient organisms in the geological record. They can help us interpret the palaeo-environment at the time when they were formed.

This type of fossil has an exceptionally long fossil record, and some of the evidence of the earliest multicellular animals is derived from trace fossils.

Trace fossils preserve the activity of an organism, not the organism itself. These traces can be regarded as fossil behaviour or evidence of the lives of the organisms that made them. These include **tracks**, **trails**, burrows, borings and excrement (coprolites).

Trace fossils are not necessarily formed by the same organism, even though they may look identical. For example, a burrow could have been inhabited by a bivalve, a crustacean or a worm, which have not been preserved. One animal may also produce many different types of trace fossil, such as walking tracks, trails or **resting traces**.

How trace fossils are made

An organism walks across soft, fine sediment and leaves the imprint of its feet. There is then a very small chance that the footprints will be filled in by sediment before they are destroyed by water currents or wind. If they are infilled, the trace fossil may be seen on the base of the infilling sediment, so may stand out.

Skolithos × 0.5
Vertical burrow

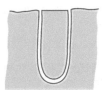

Diplocraterion × 0.5
U-shaped burrow

Thalassinoides × 0.2
Branching burrow

Rusophycus × 1
Trilobite resting trace

Cruziana × 1
Trilobite walking tracks

◔ **Figure 1** Trace fossil morphology

Types of trace fossils

Most trace fossils are found in rocks that formed in low-energy environments because higher-energy conditions would immediately destroy the traces left. The sediments themselves are generally fine grained, such as mud, clay or silt. **Terrestrial** trace fossils are much rarer than those found in marine conditions, because they are more likely to be weathered or eroded before they can be preserved.

A comparison of the main types of trace fossil is given in the table.

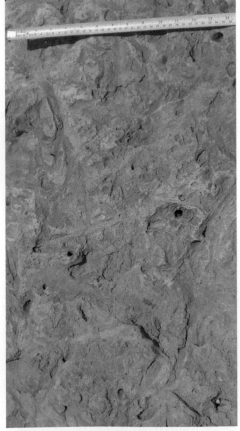

◔ **Figure 2** Burrows both vertical and horizontal with sediment mixed by bioturbation

	Description	Conditions	Example
Tracks	Found on the surface of bedding planes, formed as an animal walked across its surface	Soft or muddy fine grained sediments. Terrestrial or marine	Walking traces of arthropods, dinosaurs or even humans. Usually the imprints of legs or feet
Trails	Traces made by whole or parts of animals, at rest or travelling along the surface of the sediment	Soft or muddy fine grained sediments. Terrestrial or marine	Resting traces may show gills or leg structures. Examples are a trilobite dragging its tail behind it or a starfish impression
Burrows	Vertical, U-shaped, stacked or branching burrows of variable sizes. May be for dwelling, locomotion, protection or feeding	Substrate must be soft sand or mud to allow the organism to burrow. Burrowing mixes layers of sediment by **bioturbation**	U-shaped *Diplocraterion*, vertical *Skolithos* or branching *Thalassinoides*
Borings	Structures formed in rock or wood. May be for dwelling, protection or feeding	Evidence of a hard **substrate**	Burrowing bivalves such as *Teredo*
Excrement	Faecal pellets (<10mm) and coprolites (>10mm). Evidence of feeding	They need to be covered quickly, before they break up in currents or bacterial action breaks them down	Evidence of large animals in the environment. Dinosaur dung
Root structures	Woody-looking impressions as anchorage structures in rocks or preserved as lignite	Shallow marine, deltaic or terrestrial conditions	Roots often branching from trees or other plants

CASE STUDY

A track of a dinosaur gives us information about the animal that made it:
- Shape of the soft parts can be seen as impressions in very fine sediment. The familiar three toes are often evident, along with impressions of claws.
- Patterns of scales on the skin are occasionally preserved in very fine sediment.
- Mass can also be estimated by looking at the size and length of the feet. The extent that the feet have sunk into the sediment also gives us information about the size and mass of the animal.
- Running or walking speed can be calculated by estimating the height of the animal based on its foot length. Generally, steps further apart mean that the animal was moving faster than those close together.

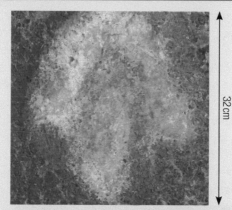

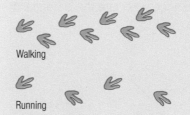

Walking

Running

🔺 **Figure 4** Dinosaur tracks and photographs showing dinosaur footprints

🔺 **Figure 3** Large U-shaped burrow into limestone infilled with clay

QUESTIONS

1 Define the term *trace fossil*.
2 Explain why some trace fossils may give us information about the type of substrate in which the organism lived.
3 Explain why trace fossils on land are rarer than those found in marine conditions.

The abbreviation **Ma** is often used to indicate a date before the present, in millions of years.

The abbreviation **myr** is often used for the duration an event in millions of years.

For example, dinosaurs existed for 165 myr. Their extinction was 65 Ma.

However, neither of these are accepted SI units.

KEY DEFINITIONS

Catastrophism the theory that changes in the Earth's crust during geological history have resulted chiefly from sudden violent and unusual short-lived events.

Gradualism assumes that change comes about gradually or that variation is gradual in nature. It is a model applied especially in evolution (Phyletic gradualism) where one species is transformed into another by slow changes.

Uniformitarianism maintains that slow, incremental changes such as erosion, created all the Earth's geological features. Uniformitarianism holds that the present is the key to the past: that the geological processes observable now were acting in the same way in the past.

CASE STUDY: SEDIMENTATION RATES

Accumulation rates of sediments	
Sediment	**Range of rates (mm per year)**
Calcareous ooze	0.01 to 0.1
Reef limestones	1 to 300
Mudstones	0.001 to 1
Sandstones	1 to 10 000
Conglomerates	50 to 50 000

The table shows the great range of rates of sedimentation. These variations are often linked to grain size but the process is also dependent on climate and tectonics which vary over time. Think about the erosion and transport of sediment from newly created mountains in comparison to the gentle accumulation of material in ancient lowlands. Contrast the sporadic and feeble sedimentation from arid continents compared to the rapid deposition at deltas. No wonder attempts to calculate the age of the Earth from the thickness of sediments produced such a range of answers.

Catastrophism!

In 1654 Archbishop Ussher had calculated that the Earth was approximately 6000 years old. The date quoted is 4004 BC on Saturday 22 October. This was based on his reading of the Old Testament in the Bible and estimating time taken for the events described. This idea constrained geoscientists; they needed to account for the formation of the geology they were observing in a very short time. **Catastrophism**, the idea that rocks and structures were formed by violent, large-scale and short-lived events, was a popular theory. After all, there *were* rapid geological events such as flood deposits, volcanism and earthquakes.

These catastrophes also neatly accounted for the extinction of species being noted in the fossil record. Georges Cuvier (1769–1832) had recognised the basics of biostratigraphy and knew that the Earth had existed for some millions of years. In establishing vertebrate palaeontology, he demonstrated that some fossils must represent animals that no longer exist.

Gradualism

In contrast to the catastrophists were those who proposed that life forms had evolved very slowly by gradual changes (**gradualism**). This meant that the rocks they were found in had also taken a long time to form.

James Hutton (1726–1797), was a Scottish geologist, often known as the father of geology, observed the processes of erosion which made sediment, which would form the sedimentary rock of the future. He realised that the driving force for mountain building and volcanism was the great heat and pressure within the planet. Published in 1788 under the modest title of *Theory of the Earth*; he called this formation and destruction of rocks 'the great geological cycle' and observed that it had occurred many times. Hutton concluded, 'that we find no vestige of a beginning, no prospect of an end'. His appreciation of the immense time spans needed to form strata and structures was enhanced by observations made of angular unconformities, such as at Siccar Point (Figure 1). He also observed that processes such as deposition and lithification were very slow and that there was no reason to suppose that such processes were any different in the past.

This view of **uniformitarianism**, 'the present is the key to the past' was popularised by Charles Lyell (1797–1875) published in *Principles of Geology*, an extensive work in many volumes. This fundamental idea in geology has meant that the age of the Earth had to be extended greatly to accommodate the potential age of the rocks observed at the surface and in mines.

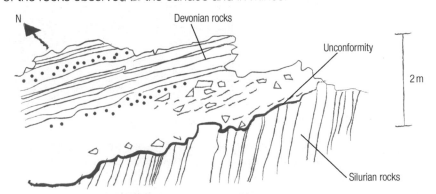

🔺 **Figure 1** A field sketch of Hutton's unconformity at Siccar Point looking NE. The lowest, steeply-dipping strata are Lower Silurian rocks which were deposited on the floor of the Iapetus Ocean around 435 Ma, folded in the Caledonian orogeny and subsequently eroded. The darker line is the unconformity and represents a time gap of some 65 Ma. Above the unconformity is a basal conglomerate containing fragments of the Silurian rocks deposited in the Late Devonian (370 Ma).

Hutton's rock cycle was limited as there was no understanding of the mechanisms due to plate tectonics nor information about the Wilson cycle.

The modern understanding of the process is not a strict cycle. The addition of basic magma from the mantle at the ocean ridges and the processes of subduction and magmatic differentiation result in the continuous generation of continental crust. The result may be that plate tectonics will eventually stop and that no new mountains will be formed after this. It could be that after a long period of erosion we will probably be left with a planet completely covered by 500 m of ocean.

In a billion years or so the Sun will increase its output by 10% and evaporate the oceans and subduction processes will cease anyway as the process of subduction requires lubrication.

British stratigraphy and geological mapping

William Smith (1769–1839) was the son of an Oxfordshire farmer. As a boy, he picked up stones containing shells in the ploughed fields. William qualified as a surveyor at a time when canals were being built to transport coal. His work took him to coal mines and many excavations for the new canals. This gave him a great opportunity to add to his fossil collection.

When he saw large numbers of fossils on a rock surface he recognised this as a collection of creatures that had lived for a short time on the sea floor. He realised that a series of layers had been deposited and that fossil groups in the lower layers were older than those in the upper layers. Arranging the fossil faunas in his collection according to the layers in which they were found, he discovered that they were not all the same. Some fossils were present in many layers but others were only found in

🔺 **Figure 2** William Smith's 1815 map. The darker colours represent the base of each of the 23 strata he mapped, showing a way-up.

a few. He was one of the first to recognise that fossil faunas succeeded one another in a definite order, the basis of biostratigraphy.

In 1815 William Smith published the first edition of his Geological Map of England and Wales called *A Delineation of the Strata of England and Wales, with part of Scotland.* It took Smith over fifteen years of single-handed mapping covering 175 000 km^2 of Britain. This gargantuan task literally broke him, as he spent some time in debtors prison! It is widely regarded as the first true geological map of any country. His brilliance was recognised and in 1831, he was presented with a medal by the Geological Society and described as 'the father of English geology'.

Uniformitarianism and catastrophe

Georges Cuvier (1769–1832) was a French naturalist and zoologist, known as the 'father of palaeontology'. He was involved in comparative studies of fossils and living animals. He was well known for establishing 'extinction' as a fact and championed catastrophism in geology.

Today geologists believe that processes including evolution occur mostly by gradual changes but occasionally by high-magnitude, low frequency events such as asteroid impacts or the eruption of super volcanoes.

HOW SCIENCE WORKS: DEVELOPMENT OF SCIENTIFIC IDEAS

New ideas build on old ones. The ideas for the age of the Earth exemplify this as follows:
- Archbishop Ussher calculated that the Earth was approximately 6000 years old based on his reading of the Old Testament.
- Later scientists, including Newton and Kepler, incorporated non-biblical historical sources into their work.
- Later still, Halley used the salinity of the oceans and Comte de Buffon used the cooling of iron balls to argue that the Earth was perhaps 75000 years old.

QUESTION

1 Draw up a simple rock cycle diagram incorporating the main three types of rocks and including the following products and processes:

Uplift	Sediment	Deposition	Extrusion
Intrusion	Magma	Lithification	Crystallisation
Transportation	Weathering	Heat and pressure	Diagenesis
Partial melting	Recrystallisation		

5.5 Radiometric dating

Measuring the age of the Earth, an historic perspective

Exactly how old is the Earth? First calculated in 1654 by Archbishop Ussher, this was a problem that inspired some brilliant geoscientists as they realised that this timescale could not be correct.

In 1862, Lord Kelvin used the rate at which the planet is cooling. He carefully measured the geothermal gradient and rock conductivity to determine the heat flow and then, assuming that the Earth had cooled from a molten state, arrived at a value of 20 to 40 Ma. This was too low because he could not know that the Earth was being continually heated internally by radioactive decay. Currently the age of the Earth is measured at around 4540 Ma.

In 1899, Joly assumed that the oceans had started as fresh water and that the salt they now contained had been added by rivers, as a result of chemical weathering of rocks. He knew the salinity of sea water (0.35%); working out the oceans' volume allowed him to calculate how much salt was contained within the oceans. An estimate of river flow and salt content gave him 160 million tonnes of sodium added each year. A simple division resulted in a time span of 90 Ma.

Measuring the age of the Earth, absolute dating

Absolute dating allows an absolute timescale to be constructed using measured ages to correlate rock units around the world. **Relative dating** allows the order of events to be established, based on stratigraphy and fossil assemblages. These were gradually 'fixed' as radiometric ages were added.

KEY DEFINITIONS

Absolute dating gives specific dates for rock units or events given in millions of years before (Ma).

Relative dating putting units or events into sequential order, by saying one event is older or younger than another.

Half-life the time taken for half the unstable parent atoms to decay and form stable daughter products.

Closure temperature the temperature at which a system has cooled, so there is no diffusion of isotopes in or out of the system.

Closed system when a mineral neither gains nor loses atoms. The higher the temperature, the more likely exchanges of atoms or ions will be.

🔻 **Figure 1** Relative amounts of parent and daughter atoms with increasing half-lives

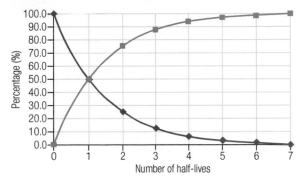

How are absolute ages measured?

Naturally occurring radioactive isotopes in rocks are unstable and break down (decay) at a statistically constant rate. This rate of decay can be measured and is often expressed as the **half-life** of the isotope; the time taken for half of the unstable parent atoms to break down to stable daughter atoms.

If the relative amounts of parent to daughter atoms can be measured, then we know how many half-lives have passed since the parent was formed. For example, if the ratio of parent to daughter is 25:75 then two half-lives have passed since the radioactive atom was created (Figure 1).

The relative amount of parent and daughter atoms can be measured using a mass spectrometer (Figure 2).

Parent isotope	Daughter isotope	Found in	Half-life (Ma)	Use
^{238}U ^{235}U	^{206}Pb ^{207}Pb	Zircon Uraninite	4470 704	Can be used to date igneous rocks older than 10 Ma
^{40}K	^{40}Ar	Muscovite mica Glauconite (mica group) Biotite mica Hornblende (amphibole)	1250	Useful for igneous (often volcanic) rocks older than 10 000 years. Also used for metamorphic and some sedimentary rocks
^{87}Rb	^{87}Sr	Muscovite mica Biotite mica Orthoclase feldspar	48 800	Used for igneous and metamorphic rocks older than 10 Ma

[Source USGS]

Potassium-Argon and Rubidium-Strontium dating

Potassium-Argon (K-Ar) dating is the most widely used method of radiometric dating. Potassium is a component in many common rock-forming minerals. There are three isotopes of Potassium; ^{40}K is the rarest and decays in two ways. Some 89% of it decays by emitting a beta particle to form ^{40}Ca. Sadly, this is no use for dating as it is indistinguishable from the calcium found in many rock-forming minerals. Fortunately, 11% decays by a different mechanism to make ^{40}Ar. This is a gas which is not formed in any other way but has the disadvantage that it can easily escape from the mineral lattice, especially when heated.

Rubidium has several isotopes, ^{87}Rb is the most useful one. As it has such a long half-life, there is some uncertainty attached to its age (quoted as 48800 Ma by the USGS, but has a range from other sources). It decays to form ^{87}Sr, which is a solid and therefore harder to lose from the rock. Rb-Sr dating is thus more useful for metamorphic rocks.

Scarcity of radioactive minerals

Radioactive isotopes are only found in small quantities in some rock-forming minerals. To obtain enough material for an accurate analysis is difficult as it may involve crushing the rock and picking out the required minerals, e.g. zircon crystals from granite.

Figure 1 illustrates another of the difficulties in obtaining an accurate age for the mineral or rock sample. As the number of half-lives increases, the amount of parent atoms approaches (but never reaches) zero. The differences between one half-life and the next become very small, e.g. some 0.8% between half-lives 6 and 7. For accuracy we must have radioactive atoms with long half-lives.

Problems with radiometric dating

Radiometric dating is routinely carried out on igneous rocks but even these need care. Major intrusions may take tens of millions of years to cool; different parts reach the **closure temperature** at different times. The most reliable igneous rocks for dating are shallowly emplaced minor intrusions and extrusive rocks that cool rapidly such as lava flows.

Sedimentary rocks are made up of fragments of older rocks. The clasts could have a range of ages, all older than the time the rock was formed. An exception is a sedimentary rock containing glauconite, a type of mica, which gives a greenish tinge to shallow marine sands. This cementing mineral includes ^{40}K and so can provide a date for the formation of the rock.

Sedimentary rocks tend to be more porous and prone to weathering which can alter the ratio of parent to daughter. Parent atoms can be added but it is more common to lose daughter atoms, especially the gas argon, leading to an incorrect younger age.

Dating metamorphic rocks causes a lot of problems. Different minerals become **closed systems** at different temperatures. Parent or daughter atoms can be lost or gained during heating events. ^{40}K–^{40}Ar has closure temperatures between 150°C and 550°C depending on the mineral analysed. During a series of metamorphic events the radiometric clock will have been reset in some minerals but not others; and so a variety of conflicting dates is possible. Dating of minerals with differing closure temperatures within the same rock can show the metamorphic history of the rock.

QUESTIONS

1 Describe why Rb-Sr dating is the most suitable method of measuring the age of metamorphic rocks.

2 Give examples of problems in obtaining accurate radiometric dates from sedimentary, metamorphic and igneous rocks.

3 Draw a half-life curve for ^{87}Rb – ^{87}Sr, taking the half-life as 50000 Ma. Using your graph, what is the minimum % of ^{87}Rb you could expect from a terrestrial rock sample?

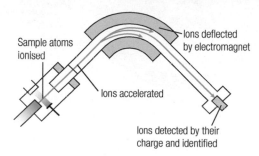

Figure 2 Atoms from the vaporised sample are ionised then accelerated towards the electromagnet in a high electric field. Heavier ions are not deflected as much as lighter ions in the magnetic field and the number of ions of each mass can be counted.

Figure 3 A zircon crystal ($ZrSiO_4$)

STUDY TIP

The half-life of ^{40}K is stated as ten times the correct value in some well-regarded text books – beware! The correct value is 1250 million years (Ma).

Be careful of the use of billions in older texts – we now always use 10^9.

STUDY TIP

^{14}C is not used by geologists because the half-life is too short (around 5730 years).

HOW SCIENCE WORKS: DECAY OF URANIUM

In 1907, Bertram Boltwood realised that lead (Pb) is one of the disintegration products of uranium and so minerals with a greater proportion of lead are older. He studied specimens whose relative geologic ages were known and found that the ratio of lead to uranium did indeed increase with age. Estimating the rate of this radioactive change, he calculated that the absolute ages of his specimens ranged from 410 to 2200 Ma. These ages were approximately 20 percent too high but orders of magnitude higher than the age of the Earth proposed by Kelvin.

The basic divisions of the time scale

What is the geological column? It is the diagram representing the relative vertical distribution of rock ages based on observations from across the globe. Originally largely based on recognition of fossil types, the **Phanerozoic eon** is divided into three eras by the most drastic of the mass extinctions. An era is a subdivision of geological time that subdivides an eon and is usually identified by large-scale extinction events.

Each era is divided into a number of periods on the basis of important changes in fossil species that are recognised worldwide. Fossil evidence has been used to further subdivide the **periods** into epochs and then even smaller divisions. The useful zone fossils ammonites and graptolites changed so rapidly that some divisions are as short as 80 000 years.

The actual length of the periods could not be known until radiometric dates were added to the palaeontological knowledge. With the help of dates from igneous rocks from all over the world and the use of cross-cutting relationships, the boundaries of the divisions of the geological column have been approximated then refined as more data became available. You will notice quite large differences in ages given in older textbooks. This book follows the timescale established by the International Commission on Stratigraphy in 2015.

Sometimes the term 'stratigraphic column' is used. It is probably best to use this term for diagrams that show rock type and thickness rather than just the relative ages of the rocks.

Relative ages using biostratigraphy

It will be necessary to be able to recognise the main fossil groups. As most of these major groups existed in some form for the entire Phanerozoic, you would have to recognise the particular morphology of sub-groups to build up a more accurate knowledge of which part of the geological column your rocks came from.

The gross morphology of some major fossil phyla are shown here with some relevant information.

Eon	Era	Period	
Phanerozoic	Cenozoic	Quaternary	
		— 2.6 —	
		Neogene	
		— 23 —	
		Palaeogene	
		— 66 —	
	Mesozoic	Cretaceous	
		— 145 —	
		Jurassic	
		— 201 —	
		Triassic	
		— 252 —	
	Palaeozoic	Permian	
		— 299 —	
		Carboniferous	
		— 359 —	
		Devonian	
		— 419 —	
		Silurian	
		— 444 —	
		Ordovician	
		— 485 —	
		Cambrian	
		— 541 —	

🔺 **Figure 1** Simplified version of the geological column with ages of boundaries given to the nearest Ma. Older versions divide the Cenozoic into the Tertiary and Quaternary. The Tertiary is now represented by the Palaeogene and Neogene, the Quaternary began some 2.6 Ma.

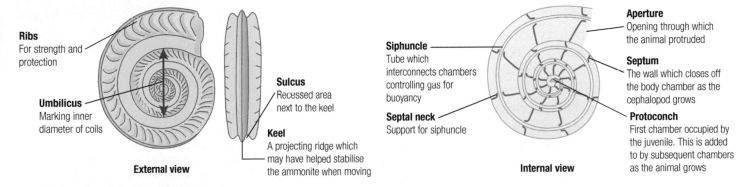

Ribs
For strength and protection

Umbilicus
Marking inner diameter of coils

External view

Sulcus
Recessed area next to the keel

Keel
A projecting ridge which may have helped stabilise the ammonite when moving

Siphuncle
Tube which interconnects chambers controlling gas for buoyancy

Septal neck
Support for siphuncle

Aperture
Opening through which the animal protruded

Septum
The wall which closes off the body chamber as the cephalopod grows

Protoconch
First chamber occupied by the juvenile. This is added to by subsequent chambers as the animal grows

Internal view

🔺 **Figure 2** Cephalopod morphology

Cephalopods are recognised by their chambered shells, they can adjust the ratio of gas to liquid in those chambers. This changes their buoyancy allowing them to move up and down in the water column. The soft parts housed in the final chamber include tentacles for swimming and grabbing prey. Some cephalopods are straight rather than coiled. Although the nautiloid cephalopods have survived from the Cambrian to the present day. Ammonites became extinct at the Cretaceous–Tertiary boundary.

KEY DEFINITIONS

Phanerozoic eon the eon of 'visible life'. The explosion of hard-shelled and thus species that could be fossilised due to the increase in oxygen levels.

Period a unit of time, a division of an era, e.g. the Carboniferous.

The example shown is a *Rugose coral* and was first seen in the Ordovician. A more primitive but colonial coral group (Tabulate) was first seen in the Cambrian. Both of these groups became extinct at the end of the Permian. Different sub-groups show variations in the septa.

Modern corals are descended from the Scleractinian group first seen in the Triassic.

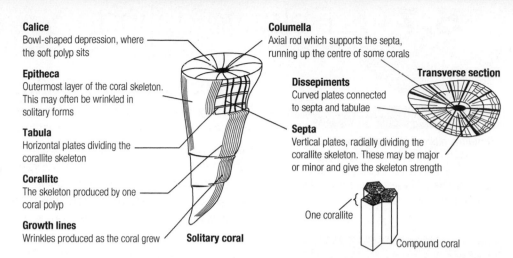

▶ **Figure 3** Coral morphology

Calice
Bowl-shaped depression, where the soft polyp sits

Epitheca
Outermost layer of the coral skeleton. This may often be wrinkled in solitary forms

Tabula
Horizontal plates dividing the corallite skeleton

Corallite
The skeleton produced by one coral polyp

Growth lines
Wrinkles produced as the coral grew

Columella
Axial rod which supports the septa, running up the centre of some corals

Transverse section

Dissepiments
Curved plates connected to septa and tabulae

Septa
Vertical plates, radially dividing the corallite skeleton. These may be major or minor and give the skeleton strength

One corallite

Solitary coral

Compound coral

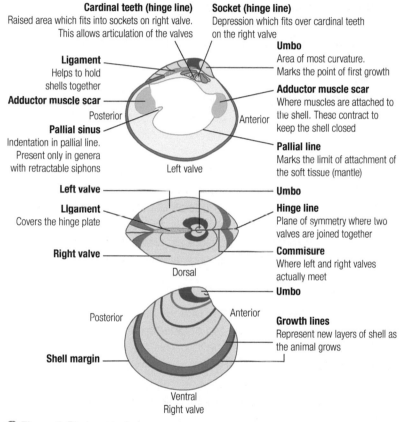

Cardinal teeth (hinge line)
Raised area which fits into sockets on right valve. This allows articulation of the valves

Socket (hinge line)
Depression which fits over cardinal teeth on the right valve

Ligament
Helps to hold shells together

Adductor muscle scar

Posterior

Pallial sinus
Indentation in pallial line. Present only in genera with retractable siphons

Left valve

Umbo
Area of most curvature. Marks the point of first growth

Adductor muscle scar
Where muscles are attached to the shell. These contract to keep the shell closed

Anterior

Pallial line
Marks the limit of attachment of the soft tissue (mantle)

Left valve

Ligament
Covers the hinge plate

Right valve

Dorsal

Umbo

Hinge line
Plane of symmetry where two valves are joined together

Commisure
Where left and right valves actually meet

Umbo

Posterior

Anterior

Shell margin

Growth lines
Represent new layers of shell as the animal grows

Ventral
Right valve

▲ **Figure 5** Bivalve morphology

Trilobites were arthropods with a chitin exoskeleton. Trilobites are divided into three sections longitudinally and laterally. They had articulated segments in their thorax, these covered the legs and gills beneath. Many had highly developed compound eyes. The pygidium or tail was made of segments fused together. They evolved in the Cambrian and became extinct at the end of the Permian period.

Cephalon

Thorax

Pygidium

Glabella part of the cephalon which may have covered the stomach

Compound eye made of many calcite crystals

Genal angle may be extended into a spine

Pleuron, the outer part of the thoracic segment

5 cm

▲ **Figure 4** Trilobite morphology

Bivalves have two similar valves and are supposed to be symmetrical through the plane of the hinge. They draw in water through a siphon extracting the nutrients and oxygen then exhale it through another siphon. They show adaptations to all manner of benthonic habitats, from burrowing to free-lying on the surface, cemented like the oyster or attached like mussels. The scallop can even 'swim' for short distances. Although bivalves have been around for most of the Phanerozoic, they really began to thrive in the Mesozoic era.

Brachiopods are outwardly similar to bivalves (their shell is made up of two valves) they are a very different organism. Much more primitive, relying on currents within the shell to feed and respire and having a complicated system of muscles to open and close. They were very common in the Palaeozoic but only just make it through the Permo-Triassic extinction.

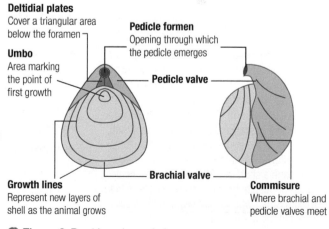

Deltidial plates
Cover a triangular area below the foramen

Umbo
Area marking the point of first growth

Pedicle formen
Opening through which the pedicle emerges

Pedicle valve

Growth lines
Represent new layers of shell as the animal grows

Brachial valve

Commisure
Where brachial and pedicle valves meet

▲ **Figure 6** Brachiopod morphology

The diagram shows a short-hinged brachiopod, the valves are different sizes and symmetrical about a line down through the foramen. The pedicle in this adaptation allowed the brachiopod to attach to the sea floor. Long-hinged species (mostly Palaeozoic) were often free-lying on the sea floor.

Applications and limitations

Relatively dating rocks means that they are put into age order, separated by events such as metamorphism and deformation.

Relative dating limits our understanding of how long it took to lay down the rocks and subject them to tectonism and metamorphism; we relied on observation of present-day processes (uniformitarianism again). It is also very difficult to sort out the order of events in highly deformed strata where sequences may be inverted. It can also be tricky when the rocks are uniform, unfossiliferous or, worse still, missing due to erosion.

Stratigraphic methods

Stratigraphy is the study of strata or layers of rock. Relative dating determines whether one layer of rock is older than another. Here we consider four methods of recognising relative dating, sometimes called 'laws' or 'principles'. Although they may seem obvious now, it was not always so:

- Original horizontality
- Principle of superposition (with way-up criteria)
- Included fragments
- Cross-cutting relationships.

Original horizontality

Most sedimentary rocks are originally deposited in water, usually shallow seas. For example, clasts carried down by rivers are deposited as beds, with breaks in deposition showing up as bedding planes. These beds are commonly laid down horizontally or very close to horizontal. It is therefore assumed that if layers of rock are tilted, then they have been moved from this original horizontal position.

The principle of superposition

This principle states that the rocks at the bottom of a sequence are always the oldest and younger rocks were laid down on top of older ones. For example, rocks at the bottom of cliffs are older than those at the top. This assumes that the rocks have not been turned upside down (inverted) which can sometimes be determined by checking the way-up criteria. These are:

- Desiccation cracks, which dry up with the cracks pointing to the oldest rocks.
- Graded bedding, where the large particles sink to the bottom first, followed by finer sediment.
- Rootlets always grow down into soil.
- Cross-bedding, these can only be truncated on the upper surface.
- Fossils can sometimes be found in life position and so indicate the way they lived.

Included fragments

Fragments from an older rock can be found within younger rock. These are:

- **Xenoliths** found in igneous rocks have to be older, as they are fragments of country rock that fell into the magma due to stoping.
- Derived fossils are also older than the sediments in which they are found. They have been eroded from older beds and redeposited in younger beds.
- Pebbles in a conglomerate are fragments of older rocks, eroded, and then redeposited.

Cross-cutting relationships

Features which cut through rocks must be younger than the rocks they cut. An example could be a dyke, which cuts through sediments. The sediments had to be there first for the dyke to be able to intrude into them. Similarly structures such as faults that cross-cut strata are, by definition, younger than the beds they cut. Be wary of sills which, when concordant, may appear to be in the sequence of deposited beds but were in fact intruded at an unknown later date.

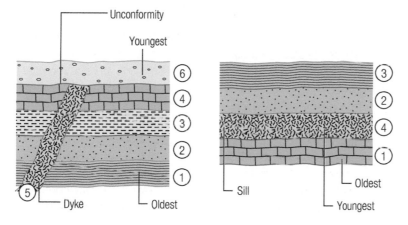

Unconformities

Unconformities represent a break in deposition and erosion of the succession. They are a gap in the geological record in which there may be time to fold and erode the older rocks before new sediments are deposited above the unconformity. There is often a greater angle of dip in the oldest beds, meaning that rocks of various ages are in contact with the young beds above. When there is no change in dip across the unconformity, it *may* be recognised by erosion fossil soils (palaeosols), rootlets or missing parts of the fossil record.

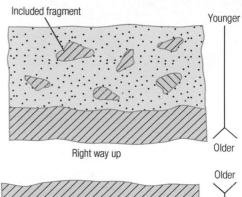

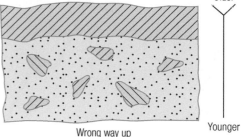

Figure 1 Included fragments and relative ages

Figure 2 Cross-cutting relations including minor intrusions and unconformity. Numbers indicate the order of deposition/intrusion. Note that there may have been many 'lost' events at the unconformity between 5 and 6.

QUESTIONS

1 Describe four methods of relative dating. Draw labelled diagrams to illustrate one example of each.

2 Make a sketch of the limestone outcrop in Figure 3. Label the sketch and write a geological history for this outcrop.

Figure 3 a shows complex deformation and thrusting in Precambrian mica schists compared to the more straightforward history seen in these Carboniferous limestones in **b**.

Relative and absolute dating can be used to work out the geological history of an area from a map or cross-section. The geological history is the order in which events occurred. It is no harder than applying the principles of superposition, cross-cutting features and included fragments, but when all the evidence is put together it results in a detailed understanding of the geology.

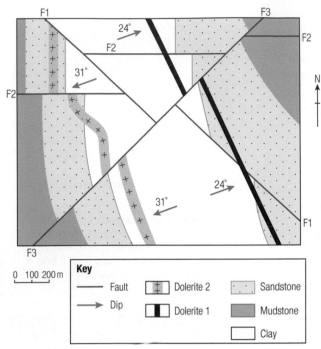

Key

—— Fault		Dolerite 2		Sandstone
→ Dip		Dolerite 1		Mudstone
				Clay

0 100 200 m

▲ **Figure 1** Geological map 1

Identifying structures on (Figure 1)

Folding type

The repeated outcrop of the mudstone and sandstone beds show that the map area is a fold. The dip arrows on the map are pointing away from each other, indicating this is an anticlinal fold. In anticlines, the oldest beds are in the axial, central part of the fold. This knowledge allows you to draw a simple section.

Folding of the beds is along a near N–S axis, with the axial plane trace lying near the centre of the clay bed between the dip arrows. The limbs have different dips with the western limb at 31 degrees while the eastern limb at 24 degrees, making this an asymmetrical fold. The western limb has a narrow outcrop pattern so the dip is steeper than the eastern limb which has a wider outcrop pattern.

Which fault came first?

Faults can also cross-cut each other and can therefore be used to determine the order of events. In this case, F2 is the oldest followed by F1, which has not been affected by F2. The final phase of faulting F3 cuts and displaces both F2 and F1, as well as the dyke and fold.

Igneous intrusions

Dolerite 2 is a transgressive sill, as it is mainly concordant to the beds, but does cross from the clay bed to the sandstone. It is younger than the beds into which it is intruded and older than all the faults as both F2 and F3 cut the sill. A sill can be recognised as the intrusion follows the bedding in the rocks.

Dolerite 1 is a discordant intrusion, or a dyke, which cuts across all the beds and structures with a straight line outcrop. It is younger than the sediments, fold and F1 and F2, but older than F3 which has displaced it.

Since all these events took place, the area was eroded to give a flat landscape and expose the older beds.

▼ **Figure 2** Geological map 2

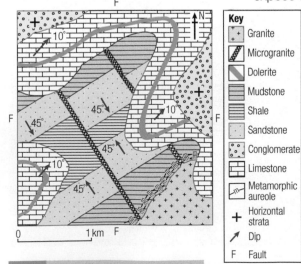

Key

	Granite
	Microgranite
	Dolerite
	Mudstone
	Shale
	Sandstone
	Conglomerate
	Limestone
	Metamorphic aureole
+	Horizontal strata
↗	Dip
F	Fault

Identifying structures on (Figure 2)

For this map, you need to start by working out the order in which the beds were deposited. We know this is a syncline as the dip arrows are pointing towards each other. The youngest beds are always in the centre of the syncline so the oldest beds are the mudstones, then the sandstone, followed by the shale.

The folding is a symmetrical syncline with the limbs dipping at 45° and the axial plane trace is NE–SW.

Two faults trending NW–SE cut across the fold. There is no horizontal displacement because the axial plane trace of the fold is not displaced. The vertical movement can be determined by looking at the outcrop pattern of the fold. The shale in the centre of the fold has a narrower outcrop on the upthrow side of a syncline so the west side has been downthrown.

Igneous intrusions

The granite batholith has intruded into the sediments and formed a metamorphic aureole affecting the mudstone. This would be metamorphosed by contact metamorphism. The microgranite dykes formed at the same time as the granite. Some of the magma was able to take advantage of the planes of weakness along the fault lines that cut the folded beds. The magma cooled more rapidly in the minor intrusion so that the crystal size is medium instead of coarse for the granite.

Unconformities

An angular unconformity is recognised by a difference in the angle of dip between the older and younger beds with beds and structures being cut off by the younger unconformable beds. Watch out for one rock type in contact with rocks of more than one age.

There must have been a period of erosion after the folding, faulting and intrusion of the batholith and then the limestone was laid down as an unconformity. This is lying on top of the older beds. The limestone has been intruded by a concordant sill and later tilted towards the NE by 10 degrees.

Another period of erosion was followed by the conglomerate being laid down unconformably as it cross-cuts the sill. It is still horizontal.

Relative and absolute dates used with maps and cross-sections

We can add palaeontological knowledge to improve our understanding of the relative ages to the cross-cutting relationships formed by faults, intrusions, folding and unconformities. It is therefore important to know the stratigraphic ranges for the major fossils studied.

See if you can write a detailed geological history for the cross-section (Figure 3) without looking at the answer below:

- The cross-section shows that the oldest bed is the lowest, which is the conglomerate with shale above. These beds are older than 170 million years as the dyke cross-cuts them.
- The dyke has been intruded at 170 million years. It has baked the rocks on either side, causing a narrow zone of metamorphism.
- A period of tilting and erosion took place.
- Then limestone was laid down unconformably. It contains ammonites, which allow it to be zoned. It formed between 170 and 38 Ma.
- A lava flow cross-cuts the limestone, which suggests that there was a period of erosion before the volcanic activity. It is likely to be a lava flow and not a sill, as there is only one baked margin below which formed a thin layer of marble.
- The siltstone contains rounded ammonites, which are likely to be derived by erosion from the limestone below.
- Tilting and erosion take place.

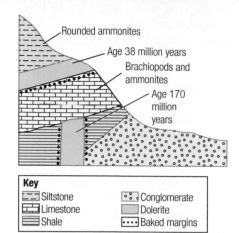

Key
- Siltstone
- Limestone
- Shale
- Conglomerate
- Dolerite
- Baked margins

Rounded ammonites
Age 38 million years
Brachiopods and ammonites
Age 170 million years

▲ **Figure 3** A cross-section with added data on fossils and radiometric ages for the igneous rocks.

QUESTIONS

1 Write a geological history for Figure 1.
2 Sketch and label the photo (Figure 4) adding measurements of apparent dip and bed thicknesses and describe the events that have formed these rocks and structures.

STUDY TIP: INTERPRETING SKETCH MAPS

In addition to simplified geological maps, you may be asked to interpret photos of outcrops. In order for this to be possible under exam conditions, these are mostly straightforward vertical sections showing structures (faults, folds and angular unconformities). Bed thicknesses, joint, dips and fault displacements can be labelled. You will be asked to draw a simplified sketch of the photo to describe the geological history and allow labelling of your interpretation. Ensure that you include some measurements, such as fault displacement or bed thickness.

2.5 m

▲ **Figure 4** An unconformity. The older beds are Devonian sandstone with Triassic conglomerates above.

5.9 Lithostratigraphic correlation

Correlation is the process of matching beds of the same age, in geographically different areas. This is often for economic purposes, such as coal or oil exploration. Most correlation is done by using borehole data. The closer the boreholes are to each other, the greater certainty there is with the data. With increased distance between boreholes there is more chance of a fault displacement or **lateral variation**, which makes correlation difficult.

Lithostratigraphic correlation

If you are very lucky, a succession may contain a **marker horizon**, a bed so easily distinguished that you instantly know where you are in the sequence. These are always unusual, such as an oxidised red bed in a sequence of dark rocks, and quite uncommon. Some of the best markers are derived from volcanic eruptions. These pyroclastic deposits are blasted tens of kilometres into the atmosphere and are laid down over enormous areas at exactly the same time; geologically instantaneously. Close to the eruption the deposits may be tens of metres thick, but further away they form a thin layer that alters over time to a clay (bentonite). These marker horizons can be recognised across the whole planet. They form deposits on land *and* in water and often have unique geochemical signatures. They can be classed under chronostratigraphy as there is a possibility of getting a K-Ar date from the minerals.

Some sedimentary rocks may be distinctive in composition with a rare mineral present in a bed. The Cretaceous–Tertiary boundary rocks are correlated by the presence of iridium, accepted to be of extra-terrestrial origin, associated with a large meteorite impact.

Some methods of lithostratigraphic correlation

Over short distances it is sometimes possible to use the relative thickness of beds, a bit like recognising a barcode. It does not matter if the whole thing is large or small, it is the sequence of thin and thick bars that gives you the information. Such a correlation might work for identifying where you are in a coalfield (Figure 2). This can have immense economic importance; is it worth drilling further to find a good thick seam such as was worked in a neighbouring field?

Another example of recognising the relative thickness of beds in a sequence for correlation is the use of **varves**. In the summer, glacial ice melts fast and the increased flows carry down silts into the lake to make a thin pale layer. During the winter, there is no meltwater and the lake itself may freeze over. The result is very low-energy deposition of clay particles and organic matter that grows under the ice to produce a thin dark layer. One year is recorded as a pair of layers on the lake bed. This means that you can count layers to find out how many years are represented in the sequence. In recent sediments you can count the years back from the present and give real chronostratigraphic ages. The correlation is provided by thicker layers resulting from hotter summers, the pattern of thick and thin bands being the same for all lakes in the same climatic region. The method is the same as that used for tree rings in very recent age dating.

The physical properties of the rocks can be correlated without ever seeing the rock itself. Downhole logging makes geophysical measurements of properties such as porosity, density and seismic velocity. Matching up properties from different logging tools allows correlation between boreholes and ties in the borehole depth data to the seismic data which extend to cover the area of exploration.

KEY FACTS: TYPES OF CORRELATION

There are three main correlation approaches that allow rocks to be compared:

Lithostratigraphic correlation relies on recognising rock types or, more usually, a sequence or succession of rock types.

Biostratigraphic correlation uses assemblages of fossils to find rocks of the same age.

Chronostratigraphic correlation relies on finding an actual age for the rocks, such as counting varve layers or determining the radiometric date of a rock using K-Ar dating.

▲ **Figure 1** Alpine glacial lake in summer. The Matterhorn in the background.

KEY DEFINITIONS

Lateral variation means there are changes in thickness or lithology in beds laid down at the same time.

Marker horizon shows a bed or bedding plane with a change of lithology easily distinguished which covers a wide geographical extent.

Varve deposits alternating light and dark layers of sediment in glacial lakes, with each pair representing one year. The summer deposits will vary in thickness according to the amount of meltwater produced and can be correlated.

Diachronous beds have the same lithology but vary in age along their extent.

KEY FACT

Succession means a series of strata or rock units in chronological order.

Problems of lithostratigraphic correlation

Lateral variation means that there is variation in the sediments because they change horizontally (laterally). Closer to the source of the sediment, the beds may be thicker, further away they may be thinner. This is clearly seen in deep sea environments (turbidites).

Worse still, the beds may change in type over an area. In a fluvial environment, point bar sands are being deposited at the same time as flood plain clays. In a shallow sea environment, there will be sandstones being laid down close to the beach, while at the same time clays are being deposited offshore in the low-energy environments.

Diachronous beds are an even greater source of confusion, a real challenge to lithostratigraphic correlation. In these cases, as conditions change with time, you may find a sequence of beds in which different parts of the bed have different ages!

Diachronous beds cut 'across time'. A good example is where a delta is building out into the sea. Over the years, a continuous layer of sands is left behind. These sands get younger from land to sea – they are definitely *not* the same age. The same is true for the other rock types in a delta cycle.

Changes in sea-level can operate in a similar way but remain a useful tool for correlation as eustatic changes operate globally affecting coastal rocks.

CASE STUDY: THE SEDGWICK AND MURCHISON CONTROVERSY

In 1835, Adam Sedgwick (England) and Roderick Murchison (Scotland) decided to name the entire succession of sedimentary rocks exposed throughout Europe. They were geology colleagues and friends, but they had a famous argument over the division between the Cambrian and Silurian in Wales.

Sedgwick's topmost Cambrian overlapped with Murchison's lowermost Silurian. Eventually the disputed rock layers were assigned the age 'Ordovician'.

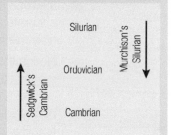

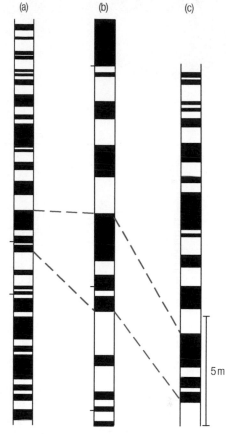

⬥ **Figure 2** Correlation of coal seams in three exploratory boreholes. In these three boreholes corresponding seams are found at different depths and are different thicknesses. It is the pattern of relative thickness that allows correlation.

◀ **Figure 3** In this outcrop of sandstone beds there are several faults evident.

QUESTIONS

1   Look at the borehole sequences of coal seams in Figure 2. Which borehole has the oldest rocks? Which borehole has the fastest rate of deposition?

2   When does lithostratigraphic correlation most likely disagree with the biostratigraphic and chronostratigraphic evidence?

3   Make a sketch or photocopy of the photograph and, having decided which beds can be matched across the faults, annotate showing the direction of movement.

If two rock units contain an assemblage of identical fossils, then the rocks are assumed to have the same relative age, even if they are separated by great distances. The likelihood of organisms having evolved into exactly the same forms at different times is very slim indeed!

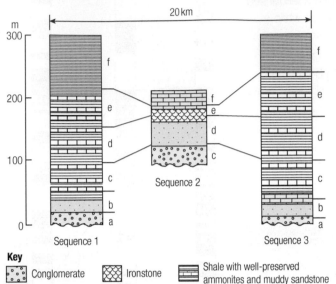

Key

- Conglomerate
- Sandstone
- Ironstone
- Shale
- Shale with well-preserved ammonites and muddy sandstone
- Limestone with broken brachiopod shells

a b c d e f fossil zones

Fossil time ranges for beds 1 to 7							
Zone fossils	1	2	3	4	5	6	7
A							
B							
C							
D							
E							

🔺 **Figure 1** A comparison of lithostratigraphic and biostratigraphic correlation. Biostratigraphic methods use first and last appearances of fossils in outcrops and in boreholes (with care). The stratigraphic ranges of the fossils in the assemblage found allow the age of the rock to be narrowed down. The lithostratigraphy would strongly suggest correlating the conglomerate/sandstone boundary across all three sequences. The biostratigraphy would show up the error of that assumption.

Obviously, some fossils are more use for correlation than others. There are many **zone fossils** including graptolites, ammonites and microfossils, which have enabled rocks to be correlated worldwide.

The basics of biostratigraphic correlation

The **stratigraphic ranges** of fossils, the time between their first and last appearance, were initially established by careful observation of their *relative* ages. Later, that relative timescale, was constrained with absolute radiometric ages.

It can be very tricky to tie down the relative age of a rock unit from just one species of fossil. If there are more species in the rock, then this **assemblage** can be used to make a more accurate assessment. The lower diagram in Figure 1 shows how this works.

Notice that if you identify fossil A, its long stratigraphic range means that you could be anywhere in the succession. In contrast, the short range of fossil B means the rock must have been laid down in the same time as bed 2. Even better, if you find B *and* C, you are restricted to the upper part of bed 2 time (a **biozone** defined by the overlap of two ranges).

The last appearance (extinction) of a fossil species is also very useful, in the example we have been using, bed 3 is uniquely defined by containing fossils A, C, D and E.

Problems of biostratigraphic correlation

- Many fossils, especially benthonic invertebrates, are restricted to one particular environment, for example lime-mud sea floor, reef or sandy sea floor. These fossils are only found in just a few rock types.
- Some kinds of fossils are very long-ranged. Their rates of evolutionary change were very slow, so are little use for establishing biozones.
- Some zone fossils such as the graptolites are delicate and are only preserved in quiet environments, being destroyed by more turbulent conditions.
- Derived fossils confuse the true sequence of beds. They will give an age much older than the rock in which they are redeposited.
- Not all sedimentary rocks contain fossils. In particular, rocks laid down in glacial, fluvial and desert environments are on land, and are unlikely to have any fossils at all.
- Changes in benthonic species may take a significant time to spread around the world, unless they have a free-floating larval stage.
- First and last appearance (extinction) of fossils can be difficult to pinpoint. When fossil groups first appear they can be difficult to find as they may be very rare at first. The same applies towards the end of a fossil's range.

What makes a good zone fossil?

Now that you understand what is needed to define a time in geological history, you could make a 'wish list' for an ideal zone fossil. This could be:

- Evolved rapidly so that each species only has a short stratigraphic range and they define a very precise part of the geological column. A fossil such as the cephalopod, *Nautilus* would be useless as it has not changed in the last 400 Ma.
- Are abundant in the rocks where they are found. They were common when it was alive *and* easily preserved as a fossil.
- Are easily identifiable. The differences between species must be obvious.
- Are found in lots of different rock types. If it is restricted to one environment of deposition it may only be found in one facies. This would restrict its use as a zone fossil.
- Have a wide geographical distribution, preferably worldwide regardless of climate or environment.
- Have strong, hard shells or skeletons which enabled them to be commonly preserved.

To match all of these criteria would be remarkable and most are not suitable as zone fossils at all. Often the mode of life of the organism is the deciding factor.

Two of the best examples of zone fossils are ammonites and graptolites. What do they have in common? Almost nothing except that they both lived in open water. This meant that they were widely distributed across the oceans and when they died, fell into a variety of marine environments.

Stratigraphic ranges – the detail

The range of fossil groups is often marked onto geological column diagrams as a vertical line. More information about the diversity of genera can be seen if the line is made into a 'kite' diagram (Figure 2) with the thickest part of the line representing the maximum diversity.

Ammonites as zone fossils

Ammonites inhabited the oceans ensuring wide distribution. Their strong calcareous exoskeletons provided an excellent prospect for preservation. They evolved rapidly and were numerous, so they tick most of the boxes for good zone fossils. They are rare in high-energy shallow waters as they would probably be destroyed.

Microfossils as zone fossils

Microfossils are found in very large numbers. Their size means that, although tricky to work with, they are preserved in the small chippings returned by drilling muds from boreholes. Some, such as Foraminifera, are calcareous. Others, such as Radiolaria, have silica skeletons, which will survive below the Carbonate Compensation Depth or can be extracted undamaged from limestone by dissolving it in HCl. A large number of microfossils consist of organic material and include spores, pollen, dinoflagellates and acritarchs which can be extracted by dissolving the rock in HF, a very strong acid.

From an economic geology viewpoint, microfossils are extremely useful for correlating boreholes, due to their abundance. Preservation is often good due to their small size.

STUDY TIP

There is often confusion and marks are lost due to misuse of the terms **pelagic** – meaning to exist within the water column rather than on or near the sea bed – and the two alternative modes of life for pelagic creatures. They are either floaters (**planktonic**) or swimmers (**nektonic**).

QUESTIONS

1 Research the use of microfossils by oil companies. Identify the microfossil groups most often used as zone fossils.

2 Using the fossils shown in Figure 2, research the numbers of genera and species alive at the time of their hiatus. What caused the decline of these organisms?

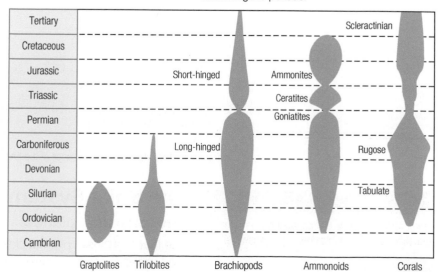

🔻 **Figure 2** Diversity and range of some fossil groups. The 'Tertiary' comprises the Palaeogene and Neogene periods.

The Solar System

The **Solar System** is made up of one yellow dwarf star which we call the **Sun**, eight **planets** and about 170 clearly defined **moons**. The planets have elliptical, co-planar orbits, moving in the same direction as the Sun's rotation.

The table below summarises our current knowledge of the planets of the Solar System.

Planet	Description	Distance from Sun (AU)*	Atmosphere	Surface temperature (°C)	Volume (Earth = 1)	Density (g cm^{-3})	Number of Moons
Mercury	Heavily cratered surface with signs of volcanism; a weak magnetic field similar to the Earth's, so has an iron-rich core	0.4	Thin, mainly helium and sodium	−183 to 427	0.05	5.43	0
Venus	Desert surface has craters, shield volcanoes and structures resembling lava flows	0.7	Dense, carbon dioxide and clouds of sulfuric acid	457 to 482	0.88	5.24	0
Earth	67% oceans; land masses with volcanoes, high mountains; extensive rivers and lakes; desert areas and ice caps; few impact craters	1.0	Nitrogen and oxygen, with variable amounts of water vapour	−89 to 56	1.00	5.51	1
Mars	Large shield volcanoes; and features which may have been formed by running water. No running water on the surface today but water is trapped in polar ice caps and may also be trapped underground	1.5	Thin, mainly of carbon dioxide	−87 to −5	0.149	3.93	2
Asteroid belt	Fragments of carbonaceous, silicate and metallic material						
Jupiter	Small rocky or metallic core. Enormous pressures in the core generate large amounts of heat, radiation and a powerful magnetic field	5.2	Hydrogen and helium cloud belts; a large red spot is a giant whirling storm of rising gas	−150	1316	1.33	4 large moons and 63 known small moons
Saturn	Rings composed of icy debris; a rocky core covered by liquid hydrogen	9.5	Hydrogen	−180	755	0.69	7 major moons and 55 known small moons
Uranus	Icy rings and a rocky core	19.2	Hydrogen, with some helium and methane	−197	52	1.27	27 major moons
Neptune	Faint ring system; a magnetic field, so probably has a rocky core	30.1	Hydrogen, with some helium and methane	−200	44	1.64	14 known moons

*AU = Astronomical Unit, defined as the distance of the Earth from the Sun = 149.6 × 10^6 km

Figure 1 The Solar System

The Solar System

The terrestrial planets

The four 'Earth-like' planets closest to the Sun – Mercury, Venus, Earth and Mars – are formed of materials similar to those of the Earth.

The gas giants

The four planets furthest from the Sun – Jupiter, Saturn, Uranus and Neptune – are made of gas surrounding a rocky or metallic core.

The asteroid belt

The asteroid belt lies between the fourth terrestrial planet, Mars, and the first gas giant, Jupiter. **Asteroids** are rocky objects left over from the formation of the Solar System, perhaps the remains of a planet that failed to form. Whilst most asteroids may be only the size of pebbles, Ceres, the largest, has a diameter of about 914 km. Collisions result in fragments which travel through the Solar System to be captured by the Earth's gravity, falling to the surface as meteorites.

Meteorites

Most **meteorites** come from the asteroid belt, but a few are thought to come from the Moon and Mars, ejected from the surface by an impact with an asteroid or **comet** and captured by the Earth's gravity. Meteorites from the Moon resemble actual rock samples brought back from the Moon. At least 130 known meteorites originated from Mars, their compositions corresponding to those of the rocks from the Martian surface and its atmosphere as analysed by the Curiosity and Viking rovers. Some Martian meteorites contain traces of water, held in the minerals' atomic structure.

Space exploration

Much of our knowledge about the planets has come from space exploration missions which began in the 1960s with early missions to the Moon, Venus and Mars. From the 1970s, fly-by missions have explored the outer planets with a few spacecraft being put into orbit, for example Juno space probe around Jupiter in 2016.

Earth's moon

The first soft landing missions were to the Moon followed by the first manned landing in 1969. The Apollo missions brought back 20 kg of rock and soil. These rocks were much older than expected, up to 4400 Ma. The Moon has a solid crust, mantle and core. The surface is made up of:

- Maria – dark areas composed of basalt lava flows, which were generated by meteorite impacts
- Highlands – light-coloured areas composed of the plagioclase-rich rock anorthosite.

Mars

So far, the exploration of Mars has occurred in three stages:

- In the 1960s, Mariner missions 3 to 7 flew by Mars, taking as many pictures as possible. They identified huge volcanoes including Olympus Mons, the largest volcano in the Solar System.
- As knowledge and technologies grew, we began putting spacecraft in orbit around Mars for longer-term, global studies. These orbital missions started in the 1970s and in 2005 the Mars Reconnaissance Orbiter was capable of taking photos showing objects just 10 cm across. This showed that sedimentary rocks were present on the surface of Mars, showing recognisable sedimentary structures.
- From 2007, spacecraft have landed on the surface to move around and explore, for example the Curiosity rover in 2012. This will tell us more about the geology of the planet, the presence of water, and maybe even clues about whether Mars was ever a habitat for life. An attempted landing in 2016 was not successful.

Venus

Venus is similar in size, mass and composition to the Earth. Spacecraft have mapped Venus using radar and have landed to take actual measurements of the temperature and pressure. Venus has no oceans and is covered by thick, rapidly spinning clouds that trap surface heat, creating a scorched greenhouse-like world with temperatures hot enough to melt lead. The clouds reflect sunlight, making Venus the brightest planet in our sky.

KEY DEFINITIONS

Solar System consists of the Sun, planets, moons, comets and asteroids.

Sun a star composed of hydrogen and helium. It is the largest object made up of more than 99.8% of the total mass of the solar system.

Planet a sizeable object orbiting a star, massive enough to have its own gravity.

Moon or natural satellite is a body that orbits a planet.

Asteroids rocky objects which failed to form a planet.

Meteorites rock fragments which fall to Earth from space.

Comet composed of ice and dust. The outer layer melts into water vapour as it gets closer to the Sun.

MATHS BOX: VOLUME OF VOLCANOES

Mauna Loa in Hawaii is the largest volcano on the Earth, with an estimated volume of 26 000 km³.

Olympus Mons on Mars is 27 km high and has a base diameter of 600 km. Assuming that Olympus Mons has a conical shape, how much bigger is Olympus Mons than Mauna Loa?

The volume of a cone
$= \frac{1}{3}\pi r^2 h$. The base diameter = 2 × radius (r).

Volume of Olympus Mons
$= \frac{1}{3}\pi \times 300^2 \times 27$
$= \frac{1}{3}\pi \times 90\,000 \times 27 = 2\,545\,020$ km³.

Olympus Mons is 98 times bigger than Mauna Loa by volume.

The figures used in this question are approximations, so to say that Olympus Mons is 100 times bigger is a sensible answer.

🔺 **Figure 2** Anorthosite. Lens cap shown for scale

QUESTIONS

1 Research the space missions to the Moon, Mars and Venus to find out the latest information that is being collected.

2 Why does the surface temperature on Mercury vary between −183°C and 427°C?

The origin of the Solar System

▲ **Figure 1** Diagram of a protoplanetary disc

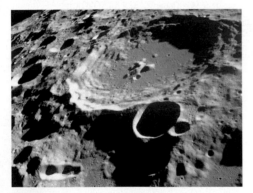
▲ **Figure 2** Craters of the Earth's moon

In the beginning

The solar nebular disc model (theory)

The Big Bang event led to the formation of the Universe about 14 billion years ago. It was the point in time when all matter and energy were created. At that moment, all matter was compressed into a space billions of times smaller than a proton. Both time and space were set to zero.

About 10 billion years later, our Solar System formed, described by the *solar nebular disc model* which was developed by Immanuel Kant in 1755. The original model was applied to the Solar System but it may work for the whole universe. By 2016, the Hubble Space Telescope had helped to discover more than 3500 planets in our galaxy outside the Solar System. The Hubble Space Telescope has photographed a number of protoplanetary discs, for example, in the Orion Nebula, a star-forming region 1500 light years away.

The NASA Kepler mission is also of importance as it is continually searching for Earth-sized and smaller planets in the habitable zone of stars in our galaxy. By 23 March 2017 it had catalogued 4496 such planets.

A giant cloud (*nebula*) of molecular hydrogen and dust collapses, possibly when hit by the shockwave from a nearby exploding star (a supernova event), becoming denser.

As the size of the cloud decreases, the rate of rotation increases, causing the hot, gaseous cloud to flatten out into a **protoplanetary disc**. Material is drawn towards the centre by gravity. It becomes so hot and dense that it triggers nuclear fusion, increasing the temperature of the surface of the star which becomes visible.

Accretion of gas into the star continues for another 10 million years.

The protoplanetary disc cools allowing the formation of small grains of rocks and ice. Electrostatic and gravitational forces bring the grains together into a vast number of small bodies called **planetesimals**, a kilometre or so across. The planetesimals accrete to form embryonic **protoplanets** about the size of the Moon.

The protoplanets nearest to the central star collide to form a small number of terrestrial *planets*. This may take up to a billion years to occur.

Giant planets may have their own spinning, flattened accretion discs of captured hydrogen and helium which is then deposited onto the planet's surface, whilst solid material within the disc accretes to form moons.

The left-over planetesimals which do not form planets are the asteroids. Meteorites, which are asteroid fragments, provide a great deal of information about the formation of the Solar System. Comets are planetesimals which form furthest from the star.

Our Moon probably formed when a protoplanet obliquely impacted with the young Earth between 30 and 50 million years after the formation of the Solar System.

The evidence

Extra-terrestrial visitors – meteorites

The assumption is made that all material in the Solar System originated from the same source. Meteorites are actual samples which indicate the chemical composition of this source. They have remained unaltered since they were formed and therefore contain a record of planet formation.

The Earth shares a common origin with the other planets and with the debris left over from the formation of the Solar System. Frequently, small (and sometimes large) fragments of debris fall to Earth as meteorites, although most burn up as

Type	Characteristics	Interpretation
Iron	Composed of an alloy of iron and nickel. 6% of known meteorites are of this type.	Thought to represent the core of a planet-like object, which formed early in the history of the Solar System.
Stony	Composed of silicate minerals including olivine, pyroxene and plagioclase feldspar. 93% of known meteorites are of this type.	Thought to represent the mantle of a small planet-like object, which formed early in the history of the Solar System.
Carbonaceous chondrites	A type of stony meteorite which contains water and organic compounds.	Probably represent an overall composition for the Earth.

they pass through the atmosphere. The meteorites are not the same as our crustal rocks, which have been influenced by weathering, metamorphism and other changes. The denser *iron meteorites* could be similar in composition to the core of the Earth; the less dense *stony meteorites* are probably similar to the mantle.

Most meteorites come from the asteroid belt. They are around 4560 million years old, which is therefore taken to be the age of the Solar System. Meteorites are divided into two groups, iron and stony. Stony meteorites which contain small inclusions (chondrules) are called chondrites.

Chrondules were originally molten droplets, for example of olivine, which formed at high temperatures in the early solar cloud.

Chondrites are chemically similar to the Sun, except for the absence of hydrogen and helium. The Sun contains 99% of the mass of the Solar System, so the composition of the Sun must be similar to the initial composition of the Solar System, which in turn is similar to the composition of chondrites.

Impact craters

Asteroid collisions with planets and moons in the Solar System have formed distinctive impact craters such as those which can be seen with binoculars on the surface of the Moon. The Earth is no less prone to impacts than the Moon, but the craters are usually destroyed by weathering, erosion and recycling by plate tectonics. One which has been preserved is the 50 000-year-old Meteor Crater in Arizona.

The craters can be shown to be the result of asteroid impact:

* They are circular and have a rim of broken rock built up of ejected material
* Quartz grains may be violently shocked and even melted
* Rock strata are tilted
* Material at depth is brecciated (broken up)
* The ejected material falls back to the surface in an inverted sequence because material closer to the surface is ejected first and falls back to the surface earlier.

The diagrams in Figure 4 show the later stages of the formation of a crater.

The meteorite which made Meteor Crater was composed almost entirely of nickel-iron. It was 45 m across, weighed 300 000 tonnes, and was travelling at 12 km s^{-1}. The explosive energy equivalent to 40 megatons of TNT would have devastated an area about 80 km in diameter.

The graph shows the relationship between the size of impacts and their frequency.

The K-T impact event of 66 Ma contributed to the extinction of the dinosaurs.

What are your chances of being killed by a meteorite impact? NASA's Near-Earth Object programme monitors asteroids which may pass close to the Earth – just in case. However, you are much more likely to be killed in a car accident in the next 2 years (1 in 5000) than by a meteorite (1 in more than 750 million!).

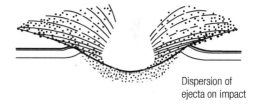

▲ **Figure 3** shows the similarity in the composition of chondrites and the Sun. Sample elements are indicated. Only those found in meteorites are shown.

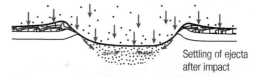

▲ **Figure 4** Ejecta around a crater (variable size, dependent on meteorite)

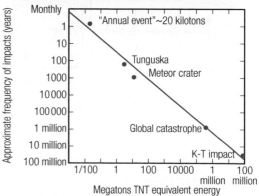

▲ **Figure 5** Meteor Crater in Arizona (measuring 60km across) and graph showing the frequency of meteorite impacts

QUESTIONS

1 The planets have elliptical, co-planar orbits, moving in the same direction as the Sun's rotation. What does that tell us about the formation of the Solar System?

2 Research the evidence for the Tunguska event of 1908.

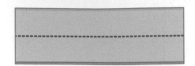

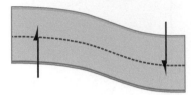
Undeformed crust

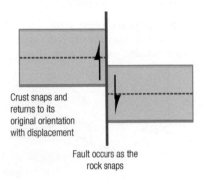
Crust deforms due to stress and is under strain

Crust snaps and returns to its original orientation with displacement

Fault occurs as the rock snaps

▲ **Figure 1** The elastic rebound theory

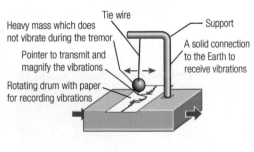

Tie wire

Heavy mass which does not vibrate during the tremor

Support

Pointer to transmit and magnify the vibrations

A solid connection to the Earth to receive vibrations

Rotating drum with paper for recording vibrations

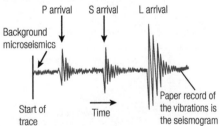

P arrival   S arrival   L arrival

Background microseismics

Start of trace

Time

Paper record of the vibrations is the seismogram

▲ **Figure 2** Seismometer and seismogram

GCSE Science tells us that the Earth is made up of a core surrounded by a mantle with a very thin outer crust. To find out how we know this, we must first take a look at earthquakes and seismology, the study of earthquake waves.

An earthquake is a vibration in the rocks of the crust and upper mantle caused by a sudden dislocation of the rocks along a fault, but sometimes from an explosion such as a nuclear bomb. The elastic strain energy released is transmitted through the rocks as earthquake or seismic waves. Particles of rock vibrate, transmitting energy from one particle to the next, away from the source or focus of the earthquake.

HOW SCIENCE WORKS: THE ELASTIC REBOUND THEORY

This theory explains how an earthquake occurs:
- Two parts of a body of solid, brittle, competent rock are under **stress** due to opposing forces acting on the rock.
- The opposing forces are often due to tectonic plates moving in opposite directions.
- The body of rock is slowly deformed and put under **strain**.
- The energy applied to the deformed rock is stored as elastic strain energy.
- The deformation continues until the stress overcomes the strength of the rock and it fractures.
- The two parts of the rock suddenly move relative to each other and there is displacement along the fracture or fault.
- The elastic strain energy stored in the rock is reduced; energy is transmitted through the Earth by seismic waves causing displacements which are felt as earthquake vibrations.

Weak or plastic (incompetent) rocks such as mudstone, shale and hot metamorphic rocks deform steadily under stress by bending or moving by plastic flow, and therefore do not fracture.

Detecting seismic waves – the seismometer

A **seismometer** detects and records ground motion. It is made up of two parts. One part is attached to a large inertial mass and does not vibrate with earth movements, while the other is allowed to move freely with the vibrations. The relative movement between the two is recorded as a **seismogram**.

Nowadays seismometers measure and record ground motion digitally. They are based on the relative movement between a coil of wire and a magnet, which induces a current which is converted into a digital recording, Modern seismometers can measure movements smaller than one nm (one millionth of a millimetre). The data from seismometers can be collected automatically and analysed by remote computers.

The seismometer combined with the seismogram is called a **seismograph**. The terms seismograph and seismometer are often used interchangeably.

A seismic station will normally have an array of seismometers arranged to pick up vibrations in the vertical and two principal horizontal directions.

P waves

P waves take their name from:
- Primary – travel fastest and arrive first.
- Push – these are longitudinal or compressional waves so the vibration of the rock particles is back and forth, like sound waves. P waves can travel through any type of material.
- Pressure – the particles alternately move together (compression) and apart (rarefaction) in the direction of travel of the wave, which is therefore a longitudinal wave.

The properties of a rock which govern how quickly the wave travels are:

- *Density (ρ)*: the denser a material, the harder it is for a wave to pass through it and the more the wave is slowed down.
- *Incompressibility (k)*: P wave energy causes rapid compression, followed by a rebound. The faster the material rebounds because of a higher incompressibility, the faster the P waves travel through it. Incompressibility is also called the bulk modulus.
- *Rigidity (μ)*: the same as shear strength or how much a material resists a bending force. A liquid has zero rigidity. Rigidity is equivalent to the Young's modulus.

The velocity of P waves depends on all three properties.

S waves

S waves take their name from:

- Secondary – travel at 60% of the speed of P waves so arrive after the P waves.
- Shear – the movement of the particles is sideways, in a shearing motion, at right angles to direction of travel of the wave, which is a transverse wave.
- Several times larger in amplitude than P waves.

MATHS BOX: THE VELOCITY OF P AND S WAVES

The formula for the velocity of a P wave is

$$V_p = \sqrt{\frac{(k + \frac{4}{3}\mu)}{\rho}}$$

The formula for the velocity of an S wave is

$$V_s = \sqrt{\frac{(\mu)}{\rho}}$$

In this example for granite (using values at the Earth's surface)

density   (ρ) = 2.7×10^3 kg/m^3 (equivalent to 2.7g cm^{-3})
incompressibility   (μ) = 32 Gpa (1 Gpa = 1×10^9 pascals): granite is very difficult to compress
rigidity   (k) = 75 Gpa: granite is very rigid

$$V_p = \sqrt{\frac{(75 + \frac{4}{3}32) \times 10^9}{2.7 \times 10^3}} = \sqrt{\frac{117.7 \times 10^6}{2.7}} = \sqrt{43.6 \times 10^6} \quad Vp = 6.6 \, km s^{-1}$$

$$V_s = \sqrt{\frac{32 \times 10^9}{2.7 \times 10^3}} = \sqrt{11.9 \times 10^6} \quad Vs = 3.45 \, km s^{-1}$$

The important points from these formulae are:

1 Seismic waves travel more quickly as the rock becomes more rigid and more incompressible.
2 Seismic waves travel more slowly as the rock becomes more dense.
3 Seismic waves travel more slowly as the rock becomes less rigid, when the temperature rises.
4 In a liquid, the rigidity (μ) is zero therefore the velocity for S waves is zero. S waves do not travel through a liquid.

L waves

L waves are surface waves confined to the surface layers of the Earth. The rock particle oscillates in a circular motion so the waves lose energy quickly with distance. Their low frequency, long duration and large amplitude make them the most destructive type of seismic wave. Surface waves include Rayleigh waves with vertical movement and Love waves (L waves) with horizontal movement.

L waves take their name from:

- Long – the wavelength of the wave is longer than for the P and S waves.
- Last – they are the slowest of the three main waves and arrive last.

QUESTION

1 Summarise the elastic rebound theory in 12 to 15 words.

CASE STUDY: MODELLING P WAVES

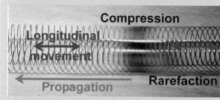

Lay a Slinky™ on a flat smooth table. Each loop represents a particle of rock. Stretch the slinky slightly and send pulses along it by pushing one end gently but sharply.

The coils alternately get closer (compression) and further apart (rarefaction) in the same plane as the movement or propagation of this pressure wave. The loops oscillate but do not move along the coil.

CASE STUDY: MODELLING S WAVES

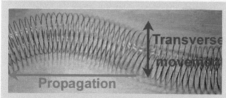

Lay out the Slinky™ again. This time move the end from side to side, at right angles to its length. Try big and small, slow and quick movements, stretch the coil a bit more. Is there any difference in the speed at which the wave travels?

S waves can be generated by P waves. Link two Slinkies™ to make a T-shape. Create a P wave in the Slinky™ forming the stem of the T; an S wave will form in the Slinky TM making up the cross bar of the T. Try different movements.

KEY FACT

Seismic waves follow curved paths through the Earth. If rocks on either side of a boundary have different values for these properties then the wave will change velocity when crossing a boundary and therefore will change direction, by refraction, just as a light ray refracts through a glass prism.

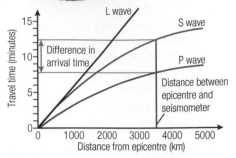

▲ **Figure 1** Distance–time curve

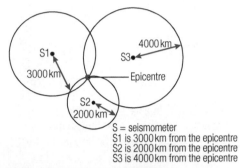

S = seismometer
S1 is 3000 km from the epicentre
S2 is 2000 km from the epicentre
S3 is 4000 km from the epicentre

▲ **Figure 2** Plotting the epicentre

KEY DEFINITIONS

Focus the point within the Earth at which the earthquake originates as movement occurs along a fault plane. Seismic waves radiate out from the epicentre in all directions.

Epicentre the point on the Earth's surface directly above the focus.

Shadow zone an area where earthquake waves are not recorded.

Intensity a measure of the surface damage caused by an earthquake.

Mercalli scale measures the intensity of an earthquake and is based on the effects that are felt in the area.

Magnitude a measure of the amount of strain energy released by an earthquake.

Richter scale a logarithmic scale that measure the magnitude (energy released) of an earthquake

Moment magnitude scale measures the magnitude and leverage on two sides of the fault in an earthquake

CASE STUDY: LEHMANN DISCONTINUITY

1936, Inge Lehmann suggested that the inner core might be solid due to the very high pressure. The phase boundary between the inner and outer core is named after her. It is a wide zone of a mixture of solid and liquid rather than a distinct boundary. She predicted that S waves travel through the inner core.

Finding an epicentre

The **focus** is the origin of the earthquake, where energy is released, whilst the **epicentre** is the point directly above the focus at the Earth's surface.

One earthquake will produce three vibrations on the seismogram, the P, S and L waves. Plotting the arrival times for these waves at many seismometer stations at increasing distances from the epicentre gives a time–distance curve (Figure 1), from which the speed of the wave can be calculated.

The time gap between the arrival of P and S waves increases with distance from the epicentre. The time gap on the time–distance curve gives you the distance of your seismometer from the epicentre. You now know how far away the epicentre is, but it could be in any direction. Using the distances from at least three seismograms, you can determine the location of the epicentre (Figure 2).

Interpreting seismograms and locating discontinuities

Analysis of many earthquakes over the last 100 years has helped geologists to build up a detailed picture of the inside of the Earth.

In 1906, Richard Oldham noticed that, for large earthquakes, the P and S waves were not recorded by seismographs located between 103° and 142° from the epicentre. This is called the **shadow zone** (Figure 3).

- Seismic waves are refracted at a boundary called the Gutenberg Discontinuity which marks the boundary of the Earth's core.
- Waves arriving up to 103° from the epicentre are not affected.
- The waves that should arrive between 103° and 142° are not detected – the shadow zone – because they are refracted at the Discontinuity.
- Beyond 142° degrees, S waves are not received and P waves arrive late as they travel more slowly through a liquid outer core.

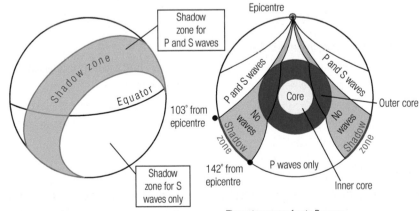

The outer core refracts P waves and absorbs S waves

▲ **Figure 3** The Shadow zones (P + S)

CASE STUDY: MOHOROVIČIĆ DISCONTINUITY

In 1909, Andrija Mohorovičić found that seismic velocities changed suddenly from 7.2 to 8.1 km s⁻¹ at a boundary now called the Mohorovičić Discontinuity, or Moho, a few kilometres beneath the Earth's surface. The rigidity, incompressibility and density of the rocks on either side of the Moho must be different.

He also found that, for one earthquake, two pulses were received at one seismograph and suggested that the waves travelled by two routes – one above the boundary and one below the boundary. The wave following the deeper, longer path arrives first, travelling through the more rigid and incompressible rocks below the Moho.

Inside the Earth

Layer	Depth	Features	Evidence
The crust	0 to average of 35 km	The nature of the crust is well known from direct observation. Oceanic crust is very different from continental crust.	
The Moho Discontinuity	Average of 35 km	A distinct and clear boundary marking a change of rock type and density.	The P and S wave velocities suddenly decrease.
The upper mantle	Average of 35 km to 700 km	• The upper mantle is solid. • The upper mantle consists of solid silicates, less dense than the lower mantle. • The main rock type is peridotite.	S waves can travel through the mantle.
The lower mantle	700 km to 2900 km	• The lower mantle is solid due to the increasing pressure, causing the rocks to become more rigid and more incompressible. • The mantle is made of the same type of silicate material as the stony meteorites.	S waves can travel through the mantle. P and S waves steadily increase in velocity with depth.
The Gutenberg Discontinuity	2900 km	A very distinct and clear boundary marking a change of material from metallic iron nickel to stony silicate material. There is also a change in state between the liquid outer core and the solid mantle.	The P wave velocity suddenly decreases and the S waves stop altogether at this boundary.
The outer core	2900 km to 5100 km	• The outer core is iron-nickel. • The liquid outer core is liquid, with reduced rigidity. • The pressure in the outer core is less than that in the inner core, allowing a liquid to exist.	S waves do not travel through the outer core which shows that it must be liquid. P waves slow down due to lower rigidity.
The Lehmann Discontinuity	5100 km approximately	A phase boundary; a zone of about 100 km where the rocks change from all liquid in the outer core through a liquid–solid transition to all solid in the inner core.	The P wave velocity increases at the boundary.
The inner core	Approximately 5100 km to the centre of the Earth, 6371 km	• The inner core is solid because of the extreme pressure, estimated at 3 600 000 atmospheres, compared to 1 atmosphere at the surface. • The composition of the core is a mixture of iron and some nickel. • Density of more than $12\,g\,cm^{-3}$.	P and S waves travel through the inner core. S waves are generated by the P waves at the Lehmann Discontinuity. Iron meteorites. The Earth has a magnetic field

HOW SCIENCE WORKS: MEASURING EARTHQUAKES

Earthquakes were originally classified by how much damage they caused. At the end of the nineteenth century, a scale of **intensity** was devised by Guiseppi **Mercalli**, later modified to a twelve-point scale. It is an arbitrary, non-mathematical scale, which depends on the opinions of the observer.

After 1904, more sophisticated seismometers allowed the energy released by an earthquake to be calculated. In the 1930s, Californian seismologists Charles Richter and Beno Gutenberg developed a scale of **magnitude** for local earthquakes. The **Richter scale** shows the magnitude of the earthquake which is calculated from the maximum amplitude recorded on a seismogram 600 km from the epicentre.

Richter's magnitude 0 was for a displacement of just 1.0 μm on a seismogram. Other earthquakes are very strong and so Richter's scale had a very great range. Gutenberg took the mathematical step of changing a value of 100 ($= 10^2$) to 2, a value of 1000 ($= 10^3$) to 3 and so on – a logarithmic scale. This makes the numbers more manageable, but we have to remember that for each increase of 1 on the Richter scale, the earthquake is some 30 times stronger. There is no theoretical limit to the energy released by an earthquake; however, the maximum recorded value is from a Chilean earthquake in 1960 at 9.5.

The seismic **moment magnitude scale**, now used by seismologists, was introduced in 1979. This is more accurate but more complicated, based on the idea of the moment or leverage developed on the two sides of the fault as they move in opposite directions. Values are similar to the Richter scale.

The term 'Richter scale' is still used by the media and understood by the general public, although values quoted are from the seismic moment magnitude scale.

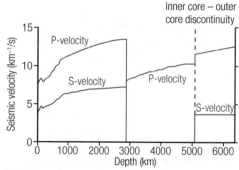

▲ **Figure 4** Change of seismic velocities with depth

CASE STUDY: ARWEN DEUSS

In 2000, Arwen Deuss confirmed the presence of S waves in the inner core.

QUESTIONS

1 Define a discontinuity.

2 Explain how the depth to the core-mantle boundary may be determined from the shadow zones.

CASE STUDY: WHY DOES THE ASTHENOSPHERE BEHAVE AS A RHEID?

The upper mantle is at a high temperature and any further increase will cause the minerals which have the lowest melting points to melt, while the high temperature minerals remain in the solid state. Only a small percentage of this molten material, between 1 and 5%, is needed to make the asthenosphere act as a rheid. As only a small part of the mantle material has melted, this process is called **partial melting**. The molten fraction is found in between the solid crystals, acting as a lubricant, reducing both the rigidity and seismic wave velocities.

Figure 1 Photo of a 20-tonne iron meteorite from Greenland (coin for scale)

Goldschmidt's classification

The 118 named elements are arranged into the pattern shown in the Periodic Table based on the number of electrons in the outer shell of atoms. Within this Table, Victor Goldschmidt (1924) grouped the elements according to how they occur in the Earth, as described in the table below.

Goldschmidt's classification	Description	Examples of elements	Other information
Lithophile 'rock-loving'	Elements which combine readily with oxygen to form low-density compounds which therefore remain near the surface of Earth. Lithophiles commonly form oxides.	Metals: aluminium, barium, calcium, chromium, potassium, lithium, magnesium, sodium, titanium, uranium. Non-metals: chlorine, fluorine, iodine, oxygen, phosphorus, silicon.	Very reactive elements. Electrolysis is needed to extract the metal from the ore.
Siderophile 'iron-loving'	High density transition elements which combine more easily with iron than oxygen, forming dense compounds. These descend to lower layers of the Earth.	Cobalt, gold, iron, iridium, nickel, platinum, tungsten.	Rare in the crust because they descend to lower levels. Ores are mined from eroded ultramafic rocks of deep origin.
Chalcophile 'ore-loving' ('chalco' originally meant copper)	Elements which combine readily with sulfur to form higher density sulfides. These occur deeper than the lithophiles, but not as deep as the siderophiles.	Silver, arsenic, copper, mercury, lead, sulfur, tin, zinc.	Crustal abundance is increased by secondary enrichment processes. Easily extracted from ores by combustion with carbon in the form of the fuel, coke.
Atmophile 'gas-loving' (including volatiles)	Occur as liquids or gases at the temperature and pressure conditions found on or above the Earth's surface.	Nitrogen, noble gases. Hydrogen (which forms water); carbon (which forms carbon dioxide).	Crustal abundance was reduced by loss from early atmosphere. Nitrogen was released by the oxidation of early ammonia as photosynthesis developed.

The layered structure of the Earth

Compared with the other planets in the Solar System, ours is medium in size and the densest. The Earth's main layers have distinct compositions and physical properties. The core and mantle make up over 99% of the Earth. The veneer of cold, solid crust is a very thin skin of the least dense rock, which cooled against the layer we call the atmosphere.

Goldschmidt's classification may help to explain how the layered structure developed. If we take the Proto-Earth as a loose collection of all possible elements, some of which reacted with each other to form compounds such as oxides and sulfides, differentiation by gravity occurred. The denser compounds moved inwards towards the centre of mass and accumulated as the core, displacing the lighter compounds outwards to form the surrounding mantle.

Density	Classification	Occurrence in the Earth	Layer of the Earth
Lower	Atmophile	Atmosphere	Atmosphere
		Hydrosphere	Oceans
	Lithophile	In rocks near the surface of the Earth	Crust
Higher	Chalcophile	In rocks below the surface	Mantle
	Siderophile	In rocks at depth	Core

Meteorites

The siderophiles within iron meteorites (Figure 1) may be equivalent to the composition of the Earth's core. It is assumed that they have not changed in composition since they were formed 4560 million years ago because they have not been subjected to weathering, metamorphism and the other changes which affect rocks on Earth. Meteorite evidence suggests that the Earth's core is made principally of iron with some nickel.

In addition, meteorites may also account for the occurrence of rare siderophiles which occur in the crust, prevented from descending further by the solid rock onto which they fell. For example, iridium is a dense siderophile, the rarest transition metal occurring within the Earth's crust as it would usually descend to a lower layer. A concentration of iridium around an impact crater indicates an extra-terrestrial origin for the object which caused the crater, for example at the boundary between the Cretaceous and the Tertiary (K-T boundary), New Mexico (Figure 2).

The stony meteorites may represent the silicate rocks of the Earth's mantle.

The upper layers of the Earth

The simple division of the Earth into the core, mantle and crust is based on the respective chemical compositions. It corresponds quite well to the Goldschmidt classification. However, seismic surveys which are based on the physical properties, including rigidity and density, suggest a different, and very significant, arrangement in the upper part of the Earth, as shown in Figure 3.

The cooler crust and uppermost mantle are rigid with similar physical properties and so are taken together as one layer called the **lithosphere** (= rock sphere). Earthquakes can be generated by faulting as the rocks are brittle and when stress is applied will respond by fracturing. Seismic wave velocities increase with depth, as expected, down to an average depth of about 100 km.

The rigid lithosphere can be divided into plates which are able to move across the surface of the Earth above the **rheid** asthenosphere.

Below 100 km seismic velocities get slower, described as a **low velocity zone**, as the rock loses rigidity because of the increase in temperature. This zone acts as a rheid, undergoing plastic flow when stress is applied. This layer is given the name of **asthenosphere** (= weak sphere). It is a rheid layer within the upper mantle immediately below the rigid upper mantle of the lithosphere. The boundary between the lithosphere and asthenosphere is taken at the 1300°C isotherm.

Below the low velocity zone, the high pressure keeps the mantle in a solid state. This part of the mantle is the same for both methods of classification.

▲ **Figure 2** Iridium layer at the Cretaceous-Tertiary boundary

Layers using chemical properties	Layers using physical properties	
CRUST	Rigid LITHOSPHERE	
Upper part of the UPPER MANTLE	Rheid ASTHENOSPHERE	
UPPER MANTLE		

▲ **Figure 3** Upper layers of the Earth (not to scale). The two columns show two different ways of classifying the same rocks. The lithosphere includes the crust and upper part of the upper mantle.

KEY FACT: GOLDSCHMIDT'S EXPERIMENTAL WORK

Goldschmidt smelted chondrites (meteorites) in a crucible, and he noticed a layering dividing the lithophile and chalcophile elements. It was this idea that led him to come up with his classification.

QUESTIONS

1. What is the difference between the crust and the lithosphere?

2. Explain how a solid rock can undergo plastic flow.

KEY DEFINITIONS

Kimberlites fine crystal size, ultramafic igneous rocks.

Xenolith fragment of 'foreign' rock included in an igneous rock, which has come from a different source.

Ophiolite section of oceanic crust and upper mantle broken off and attached to the edge of a continent during plate movement.

Peridotite an ultramafic igneous rock composed of the minerals olivine and pyroxene.

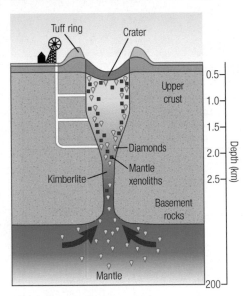

⚠ **Figure 1** Kimberlite pipe.

⚠ **Figure 2** Peridotite – polished surface and orange-coloured weathering surface, Newfoundland (coin for scale)

We can find out about distant planets and moons more easily than we can find out about the composition of the Earth's interior. First-hand observation and sampling of rocks and structures allows us to touch rocks from a maximum depth of just 250 km. As it is 6371 km to the centre of the Earth, the only part of the Earth that can be observed directly is the top 3.7%!

The crust beneath our feet

Most of the land surface of the Earth is now very well mapped, even though there may be some debate about detail and interpretation. Analysis of surface geology maps suggests what lies below the surface, especially in areas where older crustal rocks have been brought to the surface during earth movements and exposed by erosion. From this we know that the upper continental crust is varied, showing a range of sedimentary, igneous and metamorphic rocks, with an overall granitic composition.

Mines and boreholes

We have direct access to the higher levels of the crust by means of mines for coal, metal ores and diamonds. Problems of drainage, ventilation, high working temperatures and lifting the ore to the surface limit the depth of a mine. The deepest mines, for gold in South Africa, penetrate nearly 4 km of the crust.

Boreholes penetrate a little deeper. Samples of rocks and microfossils can be brought up from boreholes and remote sensing undertaken. The deepest oil well penetrates 10 km into the crust, the depth below which oil is not found. The Russian Kola Deep Borehole in Siberia reached a depth of almost 13 km.

Volcanoes bring magma from depth

The magma that feeds volcanoes originates in the lower crust or upper mantle and carries up samples of the rocks from these layers to the surface. Basalt lavas which erupt at mid-ocean ridges, are formed by partial melting of the upper mantle. We can estimate the composition of the upper mantle from the chemical analysis of these basalts and their volatiles.

Occasionally, the igneous material in a volcanic pipe includes diamonds and other minerals which have a compact crystal structure. Diamonds crystallise under the high-pressure conditions of the upper mantle, at depths of up to 250 km. The enclosing igneous rock, called **kimberlite** after the Kimberley diamond mine in South Africa, also includes mantle **xenoliths** of peridotite. The magma carries up fragments of country rock torn from the walls of these very deep volcanic vents. Kimberlite pipes are the result of explosive volcanism from deep mantle sources. Within 2 km of the surface, the highly pressured magma explodes upwards and expands to form a conical vent a few hundred metres to a kilometre across.

Ophiolite suites – ocean crust on land

During the collisions between lithospheric plates, sections of oceanic crust may be broken off and thrust onto the edge of the continental plate. These **ophiolites** may then be exposed by erosion. The section of ancient oceanic crust can be examined on land without the need to drill a borehole through the ocean floor. The **peridotite** at the base of an ophiolite sequence is from the upper mantle. If the structure of an ophiolite sequence is returned to its original undeformed orientation, its total thickness is usually about 7 km, the thickness of the oceanic crust.

Ophiolites can be seen in the Lizard Peninsula in South-West England, Anglesey and the Troodos Mountains in Cyprus. These ophiolite suites are millions of years old, but they are similar to the modern-day oceanic crust and upper mantle.

Deep boreholes

Project Mohole (1961–66) was an American enterprise which aimed to drill through the thin oceanic crust into the mantle. Drilling took place from a ship in 3660 m of deep water, off the coast of Mexico. It reached a depth of 183 m from which core samples of basalt were recovered. This remarkable achievement was abandoned due to increasing costs, but the new technology was later used for off-shore drilling for oil. It was a failed attempt to reach the Moho.

The **Kola Superdeep Borehole** (1970–1994) was a Russian project which aimed to drill 15 km through the thicker continental crust on the Kola Peninsula. One of the boreholes branching out from the main bore reached 12 262 m, a third of the way through the continental crust and the deepest, but not longest, in the world. At that depth, the 300°C temperature was much higher than expected and the rock was too plastic for drill bits to be effective.

At the time, seismic surveys had suggested that the granite of the upper crust would give way to basalt in the lower crust at a depth 7000 m. Instead, metamorphosed granite was found. The granite was heavily fractured and saturated with water which probably came from the minerals of the lower crust.

It took the same time to drill 12 km through the Earth's crust as it did for the Voyager 1 spacecraft to reach the edge of the Solar System!

At 3900 m deep, the Western Deep Levels gold mine in South Africa is the deepest mine in the world. The mine today has 800 km of tunnels and employs 5600 miners. The mine is a dangerous place to work and an average of five miners die in accidents each year. The journey to the rock face can take one hour from the surface. Temperatures can rise to dangerous levels due to the geothermal gradient. Giant air conditioning equipment sends cool air down to reduce temperatures from 55°C to a more tolerable 28°C.

A lucrative gold deposit has been located at a depth of 5 km where the temperature is a blistering 70°C. Mining could be carried out by robotic machines controlled by computers operated from the surface.

The upper layers of the Earth

Using information from rocks which we are able to collect at the surface, from mines and boreholes, and from xenoliths we can be certain about the nature of the upper layers of the Earth.

Layers using chemical properties		Oceanic areas	Continental areas	Layers using physical properties
CRUST	Composition	Rich in Fe and Mg Basalt (pillow lavas) Dolerite (dykes) Gabbro in layers	Rich in Al and Si Granitic rocks Igneous, metamorphic and sedimentary rocks	Rigid LITHOSPHERE
	Density	2.9 g cm^{-3}	Average 2.7 g cm^{-3}	
	Age	Oldest oceanic crust is 200 Ma	All ages, up to 4000 Ma	
	Thickness	5–10 km; average 7 km	Up to 90 km under highest mountains; average 35 km	
——Moho——				
Upper part of the UPPER MANTLE	Composition	Peridotite of lithosphere upper mantle		Moho
	Composition	Peridotite of asthenosphere upper mantle		Rheid ASTHENOSPHERE
	Thickness	7–200 km deep	35–300 km deep	
UPPER MANTLE		UPPER MANTLE		UPPER MANTLE

QUESTIONS

1 The crust is thin under the oceans. Why not drill boreholes there to sample the full crustal sequence?

2 How do we know that the upper mantle is made of peridotite?

3 Using the information in the case study, construct a graph to show the thermal gradient of the Western Deep Levels gold mine.

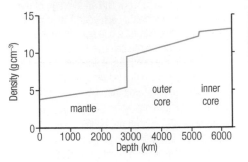

▲ **Figure 1** Density of the Earth

Density

To have validity, the proposed model of a layered Earth with a dense core must agree with known data such as the mean density for the whole planet.

MATHS BOX: MEAN DENSITY OF THE EARTH	
The mass of the Earth is	5.97×10^{24} kg
The volume of the Earth is	1.08×10^{12} km^3 = 1.08×10^{24} m^3
The mean density of the Earth	= 5.5 kg m^{-3}

Direct measurements show that the density of the rocks of the continental crust is 2.7 g cm^{-3} and that of the oceanic crust 2.9 g cm^{-3}. The density of the core must therefore be high to balance the lighter crust. On the surface, the density of iron is 7.9 g cm^{-3} and nickel is 8.9 g cm^{-3}; these values would be substantially higher, at least 12 g cm^{-3} under the extreme pressures within the core.

By calibrating seismic wave velocities in rocks of known physical properties, we can suggest densities for the various layers of the Earth's interior, taking into account the expected conditions of temperature and pressure. The graph in Figure 1 shows the changes in density with depth, with a clear increase at the mantle–core boundary where the composition is believed to change from silicates to iron and nickel.

Is there enough iron available to make up the core?

The stable isotope of iron (^{56}Fe) is the most common endpoint of nuclear fusion processes inside massive stars, and is therefore the most abundant element in their cores. Iron is the sixth most abundant element in the Universe, and the most common refractory (resistant to heat) element. There is no problem in finding enough for the Earth's core.

MATHS BOX: MASS OF THE EARTH
The mass of the Earth may be estimated from the length of the lunar month and the universal gravitational constant G = 6.674×10^{-11} N m^2/kg^2.
When an object (the Moon, mass = m) is moving (velocity = v) in a circle (radius of orbit = r) the force (F) which keeps it in the circle = $\dfrac{mv^2}{r}$.
The force (F) between two bodies, the Earth, (mass M) and the Moon, (mass m) which are distance r apart = $G\dfrac{Mm}{r^2}$.
Combining the two equations gives the mass of the Earth M = $v^2 \dfrac{r}{G}$

One orbit of the Moon (T) takes 27.3 days	= 2.36×10^6 seconds
The average radius of orbit (r) of the Moon	= 3.844×10^8 metres
The velocity (v) of the Moon in circular orbit	= $\dfrac{2\pi r}{T}$ = 10^{24} m s^{-1}
The mass of the Earth (M)	= $v^2 \dfrac{r}{G}$ = 6.04×10^{24} kg
The accepted mass of the Earth is 5.97×10^{24} kg	

Gravity surveys

GCSE Physics tells us that the acceleration due to the Earth's **gravity** is 9.81 m s^{-2} and that the gravitational attraction between two bodies depends on their respective masses and the square of the distance between them. This implies that the value of gravity measured over an area of dense rock, such as gabbro, is very slightly different from the value over a less dense rock, such as granite or rock salt. The value of 9.81 is only an average value.

KEY DEFINITIONS

Gravity measured in Gals (after Galileo). 1 Gal = acceleration of 1 cm s^{-2}.

As variations in gravity are very small, the unit used is the milligal, or mgal.

The average value for gravity on Earth is 981 000 mgal, equivalent to the well-known 9.81 m s^{-2}

Isostasy the theoretical state of equilibrium between Earth's lithosphere and asthenosphere such that the lithosphere 'floats' at an elevation that depends on its thickness and density. The less dense continents rise to a higher elevation than the denser ocean floors, although the pressure exerted on the underlying asthenosphere is the same.

Isostatic rebound the rising up of land masses that were once depressed by ice sheets.

CASE STUDY: PIERRE BOUGUER

Pierre Bouguer was a mathematician, astronomer and surveyor who worked in the Andes. Bouguer knew that the mass of a mountain would exert a gravitational pull on the mass of a plumb line. In 1735, he made measurements on either side of the large volcano of Chimborazo and found that the plumb line was not deflected by as much as he calculated. The mountain wasn't 'pulling its weight'. The idea that a volcano just sat on the crust must be wrong. Instead, the less dense rock of the volcano extends deeper into the mantle.

Gravity anomalies

The value of gravity itself cannot be used to identify a particular rock type because too many variables are involved. It is more useful to plot the variations in the values from place to place. These values which are different from the average are called *gravity anomalies*.

Measured value	Anomaly	Mass	Rock density
greater than expected	positive	more than average	relatively high
less than expected	negative	less than average	relatively low

Corrections must first be made to allow for non-geological effects.

1 Latitude. Due to its spin, the Earth is slightly 'flattened' at the poles and it bulges slightly at the Equator. The poles are slightly closer to the centre of mass, the centre of the Earth, and the value for gravity there is slightly higher (983 200 milligals) than at the Equator (978 100 milligals).

2 Altitude; the *free air anomaly*. If a location is at a height of, say, 100 metres above sea level, it is 100 metres further away from the centre of the Earth. The value is adjusted to sea level by adding 0.31 milligals per metre.

3 *Bouguer Anomaly*. The final value must account for the extra 100 metres of mass which is above sea level; 1 milligal per 10 metres is subtracted from the free air anomaly to remove the effect of the mass.

Gravity anomalies are used to investigate the interior of the Earth and also in the search for ores and geological structures which may hold hydrocarbons. Typical variations in the value of gravity are 20 milligals, although over the deep ocean trenches they exceed 300 milligals.

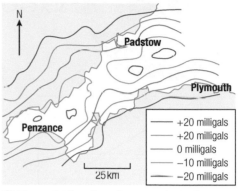

▲ **Figure 2** Map of gravity anomalies in SW England

Isostasy

Strong negative gravity anomalies across mountain ranges indicate that rocks below them appear to be lighter than expected. In 1855, Sir George Airy suggested that mountains, made of less dense continental crust, were rather like icebergs, sinking down into the mantle until they were in a state of balance, supported by the denser rocks of the mantle. This balance is called **isostasy**.

Isostasy is most interesting when the balance is not achieved, as shown by gravity anomalies. There is a negative gravity anomaly over Scandinavia where the continental crust extends further into the mantle than expected. This is a result of the overloading of the area by thick continental ice sheets during the Ice Age. The ice melted relatively quickly, leaving the crust to rise slowly back to its original level by **isostatic rebound**. Evidence of this uplift is shown by raised beaches which are now perched up above present sea level, seen around the coast of Scotland and Norway. These vertical movements of land masses show that the mantle must be able to flow, as a rheid. The rate of uplift allows us to calculate the rate of flow and therefore the viscosity of the mantle.

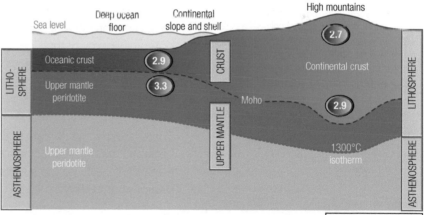

▲ **Figure 3** Isostatic balance

QUESTIONS

1 Calculate the volume of the Earth's core.
The volume of a sphere is calculated using Volume = $\frac{4}{3}\pi r^3$.
Take the radius of the Earth to be 6370 km and the radius of the core to be 3470 km.

2 Find out how a pendulum can be used to determine a value for gravity.

3 Describe how a sensitive spring balance can be used to measure gravity anomalies.

▲ **Figure 4** Model of isostasy using wooden blocks. The longer pieces of wood sink deeper into the water.

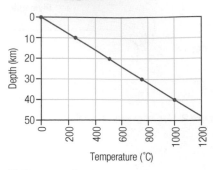

Figure 1 Geothermal gradient

Why is the interior of the Earth hot?

We know that the interior of the Earth is hot because volcanoes erupt molten lava and the temperature in boreholes and deep mines increases with depth. The highest temperatures are in the core but the mantle stores the greatest amount of heat energy because silicates are poor conductors of heat.

The increase in temperature with depth is called the *geothermal gradient*. The gradient is measured by taking the temperature of rocks at the top and bottom of boreholes. In most parts of the world, the gradient near the surface is 25 °C km^{-1}. Where does this heat come from?

Heat and the formation of the Earth

Around one third of the Earth's geothermal energy results from the processes that formed the Earth. There are three main processes which contribute to this heat:

1. Formation of the proto-Earth

2. During the process of segregation into core, mantle and crust

3. Change in state at the inner outer core boundary, from liquid to solid.

The proto-Earth formed by planetoids coming together in collisions, further catastrophic collisions occurred in the giant impact that formed the Moon. Kinetic energy which was transferred to the Earth resulted in an increase in its thermal energy. Early in Earth history the temperature may have been high enough for the Earth's surface to be a magma ocean.

During the process of segregation into the crust, mantle and core the gravitational potential energy of the siderophile elements is transferred to kinetic energy as they move down to the core. To conserve angular momentum gravitational potential energy is also transferred to thermal energy. The amount of energy transferred as the proto-Earth grew by accretion was enough to melt the higher density siderophile metals, resulting in the formation of the molten core.

There is a change of state in the Earth's core, from molten to solid at the boundary between the outer and inner core. This change from molten to solid leads to a transfer of heat into the surrounding material. The inner core continues to grow at a rate of about 0.5mm per year.

Heat from radioactive decay

The remaining two-thirds of the Earth's geothermal energy results from radioactive decay. As unstable parent atoms change to stable daughter atoms, thermal energy is transferred to the surrounding Earth. Radioactive isotopes of potassium, uranium, and thorium are lithophiles and occur in the minerals of the mantle and the crust.

Isotopes with short half-lives, now decayed, originally contributed to the geothermal energy of the young Earth, resulting in higher temperature gradients and higher rates of mantle convection. Temperatures were then high enough to form ultramafic lavas; such eruptions are unknown today.

How heat is transferred within the Earth

Heat flow

Heat is constantly transferred from its source within the Earth to the surface. This gives rise to the geothermal gradient and ideas about geothermal flux, or heat flow.

Thermal flux is the rate of heat energy transfer through a given surface per unit time. At the Earth's surface, the mean heat flow is 65 mW m^{-2} over the thicker continental crust and 100 mW m^{-2} over the thinner, younger oceanic crust. The Earth's crust is a very effective insulator so where there are areas of high heat flow, it suggests that:

- Magma or hot water is rising through the crust transferring thermal energy by a process called **advection**.
- The underlying mantle is moving upward, at mid-ocean ridges and hot spots, by a process called **convection**.

If the geothermal gradient of 25°C per km were constant, temperatures would soon be reached at which all rocks would melt. The gradient in fact decreases with depth (although the actual temperature continues to increase):

- Heat production by radioactive decay is greater in the crust which has the highest concentrations of uranium, thorium, and potassium.
- Thermal transport changes from **conduction** in the lithosphere to advection in the mantle.

So the geothermal gradient within the bulk of Earth's mantle is much lower than 25 °C per kilometre.

Convection cells

In 1928 Arthur Holmes proposed that convection in the mantle drives the Earth's geological processes. Holmes envisaged that convection in the mantle moving carried the overlying rigid crust with them. In the 1990s it became apparent that this model was incorrect and that convection is driven by the transfer of heat thermal energy from the lithospheric plates to the atmosphere and the sinking of cold lithospheric plates by negative buoyancy into the hot mantle by subduction.

In the mantle plumes and immediately below mid-ocean ridges there is **convection**; however, these only represent a fraction of the **geothermal flux**.

Current thinking

The latest theory is that mid-ocean ridges (MOR) do not represent the rising limb of convection currents but rather the passive drawing up of mantle material as the lithospheric plates move apart. We once thought that an MOR marked the place where a convection current rose to the surface building up a linear chain of volcanoes. But it is only in places such as Iceland where the MOR overlies a mantle plume that it is part of the convection cell.

Characteristics of MORs and subduction zones

At the MORs, there are:

- High heat flow
- Rising magma
- Eruption of lava.

At subduction zones, there are:

- Deep ocean trenches
- Low head flow
- Evidence of compression.

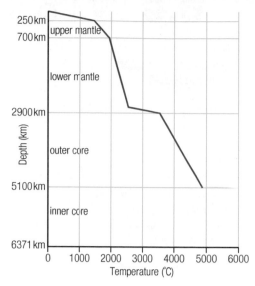

▲ **Figure 2** Geothermal gradient through the Earth

QUESTIONS

1. Using information from the graph in Figure 2, describe the variation in the geothermal gradient from the Earth's surface to the outer core. Estimate the temperature at the centre of the core and assess the accuracy of your answer.

2. Explain why radioactive decay releases thermal energy.

3. In 1862 Lord Kelvin (the eminent physicist William Thomson) estimated the age of the Earth to be between 20 and 100 million years. Research his methodology and explain why his date is so different from modern estimates.

Electromagnetic surveys

Electromagnetic (EM) surveys work in much the same way as transformers in GCSE Physics.

- An alternating current is passed through the primary, transmitting coil.
- A magnetic field is induced which spreads into the ground.
- The magnetic field induces a secondary current in any underground material which can act as a conductor, such as water or iron-bearing minerals.
- This current sets up an alternating secondary magnetic field.
- This field is detected by the secondary, receiving coil and the information is sent to a data logger.

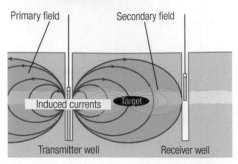

▲ **Figure 1** The principles of EM surveying

On a small scale, this would be called a metal detector, used at the surface. The depth of penetration into the ground can be increased by setting the two coils further apart. The instruments can be used on the ground, in ships or in aircraft, for example to locate sulfide ore deposits.

EM surveys are used to detect the presence of fluids in subsurface rocks. For example, partial melting in the upper mantle and below mid-ocean ridges means there is an increase in conductivity of the rock. This conductivity can be mapped.

Partial melting of only 1% to 2% can be detected, especially if the magma is hydrated. Surveys show where the melting takes place under the mid-ocean ridge.

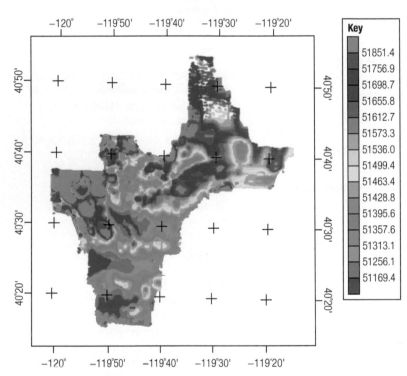

▶ **Figure 2** Example of a map produced by an EM survey

Seismic tomography

Tomography is a technique in which a virtual 3D model is built up by analysing the behaviour of seismic waves as they pass through sections (tomo = section) of a body.

Seismic tomography involves looking for variations in P wave, and sometimes S and L wave velocities as the waves pass through rocks which have different physical properties. The useful information comes from the identification of anomalies where the waves do not arrive at the seismograph at the expected times. The waves may be faster or slower than the average for the Earth. Waves travel faster in colder rocks than in hotter rocks of the Earth's interior. For example, mid-ocean ridges, which are areas of younger, hotter material, are 'slow areas' and the old, colder continents are 'fast areas' on the velocity map. The method involves computer analysis of a large amount of seismic data and is more successful where there are lots of earthquakes and seismographs.

Seismic tomography is used to look at a number of geologically active situations.

Continental roots

3D seismic tomography models of the continents show that at depth, under mountain ranges, the seismic velocity is higher than expected. This supports the theory of isostasy which suggests that mountains have cold 'roots' which extend into the mantle.

Subduction

Associated with ocean trenches, for example along the west coast of South America, there are high velocity zones indicating the presence of cold rock beneath the coastline. This shows that sections of cold lithospheric plates have descended into the mantle by subduction.

Hot spots and mantle plumes

Hot spots occur where there is a build-up of magma beneath the surface causing volcanic centres at the surface. The source of this magma can be investigated by tomography. Analysing the mantle, section by section, shows that low velocity and therefore hotter regions extend beneath hot spots deep into the mantle. Changes in S wave velocities indicate that the plumes extend to the core–mantle boundary. Magma is generated at great depth and rises towards the surface as a **mantle plume**. Beneath Hawaii, the plume is about 500 km wide and up to 2000 km deep. Worldwide, 32 regions in the mantle have been identified as mantle plumes.

KEY DEFINITIONS

Seismic tomography a technique for 3D imaging of the subsurface of the Earth using seismic waves.

Mantle plume a stationary area of high heat flow in the mantle, which rises from great depths and produces magma that feeds hot spot volcanoes.

STUDY TIP

The terms hot spot and mantle plume are often used synonymously. Technically, the plume refers to the hot rising mantle material, which results in the high heat flow values, which can be recorded as a 'spot' on a map.

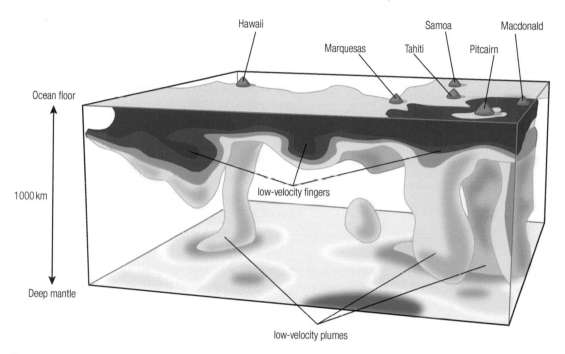

Figure 3 Mantle plumes below the Pacific Ocean

KEY FACT: MELT

Electromagnetic and Tomography (MELT) experiment in the 1990s on the East Pacific Rise MOR transformed our knowledge of the processes that operate in these areas. A broad partial melting zone (of 1 to 4%) was detected up to 70 km below the MOR using EM and seismic surveys. The melt appeared to migrate upwards to a narrow zone below the MOR rather than into a large magma chamber.

CASE STUDY: THE MID-ATLANTIC RIDGE

Although the Mid-Atlantic Ridge is a large volcanic mountain range marking the position of a divergent plate margin, it does not normally reach above sea level. Where islands do appear above the surface, the amount of lava produced is much higher than normal, suggesting the presence of a hot spot and mantle plume.

Iceland lies on a MOR and is also a hot spot, two reasons for the large amount of volcanic activity. The Icelandic magmas are said to be 'evolved', implying there has been plenty of time for them to develop as they rise through the mantle, producing silicic and intermediate lavas as well as mafic basalts.

QUESTION

1 Using the British Geological Survey website, find an EM survey map for an area known to you. Explain the variations on the EM survey map.

The Earth is magnetic

Ancient mariners used needles of a material they called lodestone ('leading stone'), known to us as the mineral magnetite (Fe_3O_4), as compasses because it aligns itself in a north–south direction. As William Gilbert observed in 1600, the Earth behaves 'like a giant magnet'. Its field is represented by a bar magnet lying approximately along the axis of the Earth's rotation. Following the lines of the Earth's magnetic field, a freely suspended compass needle lies exactly vertical at the magnetic poles. This is how the magnetic poles are located. At the magnetic equator, the needle is exactly horizontal.

Origin of the Earth's magnetic field

We know that the core of the Earth is made of iron and nickel, two of the three main magnetic elements.

The inner core is solid with a temperature of about 5700°C. The outer core is liquid with a temperature at the mantle-core boundary of about 3500°C. This temperature is well above the **Curie point** at which materials lose their magnetism. So the Earth's magnetism cannot be permanent and must be constantly generated.

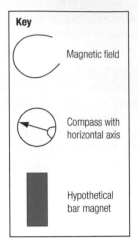

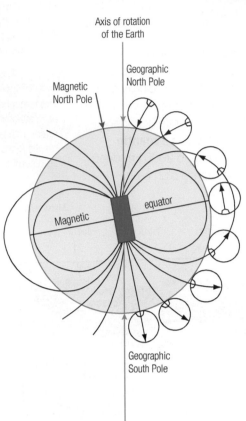

Key

Magnetic field

Compass with horizontal axis

Hypothetical bar magnet

Axis of rotation of the Earth

Geographic North Pole

Magnetic North Pole

Magnetic

equator

Geographic South Pole

🔺 **Figure 1** The Earth's magnetic field and inclination at different latitudes

The temperature difference sets up convection currents within the liquid outer core. A convecting mass of molten iron generates electricity. The generation of electricity induces magnetism, which generates more electricity, and more magnetism, and so on. The balance between generation and destruction allows the Earth to show a continuous, if weak, magnetic field. The effect is called the 'self-exciting dynamo'. The circulation of the convection currents is affected by the rotation of the Earth and therefore the position of the magnetic poles is close to the geographic poles. The circulation is also controlled by the presence of the solid inner core. Without a solid, very hot inner core, convection patterns within the outer core would not be possible.

Variations in the field – magnetic reversals

The 'self-exciting dynamo' sometimes runs down as convection in the core changes. The magnetic field gradually fades away over a period of several thousand years. It then increases again but with the poles the opposite way around. The evidence for these reversals is found in the rocks as **remanent magnetism**.

Palaeomagnetism

The iron-rich magnetic minerals in lavas align themselves with the Earth's magnetic field and as they cool through the Curie point, retain this **palaeomagnetisation** permanently. They act like 'frozen compasses', showing the direction to the poles at the time of their formation. Minerals in lavas erupted when the polarity was normal are aligned north–south, and in lavas erupted when polarity was reversed are aligned south–north. The age of the lavas, and therefore the date of each magnetic reversal, is known from radiometric dating and from fossils in interbedded sediments.

KEY DEFINITIONS

Curie point the temperature above which magnetic materials lose their permanent magnetism. For magnetite, this is 585°C.

Remanent magnetism recorded in rocks due to the alignment of their magnetic minerals according to the Earth's magnetic field at the time of their formation.

Palaeomagnetism ancient magnetism preserved in the rocks.

Magnetometer an instrument which detects the strength and direction of the magnetic field.

Magnetic inclination the angle of dip of the lines of a magnetic field. It is the dip angle made with the horizontal and the Earth's magnetic field lines, measured with a compass.

Sensitive **magnetometers** show normal field strength over the zones of normal polarity. Above the zones of reversed polarity, measurements are weaker because of the opposing effects of the reversed remanent in opposition to the normal modern polarity. Aerial surveys show many reversals in the remanent magnetism in basalts of the ocean floor.

Magnetic inclination

We can also find the latitude of a volcano at the time that it erupted its lavas by using the **magnetic inclination** of these 'frozen compasses'. Careful measurement of the magnetised minerals shows the angle that they make with the horizontal or ground (adjusted to take later earth movements into account) and from this the palaeolatitude can be calculated.

HOW SCIENCE WORKS: THE EARTH'S MAGNETIC DECLINE

By sifting through ships' logs recorded by Captain Cook and other mariners dating back to 1590, researchers have been able to study the variations of the magnetic field. These data show that the overall strength of the planet's magnetic field was virtually unchanged between 1590 and 1860. Since then, the field has declined at a rate of roughly 5% per 100 years.

Every 300 000 years on average, the north and south poles of the Earth's magnetic field swap places. The field must weaken and go to zero before it can reverse itself. The last such reversal occurred about 780 000 years ago, so we are long overdue for another magnetic flip. Once it begins, experts believe that the process of reversing will take less than 5000 years.

The field's strength is now declining at a rate that suggests it could virtually disappear in about 2000 years. Researchers have speculated that this ongoing change may be the prelude to a magnetic reversal.

HOW SCIENCE WORKS: MEASURING MAGNETIC FIELDS

The simplest way to measure a magnetic field is to use a compass, a suspended magnetised needle. On a vertical axis, it shows the horizontal component of the field. On a horizontal axis, it measures the angle of inclination. The magnitude of the horizontal field is determined by accurately measuring the very small oscillations of the compass needle. Over the last 100 years, recordings from fixed stations have been used to construct magnetic maps of the world, used as a basis for navigation and magnetic surveys.

In the proton-precession magnetometer, a strong magnetic field is produced by an external coil. Protons, positively charged sub-atomic particles, in a fluid are aligned in the direction of the field. When the magnetic field is suddenly turned off, the protons try to re-align themselves with the Earth's field. The field makes the protons 'wobble' by precession of the spin axis which depends on the magnitude of the field. The wobble induces a weak voltage which is detected by the coil and analysed electronically. These devices used for mobile surveys (foot, ship, aircraft and spacecraft) to identify anomalies in the magnitude of the local magnetic field.

The flux-gate magnetometer uses three sensors to measure the three components of the field rather than the strength. Each sensor is a transformer. A high frequency current is applied to the primary coil. The induced voltage in the secondary coil depends on whether the sensor is parallel with or at right angles to the magnetic field. So the direction of the field is determined. These are generally used on spacecraft.

Fixed magnetic observatories have both types of magnetometers, feeding the digitised data to a worldwide system of computer databases and used for research into the Earth's magnetic field and how it is affected by the Sun.

The unit for the strength of a magnetic field is the tesla (T), defined as one weber per square metre. The Earth's weak magnetic field is measured by the nanoTesla (nT) = 1×10^{-9} T.

QUESTIONS

1   Describe how rocks become magnetised and explain how instruments can measure this magnetism.

2   Describe how palaeomagnetism can be measured in rocks and explain what this can tell us about global positions of rocks at the time of formation.

3   Latitude can be determined using palaeomagnetism. Why is it not possible to find palaeolongitude?

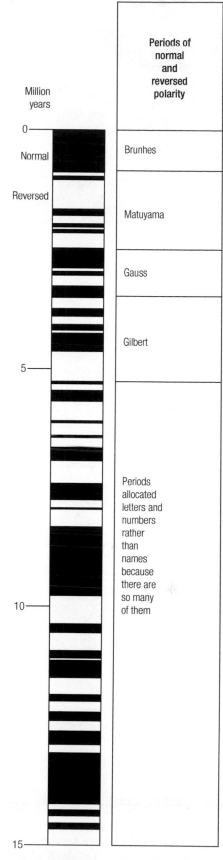

Figure 2 Magnetic reversals over the last 15 million years. The most recent of these is known as the Brunhes–Matuyama reversal, approximately 781 000 years ago.

Early ideas about the Earth

Geosyncline model was developed in the nineteenth century by both American and British geologists who worked in the Appalachian and Caledonian mountains. They believed that the continents were fixed through time and that lateral compression, of a few tens of kilometres, could explain *orogeny*. They thought that compression could be explained by *contraction theory*, the idea that the cooling had caused the radius of the Earth to shrink. This model was the dominant theory from mid-nineteenth century until the late 1960s.

Continental drift is the theory that the position of the continents has changed through time, proposed by Wegener in 1915, and no mechanism was proposed. This was based on distribution of rocks and fossils in different continents.

Arthur Holmes, who pioneered radiometric dating, developed the idea that heating by radioactive decay in the mantle caused mantle convection currents that carried the contents like rafts. Holmes popularised the theory published in a set of influential textbooks between 1944 and 1966.

Passive plate tectonics Evidence from ocean basin research after WWII and the global seismic network set up to monitor Cold War nuclear tests gradually provided mounting evidence that the continents were not static, and that lateral displacements of hundreds of kilometres had occurred during orogeny. These ideas were accepted only in the mid 1960s.

Active plate tectonics – through the 1980s and 90s, evidence mounted that the passive model was neither mechanically nor thermodynamically robust. These ideas are discussed in later chapters in this book.

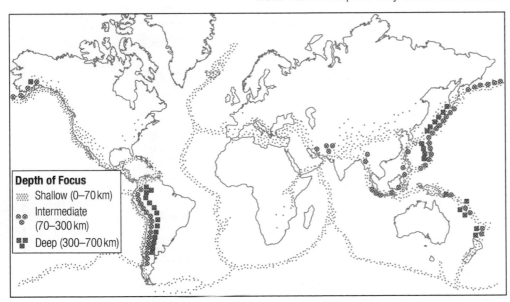

Depth of Focus
- Shallow (0–70 km)
- Intermediate (70–300 km)
- Deep (300–700 km)

▲ **Figure 2** Simplified map of world earthquake zones divided into shallow, intermediate and deep

STUDY TIP

Make sure you know the differences between continental drift, seafloor spreading and plate tectonics. Evidence for continental drift is the evidence found on land. Evidence for seafloor spreading is found in the oceans. Plate tectonics describes what is happening at plate boundaries. Modern plate boundaries are active areas with earthquakes and volcanoes.

HOW SCIENCE WORKS: DEVELOPING THE THEORY OF PLATE TECTONICS

In the early 1960s, Harry Hess proposed the hypothesis of seafloor spreading, in which basaltic magma from the mantle rises to create new ocean floor at mid-ocean ridges. Robert Dietz suggested that the surface of sliding was at the base of the lithosphere, not at the base of the crust. As continents drift apart, the oceans become wider. Hess and Dietz succeeded where Wegener had failed, by providing a mechanism for the movement of continents on the soft, plastic asthenosphere, with convection currents as the driving force.

Hess recognised that while the sea floor is created at the MOR, it is consumed at the oceanic trenches. The ocean floor acts as a 'conveyor belt' of oceanic crust from the MOR to the trench. The Pacific Ocean and South American plates are colliding with each other so strongly that fold mountains are formed.

In 1965, Tuzo Wilson suggested that the surface of the Earth is divided into a small number of large sections which are much thicker than just the crust. Some of these tectonic plates are capped by oceanic crust, some by continental crust, but most by both.

A unifying theory – plate tectonics

The plate tectonic theory gives a coherent explanation for many of the major global geological events such as earthquakes, volcanoes, faults, folds and fold mountains. It brings together a number of the earlier hypotheses of seafloor spreading, continental drift and subduction zones.

The uppermost layer of the Earth is divided into a number of sections, which are constantly in motion relative to each other, carried by moving material beneath. The sections are the *rigid lithosphere plates*.

The moving material is the *plastic asthenosphere*, which is partially melted and acts as a rheid.

What is a lithospheric plate?

Most plates are a combination of oceanic and continental crust, so the term 'oceanic' or 'continental' applies to that part of the plate which occurs at a plate margin. The Eurasian Plate starts at the Mid-Atlantic Ridge where it is oceanic, then extends as a continental plate across Europe and Asia.

	Oceanic plate	Continental plate
Thickness of lithosphere	About 5 km thick under the MOR, thickening to 200 km away from the ridge	Up to 300 km thick under mountain ranges; about 75 km under the continental platform
Density (g cm^{-3})	2.9	2.7
Age	0–200 Ma	0–3960 Ma
Average composition	Lithospheric mantle (peridotite of the upper mantle) capped by basaltic oceanic crust with a thin layer of sediments at the top	Lithospheric mantle (peridotite of the upper mantle) capped by granitic crust, which is deformed and contains igneous, metamorphic and sedimentary rocks
Examples	Pacific, Cocos and Nazca Plates	Arabian Plate is almost entirely capped by continental crust. Most other plates carry continental crust

What is a plate boundary?

New crust is formed along the MOR because plates are moving apart, and destroyed along subduction zones because plates are coming together. This displacement causes earthquakes which can be used to identify the boundaries between the plates.

The interior parts of the plates are not involved in any interactions and are *aseismic*.

The map of plates shows that earthquake zones mark plate boundaries.

There is plenty of other evidence to help in mapping the plate margins. These are: changes in heat flow, chains of volcanoes, trenches and rift valleys, changes in gravity and fold mountain belts.

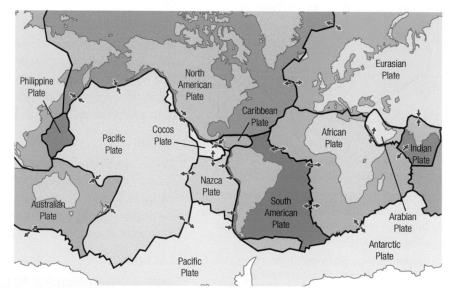

▲ **Figure 3** Present-day plate tectonic map of the world

QUESTIONS

1. Outline the theory of plate tectonics in 20 words.
2. What is the difference between plate tectonics and continental drift?
3. Explain how seafloor spreading causes continents to move (drift). Use the terms convection, asthenosphere, lithosphere and partial melting.
4. Apart from earthquakes, what other geological feature marks the position of plate boundaries?

Moving plates

The evidence for continental drift

The plate tectonic theory suggests that lithospheric plates have moved to their present positions over a period of time and that they are still moving. It is believed that a large southern land mass called *Gondwanaland* existed about 250 Ma made up of what are now South America, Africa, Antarctica, India and Australia. As a single continent, it had the same rocks and fossils and was glaciated at the same time. Gondwanaland began to break up in the Jurassic about 170 Ma, the fragments drifting apart to form the present arrangement of continents.

What evidence supports these ideas?

Fit of the continents

Wegner's idea of the fit of the continents, using the present coastline of Africa and South America, does not give an exact jigsaw fit. This is not surprising because:

- Sea level is constantly changing, so a coastline is a temporary feature.
- Deposition and erosion have occurred since the two continents drifted apart 167 Ma.
- Where there has been erosion of the continents, there is a gap.
- Where there has been deposition of sediment, there is an overlap.

There is a much better fit if you use the edge of the continental shelf at a specific depth, such as 1000 m, or 500 m.

Rock types

To prove that two rocks on either side of an ocean were once part of the same outcrop, they must have the same:

- Distinctive characteristics of mineral composition and physical features
- Age, as determined by radiometric dating.

Examples of matching rocks include Precambrian cratons, Carboniferous coals and **tillites**, Permian red sandstones and **evaporites** and Upper Triassic flood basalts.

Mountain chains

Fold mountain chains are linear features hundreds of kilometres long. The map of Gondwanaland shows how one Precambrian fold mountain chain crosses from Africa to South America and back to Africa as a continuous belt – so the two continents must have been joined together in the Precambrian. The trend of fold mountains provides a way to match geology across continents.

Fossils

If Africa and South America have always been separated, they should have a different fossil record, especially for animals and plants which lived on land or on shallow sea floors. Such animals and plants would be unable to spread across a wide ocean. During the Carboniferous, land-based reptiles (*Mesosaurus*) and plants (*Glossopteris*) are found in both Africa and South America.

Glaciation

In both South America and Africa, there are sedimentary deposits of angular, poorly sorted and scratched pebbles (clasts) in a fine grained matrix. This is a fossil boulder clay or tillite deposited by an ice sheet during the Carboniferous about 300 Ma. Glacial striations are used to trace the movement of the glaciers from one common source in central southern Africa.

Gondwanaland probably occupied a position near the South Pole, as ice sheets cannot extend to the Equator. Africa and South America are now much further north, so this is clear evidence that the continents have moved.

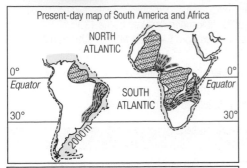

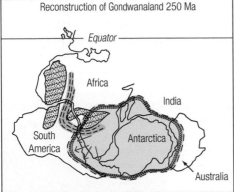

Trend of fold mountains · Extent of Carboniferous polar ice sheets · Shield area 2000 Ma → Glacial striations

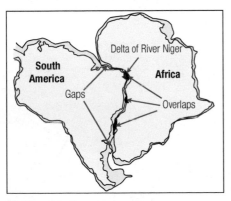

Figure 1 Map of Gondwanaland and fit of Africa with South America

Figure 2 Gaps and overlaps

KEY DEFINITIONS

Tillites ancient glacial deposits preserved within a rock sequence.

Evaporites minerals formed by the evaporation of hypersaline water, for example halite (rock salt), gypsum and anhydrite.

Global positioning system (GPS) – measuring the width of the oceans

Satellites are used to measure the exact location of a series of fixed receivers on the Earth's surface. From these positions, the relative movement of the plates is calculated. In Iceland, GPS measurements prove that the two halves of the island are moving apart at a rate of 2.5 cm per year, well within the capabilities of the current GPS system to measure accurately.

A mechanics problem for plate tectonics is that the plates move over the curved surface of the Earth. They therefore sometimes slightly rotate around a 'pivot' as they move. As with all circular motion, the further the GPS station is from the pivot, the greater its speed. Accurate GPS measurements allow this rotational movement to be analysed.

GPS measurements show the rate and direction of the current plate movement. Geology tells us that information for times in the past.

Geodesy

Geodesy ('division of the Earth'), or geodetics, is a field of science and mathematics which concerns itself with the size and shape of the Earth and the precise location of points on its surface.

The ellipsoid and geoid

The *mean (Earth) ellipsoid* is the basic theoretical shape of the Earth, used for work on a global scale. It can be completely smooth and does not include mountains or valleys because it shows the average elevation.

The *geoid* is the mean sea level due to the Earth's gravity and rotation alone; it extends through land masses as well. The geoid is the true shape of the Earth and is used when additional detail is needed. The geoid is an ellipsoid: the short polar axis is aligned with the rotational axis of the Earth; the equatorial axis is 21 km longer.

All points on the geoid have the same *isostatic balance*. The force of gravity is perpendicular to the geoid; plumb lines hang perpendicularly and water levels are parallel to the geoid.

For regional mapwork, a *reference (Earth) ellipsoid* is used and takes local variations into account. For example, the surface of the mean geoid is higher than the reference ellipsoid where there is a positive gravity anomaly and lower where there is a negative gravity anomaly.

Setting the co-ordinates

Geodesists allocate co-ordinates to set points all over the Earth. The height, angles, and distances between these points gives a *spatial reference system* so that we can quickly and accurately find out where we are.

Around 200 years ago, land-based measurements were used, measuring the length of a base line along a line of latitude and then taking angles to other fixed points using triangulation. In Britain, the old concrete 'trig points' are still useful hilltop landmarks. Ross Clarke (1887) and John Hayford (1910) used this method to find the ellipsoid shape and size of the Earth. Later, other Geodetic Reference Systems were adopted: GRS-67 in 1967 then GRS-80 in 1980. The World Geodetic System (WGS) is now the standard coordinate system.

With the arrival of GPS, surveying methods, reference systems and *accuracy* have all improved. Fixed points are set by satellite observation. With high resolution measurements, we can determine exactly by how much the Earth's surface changes over very short and very long periods of time. For example, the Earth's surface rises and falls about 30 cm every day due to the gravitational pull of the Moon and the Sun.

The first satellite, Russia's Sputnik in 1957, paved the way for advances in communications during the Cold War. In 1960, five US Navy satellites allowed their ships to fix their position once an hour. In 1967, a satellite showed that highly accurate atomic clocks (based on the natural oscillation of an atom of caesium) could be used in space. GPS then developed quickly for military purposes. When an airliner was shot down in 1983, GPS was made available to civilian aircraft and shipping to stop them straying into foreign territory.

In 1995 the modern (American) GPS network of 24 satellites was fully operational. Now there are around 30 active solar-powered satellites in use, orbiting at approximately 20 000 km. A receiver anywhere in the world can detect at least four of them.

Each satellite sends out a microwave signal carrying a time code and geographical data point. The receiver's computer measures the time that it takes for the microwave signal to travel from three satellites, calculates the distances and triangulates the position.

The *accuracy* of most GPS receivers is ±3.5 m, i.e. your measured position is within 7 metres of your actual position.

KEY FACT: GEODETIC SURVEYING TECHNIQUES

Using geodetic surveying techniques and analysis it is possible to measure your position within a centimetre of your actual position. This is an example of accuracy

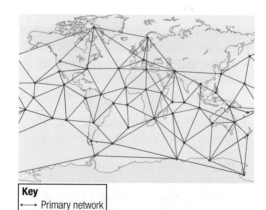

Key
→←→ Primary network
→←→ Baselines

🔺 **Figure 3** Worldwide geometric satellite triangulation network

QUESTION

1 Which lines of evidence was Alfred Wegener unaware of when formulating his ideas of continental drift?

7.3 How fast does a plate move?

KEY DEFINITIONS

Apparent polar wandering curve a line on a map which joins up the apparent positions of the magnetic north pole over time.

Orogeny a period of mountain building.

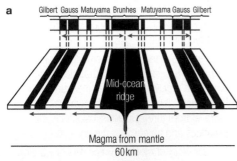

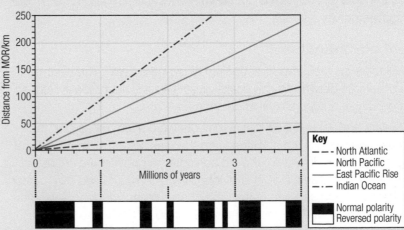

▲ **Figure 1** Magnetic reversals over the last 5 Ma. **a** theoretical model **b** actual map showing localised variation in lava produced across 500 km of ocean floor

STUDY TIP

- If movement is 1 cm per year on both sides of a ridge, the rate of spreading of the ocean is 2 cm per year. Remember to double your answer.
- A change in gradient of the line on a graph shows a change in the rate of spreading for the ridge.
- Make sure of the zeros! There are 6 zeros in a million and 5 zeros to convert km to cm.
- Rates of change of a value are found from the formula:

$$\text{rate of change} = \frac{\text{the amount of change}}{\text{time}}$$

For example:

velocity is the rate of change of distance with time: $\text{velocity} = \dfrac{\text{distance}}{\text{time}}$

$$\text{rate of sedimentation} = \frac{\text{thickness of sediment}}{\text{time}}$$

The units follow the same formula:

$$\text{velocity} = \frac{\text{metres}}{\text{time}} = \text{m/s or m\,s}^{-1}$$

The fit of the coastlines of South and North America with Africa and Europe and the occurrence of the Mid-Atlantic Ridge halfway between suggests that seafloor spreading took place by the splitting apart of a larger continental land mass (Gondwanaland). The date when the splitting began and the rate of spreading can be estimated from:

- The age of the youngest rocks affected by the split and the distance between them now on either side of the Atlantic
- The age of the basalt lavas at known distances from the MOR
- The age and thickness of the sediments at known distances from the MOR.

Magnetic anomalies

The Earth's magnetic field undergoes complete reversal, so that north becomes south, and vice versa, up to four times in a million years. As new magma is erupted at a mid-ocean ridge (MOR), the iron particles line up parallel to the Earth's existing magnetic field. As the rocks cool down, the particles remain permanently magnetised. The result is a striped magnetic anomaly pattern on the sea floor, shown as normal (black) and reversed (white) polarity. The pattern is symmetrical, parallel to the MOR. These zones vary in width according to the length of time during which the magnetic field stayed in one direction and with the amount of lava erupted during any particular period.

The only possible interpretation for this symmetry is that the crust on the two sides of the ridge is moving apart. Basalt erupts at the ridge to form new oceanic crust, which then spreads away from the ridge equally on both sides. Radiometric dating of the basalts give absolute dates from which the rate of spreading can be calculated. This is proof for seafloor spreading.

MATHS BOX: RATE OF SEAFLOOR SPREADING

The graph below shows a magnetic anomaly pattern on one side of an MOR. For the Mid-Atlantic Ridge, basalts erupted 3 Ma are now 30 km from the middle of the ridge. It has taken 3 000 000 years for the basalt to move 3 000 000 cm at an average rate of 1 cm/yr.

$$\text{Annual rate of spreading (cm/yr)} = \frac{\text{Distance moved (cm)}}{\text{Time taken (years)}} = \frac{3\,000\,000}{3\,000\,000} = 1 \text{ cm per year}$$

▲ **Figure 2** Spreading rates of oceans

The fastest rate of spreading currently is 16.8 cm per year at the East Pacific Rise between the Nazca and Pacific Plates. Rates can also vary along the same boundary. The South Atlantic is spreading at about 4 cm per year and the North Atlantic at only 2 cm per year along the Mid-Atlantic Ridge.

Figure 2 axis: Distance from MOR/km (0–250); Millions of years (0–4). Key: North Atlantic, North Pacific, East Pacific Rise, Indian Ocean; Normal polarity, Reversed polarity.

Figure 1a labels: Gilbert Gauss Matuyama Brunhes Matuyama Gauss Gilbert; Mid-ocean ridge; Magma from mantle; 60 km.

Figure 1b: Mid-ocean ridge. Example of a symmetrical magnetic anomaly pattern across a mid-ocean ridge.

Palaeomagnetism and apparent polar wandering curves

Iron-bearing minerals in some rocks hold a record of the Earth's magnetic field at the time of their formation. A large number of rocks are collected, dated, and the direction of palaeomagnetism measured. This data is then plotted as an **apparent polar wandering curve**. The curves for South America and Africa suggest that, before 160 Ma, one North Pole was in two positions at the same time. However, we assume the North Pole is fixed and it is the continents that have moved. If the two continents are re-positioned next to each other, the two curves match up, and there is then one position for the pole. The curves diverge only after the continents started to drift apart, 167 Ma.

KEY FACT:
MAGNETISING ROCKS AT THE MOR

The unpaired electrons in the outer shell of Fe atoms forming Fe minerals within the magma spin parallel to the Earth's magnetic field and acquire magnetism. Below the Curie temperature the minerals remain permanently magnetised, preserving a record of the geomagnetic field where they crystallised.

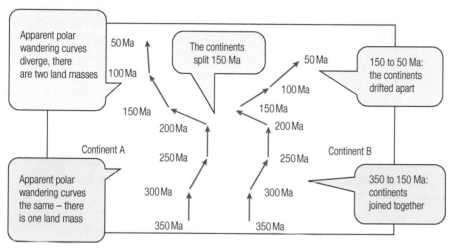

Figure 3 Apparent polar wandering curves

The curves for North America and Europe are the same between 400 Ma and 100 Ma when the continents were joined.

HOW SCIENCE WORKS: CAN THE POLES WANDER?

Geologists studying palaeomagnetic pole positions have found evidence suggesting that the magnetic poles appear to have wandered all over the globe. This is called apparent polar wandering. Polar-wandering is a topic that has been viewed differently by geologists over the years. Some geologists believed that the continents moved, and some believed that the poles moved. Rocks of the same age in Europe and North America suggested that the North Pole was in two positions at the same time. The Earth's magnetic pole now is close to the geographical poles and it is likely that this has always been the case. There must be another explanation, that is, the continents have moved across the surface of the Earth.

CASE STUDY: BRITAIN'S PLACE IN THE WORLD – 1

Britain has been involved in its share of tectonic activity. 500 Ma, Scotland and England/Wales were on the former plates of Laurentia and Baltica respectively, separated by the Iapetus Ocean. The plates came together during the Caledonian **Orogeny** about 400 Ma forming Laurasia with an orogenic belt extending from the Appalachians, through Newfoundland and Britain to Greenland and Norway. Now separated by the young Atlantic Ocean, the Appalachians and the Scottish Highlands have the same rock types and structures of the same age, showing they were once part of the same mountain range.

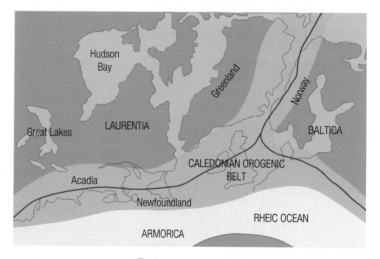

Figure 4 The Caledonian orogenic belt

CASE STUDY: BRITAIN'S PLACE IN THE WORLD – 2

In Britain, coal, coral limestones, desert sandstones and evaporite deposits all suggest tropical or equatorial climates during Carboniferous and Permian times. We assume that the world's climate belts have followed the same pattern throughout geological time as they are all created by the relationship of the Earth to the Sun and by atmospheric circulation. Therefore Britain must have moved from tropical latitudes 250 Ma to its present-day temperate climate belt.

QUESTIONS

1 The age of the oldest oceanic sediments at the edge of the Atlantic is 190 Ma. If the distance across the Ocean is 3800 km, what is the average rate of seafloor spreading?

2 Using the graph in Figure 2, calculate the average rates of spreading for the four oceans. Which ocean is spreading fastest?

Figure 1 The San Andreas Fault

One reason why the Plate Tectonic theory was developed relatively recently is that plate boundaries are rarely seen on land. Those that are visible were thought of as ordinary fault lines. The San Andreas Fault in California, along with its many associated faults, is now interpreted as a transform plate boundary or a **transform fault**. These boundaries are also called conservative plate margins. Plates slide past each other with a horizontal movement. There is no subduction, no volcanoes and no creation or destruction of crust. Plate movement is an unremitting 5 cm a year, so California will continue to suffer from severe shallow-focus earthquakes well into the future.

CASE STUDY: THE SAN ANDREAS FAULT

The San Andreas Fault and its movement was well studied for many years before the plate tectonic theory was developed, owing to its frequent and sometimes strong earthquakes. Initially it was designated as a transcurrent (or **tear** fault), a type of strike-slip fault. When it was realised that the fault marks a plate boundary, one of the few which can be seen on land, it was called a **transform plate boundary** because no crust is created or destroyed. The crust moves horizontally on either side of the fault. Now it is referred to as a transform fault because it links two other plate boundaries.

The San Andreas Fault zone has seismically active sections, where small earthquakes are frequent, and locked sections where there is no known recent movement. Strain builds up over many years in the locked sections, until there is a sudden 'break'. The last big earthquake on a locked section was at San Francisco in 1906. It registered around 8.3 on the Richter scale (different sources give different figures). Many parts of the city were completely flattened and approximately 700 people were killed. This part of the fault has not moved since. If the fault had become locked between 1906 and 2016 but then suddenly released, a devastating displacement of 5.5 m would be possible. This will happen at some point in the future.

Deformation on either side of the fault

You can demonstrate the effects of this transcurrent movement by placing two heavy books next to each other on a tablecloth covering a table, and moving the books in opposing directions past each other.

Opposing movement of the plates on either side of the fault causes shear stress in the rocks. The Ventura anticline, between Los Angeles and San Francisco, has developed in sediments which are only about 500 000 years old. It is still rising at the rate of 5 mm/year, and would have reached a height of 2500 m if the unconsolidated sediments were not being eroded rapidly at the same time.

Small thrusts are also developed. The fracturing of an oil well drilled through one thrust shows that the displacement along the thrust plane would be 15 km in one million years.

CASE STUDY: SAN ANDREAS FAULT ZONE

The San Andreas Fault is a transform fault which occurs on land. It connects two volcanically inactive oceanic rift systems in the Gulf of California to the south and in the Pacific to the north of San Francisco. San Francisco is actually on the Pacific Plate and not the North American Plate!
- The length is over 1300 km.
- The Pacific Plate is moving north-west and the North American plate west-north-west but at a slower rate. Plates move past each other at a relative rate of 5 cm per year.
- In the last 140 million years, the displacement (of Jurassic rocks) has been 560 km.

The Plate tectonic setting of the San Andreas Fault has changed over its 200 million year history. As a transform fault, it was associated with an ocean ridge which was once nearer the middle of the Pacific than it is now. The original MOR has moved towards the American Plate due to rapid subduction of the Pacific Plate beneath the continent and so the San Andreas Fault is a transform fault which, unusually, affects the edge of a continent.

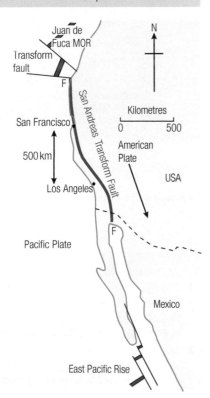

Figure 2 Map of the San Andreas Fault

Stress transfer

When a section of crust moves during an earthquake, stress is released. For large earthquakes, the stress may not be totally used up in the displacement of the crust and the remaining stress may be transferred to another section of the fault, triggering other earthquakes.

The difficulty when trying to predict this movement is that some parts of the fault are creeping, making many small movements, while other sections are stuck. In these locked areas there are *seismic gaps* where no earthquakes have occurred to release the stress. Crustal failure will occur once the stress exceeds a certain value, which depends on the shear stress, normal stress, pore pressure and the friction of the fault plane. If this value can be calculated for a section of the fault line, the probability of a subsequent earthquake can be assessed.

CASE STUDY: STRESS TRANSFER

In 1992, the magnitude 7.2 Landers earthquake in California was followed by the 6.5 Big Bear earthquake 40 km away; 20 000 aftershocks were recorded within a month and within a 5 km radius of these two locations. Over 75% of these were in areas where calculations showed that stress had increased as a result of the displacement along the faults.

Along the North Anatolian fault system in Turkey, 11 of the 13 earthquakes between 1939 and 1999 occurred in areas of increased stress caused by a previous rupture. The 13th was successfully predicted before the earthquake occurred, allowing the evacuation of unstable buildings. The probability of another earthquake along the Anatolian fault system is estimated at 60% over the next 30 years, probably close to Istanbul.

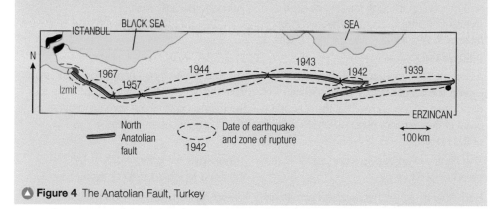

▲ **Figure 4** The Anatolian Fault, Turkey

KEY DEFINITIONS

Transform fault a strike-slip fault which:
ends at the junction of another plate boundary or fault;
shows the same amount of deformation across the fault zone;
can form a tectonic plate boundary.

Tear fault a strike-slip fault which:
dies out without a junction with another fault;
has more displacement in the middle of the fault zone;
cannot form a tectonic plate boundary.

Transform plate boundary a fault where no crust is created or destroyed.

STUDY TIP: TRANSFORM FAULTS

Working out the movement along a transform fault can be confusing because the crust spreads away from the MOR as well as being displaced by the fault. Look at the relative movement along the faults at various points along each fault shown in Figure 3.

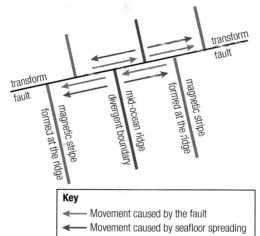

▲ **Figure 3** Displacement along a transform fault

MATHS BOX: MOVEMENT ALONG THE SAN ANDREAS FAULT

The amount of movement along the fault is recorded by the displacement of rocks of known ages.

The rate of displacement $= \dfrac{\text{distance moved}}{\text{time}}$

Complete the table below to show how the rate of movement has changed over time. The first one has been done for you. Is the rate likely to be constant over a long period of time?

Age (Ma)	Cumulative displacement	Displacement since previous date	Displacement per million years
140	560 000 m		
100	510 000 m	50 000 m in 40 million years	1250 m
50	360 000 m	150 000 in 50 million years	
12	10 000 m		
1	15 000 m		
Present		5 cm per year	50 000 m

QUESTIONS

1   What evidence would be considered when the San Andreas Fault was re-classified from a tear fault to a transform fault?

2   Research how strain is measured in rocks.

7.5 Divergent plate boundaries

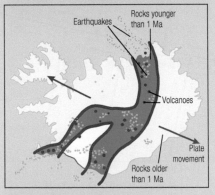

▲ **Figure 1** Map of Iceland and photo across the MOR. Arrows show plate movement.

KEY DEFINITIONS

Sheeted dykes form from magma which usually does not reach the surface, and usually composed of dolerite.

Pillow lava form when lava erupts underwater, and cools rapidly in cold sea water, usually composed of basalt. The rounded masses formed are called pillows.

The Plate Tectonic model incorporates Hess's idea of seafloor spreading, that the ocean floor moves away from a central line, the mid-ocean ridge (MOR) where the lithosphere splits apart by extension.

What is the evidence that the sea floor is diverging in this way and what is the mechanism by which it happens? The present-day Mid-Atlantic Ridge (MAR) is used as an example.

Features of the Mid-Atlantic Ridge and the ocean floor

Description	Interpretation
The 7000 km length of the Mid-Atlantic Ridge (MAR) is a submarine mid-ocean ridge (MOR).	The MAR is one of the largest mountain ranges in the world.
MORs are made of mafic and ultramatic igneous material.	The igneous rocks are derived from the ultramafic magma of the upper mantle.
The MAR is mostly submarine but Iceland is above sea level.	Iceland is located above a hot spot as well as lying on the MAR.
Magma accumulates in shallow magma chambers below the fast-spreading MOR and moves up through feeder **sheeted dykes** (dolerite) to erupt on the sea floor as fissure lava flows, often forming basalt **pillow lavas**.	When the plates move apart, pressure is reduced, allowing the ultra-mafic peridotite mantle to partially melt.
Along MORs, igneous rocks are new. Dating the basalts and the sediments deposited on top shows that the age of the crust increases with distance on either side of the ridge.	The two plates move apart, spreading away from the MOR. Mafic material adds on to the trailing edge of the spreading plate, creating new oceanic crust.
The centre of spreading along an MOR may be an axial rift valley.	This median rift is formed by extension of the crust which is under tension.
Heat flow is higher than average across an MOR, peaking at the median rift.	Upwelling of magma beneath the MOR and expansion of the hot rock may help to increase the height of the ridge.
Shallow-focus earthquakes occur along an MOR.	As the magma moves up, it makes harmonic tremors.
Shallow focus earthquakes also occur along transform faults which extend at right angles to an MOR.	Displacement occurs along transform faults which allow the plates to move away from the rift, section by section.
There is no sediment on an MOR. The thickness increases with distance from the ridge up to 2000 m thick at the furthest point. This pattern is symmetrical on either side of an MOR.	The thicker the sediments, the longer they have had to be deposited, and the older the crust. Sediment accumulates very slowly (1 mm per 1000 years) from microscopic plankton such as radiolaria or foraminifera.
Thin oozes and other fine grained sediments on the ocean floor, collected as core samples, can be dated using pelagic microfossils. The age of the sediment is greatest at the furthest distance from an MOR dated at 200 Ma.	The sediment at the base of the core was deposited first and therefore shows the date when sedimentation started on that part of the newly formed basalt sea floor. The symmetrical pattern shows that the plates are moving apart.

Rifting and extension

A rift valley, such as the median rift, is a linear strip of crust that has slipped down along normal faults which dip towards the valley. Normal faults are formed by extension of the crust, which is either pulled apart by tension or gently arched upwards by rising magma. The strip of crust subsides between the fractures, due to gravity, to form the valley. Shallow-focus earthquakes are common along the fault lines.

A new divergent boundary

The East African Rift Valley may become a new ocean in time. The northern part of this rift system, in the Red Sea, is now a divergent plate boundary where new basaltic ocean crust is being formed. This split started about 40 Ma and accelerated about 23 Ma. The Red Sea is likely to become an ocean in another 50 million years.

An old divergent boundary

South America separated from Africa about 170 Ma along a major rift system in the centre of Gondwanaland, probably similar to the East African Rift today. Widespread extrusions of plateau basalts accompanied the split. The rift valley widened and subsided below sea level, allowing the ocean to flood in. The continental crust stretched and became thinner, leaving wide continental slopes on the edge of each plate. The thinned continental crust finally split completely and the gap was filled with basalt, forming the beginnings of the oceanic crust. One plate had become two.

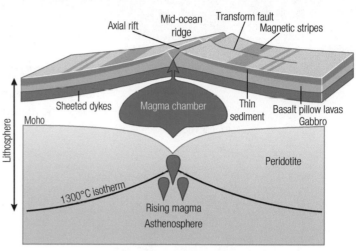

▲ **Figure 2** Section across a divergent plate margin

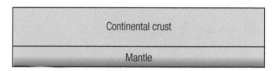

a   Original crust and mantle

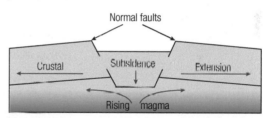

b   Tension caused by extension and gentle uplift due to rising magma underneath. Normal faults develop.

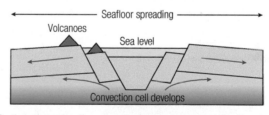

c   Continued extension. Formation of volcanoes and perhaps flood basalts. Incursion of the sea. Seafloor spreading starts.

▲ **Figure 3** Formation of a rift valley

HOW SCIENCE WORKS: OCEAN CORE COMPLEXES

In the 1990s, unusual dome-like structures were discovered on the ocean floor next to the MAR. These *ocean core complexes* are up to 30 km across, with conspicuous grooves on the top. This discovery provided an opportunity to revise our understanding of what happens at MORs.

The slow rate of spreading at the MAR suggests that the supply of magma is low and so faulting becomes more important. Extension causes normal faults along the side of the median rift. Uplift of the footwall brings the ductile lower crust (gabbro) and upper mantle (peridotite) up to the sea floor. As uplift continues, new gabbro from the magma chamber is added on to the footwall. Therefore, heat and material are transferred from deep (hot, ductile) to shallow (cool, brittle) levels.

The fault starts as a high-angle normal fault. Lateral plate movement causes the dip of the fault plane to flatten to about 30°. The fault plane is lubricated by seawater and by the formation of talc (an easily cleaved, hydrated silicate mineral) so displacement of up to 100 km continues for a million years or more in a direction parallel to the extension, at right angles to the MOR, creating significant regions of new sea floor. The grooves on the top of the domes are left over from the faulting.

These oceanic core complexes of uplifted footwalls of low-angle normal faults become inactive when a renewed supply of magma overwhelms the fault.

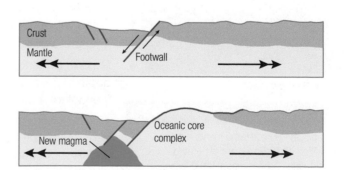

▲ **Figure 4** Formation of an ocean core complex. Length of section 30 km.

QUESTION

1   How would you demonstrate that a plate margin is diverging?

Lithosphere in the ocean basins

The Plate Tectonic theory identifies divergent margins as the place where new lithospheric material is made. What is the evidence?

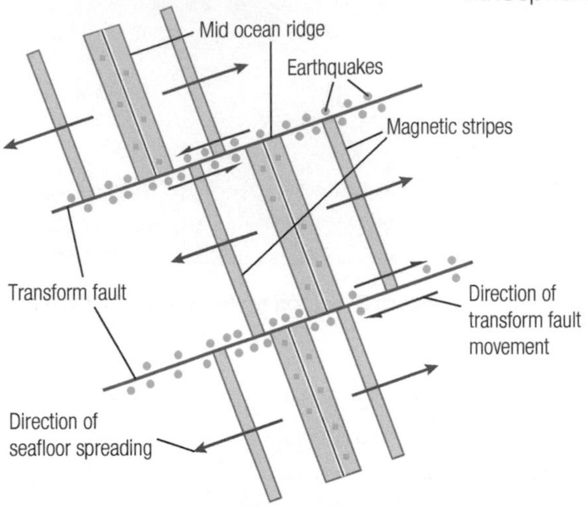

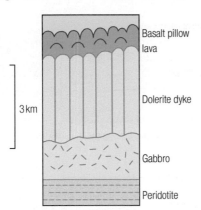

▲ **Figure 1** Transform faults displacing an MOR

▲ **Figure 2** Structure of the ocean crust and lithosphere

Mid-ocean ridges (MOR)

Over the mid-ocean ridges there is high heat flow, which indicates rising magma.

Generation of magma is shown by the eruption of basaltic pillow lavas.

Basalt lava is derived from the partial malting of peridotite.

Earthquakes have a shallow depth of focus caused mainly by harmonic waves. When magma moves, it vibrates to produce harmonic waves (similar to the wave patterns in some musical instruments), which are detected as small, shallow-focus earthquakes. It is the same phenomenon that makes old pipes rattle and bang as water flows through them.

The absence of deep earthquakes suggests that the rock is ductile. The lithosphere is, by definition, rigid, so there is no, or very thin, lithosphere beneath the MOR.

Some shallow-focus earthquakes are associated with faulting along the axial rift valley. These rifts are formed by normal faults, a sign of extension of the crust which allows mantle material to rise towards the surface.

New molten material is brought to the surface at the MOR, then cools and differentiates into a mafic crust underlain by ultramafic mantle which makes up the rigid lithosphere.

Shallow earthquakes occur in lines across the ocean floor at right angles to the MOR, indicating horizontal movement along the transform faults which divide the lithosphere into sections. This makes it easier for plate movement to occur away from the MOR. There are hundreds of these earthquakes daily, measured as less than 2 on the moment magnitude scale. Sometimes there are larger earthquakes of around magnitude 5.

Lithosphere of the ocean basins

On either side of the MOR, core sampling of the new crust and of ocean core complexes shows a consistent structure for the uppermost part of the lithospheric plate.

- Basaltic pillow lava has small crystals which cooled rapidly under water.
- Dolerite dykes are the vertical feeder pipes from the magma chamber supplying the volcanoes. Dolerite has medium sized crystals meaning it cooled fairly slowly.
- Gabbro cools very slowly, forming coarse crystals in the magma chamber.

Seismic evidence shows a change in physical properties at the Moho at an average depth of 7 km where the mafic (basalt and gabbro) crust changes to the ultramafic (peridotite) mantle.

As lithosphere moves away from the MOR, it cools, contracts and becomes more dense. At depth, the rigid lithosphere changes to the asthenosphere where the rock is more ductile. Calculations suggest that this happens at about 1300°C and this is taken as the base of the rigid lithosphere. The ductile asthenosphere moves steadily without sudden displacements and does not produce earthquakes. The depth of the deepest earthquake foci indicates the depth of the base of the lithosphere.

▲ **Figure 3** Pillow lava

STUDY TIP

Remember the Benioff zone marks a plane along which earthquakes are generated. The subduction zone is where the downward movement of an oceanic lithospheric plate, beneath another plate, is 'destroyed' by partial melting as it descends into the mantle.

Subduction zones

When an area of seafloor spreading reaches another plate, it descends into the mantle.

If the other plate is continental, the denser oceanic plate is dragged down under the less dense continental plate by the force of the collision. The boundary is marked by a deep ocean trench and fold mountains with andesite volcanoes.

If the other plate is oceanic, the colder, denser and larger of the two plates descends under the smaller plate. The boundary is marked at the surface by an **island arc**.

The descending plate moves down by subduction. The inclined boundary between the plates, starting at the trench and ending where the descending plate melts to magma, is an active subduction zone of stress and displacement, causing progressively deeper earthquakes. Rising magma under fold mountains and island arcs causes shallow-focus earthquakes.

Earthquakes are actually quite shallow under the **deep-ocean trenches** and along the oceanic sides of fold mountains, such as along the Pacific coast of the Andes.

Moving towards the continents, the earthquake foci get progressively deeper along a zone that slopes away from the trench at about 45°. This marks the top of the **Benioff zone** between the two plates where one moves beneath the other.

This active zone is present underneath the full length of the mountain range. Peru and Chile experience many devastating earthquakes. The Benioff zone is also present under island arc systems which experience earthquakes which may cause tsunamis.

Earthquakes are due to friction between the two plates as they slide past each other or get stuck and then suddenly move. Below a depth of 720 km, the hot rocks are too ductile to cause earthquakes.

Collision zones

Where two continents collide, such as India and Asia, there are shallow and medium-focus earthquakes along faults. Compression is still taking place as the Indian Plate is being pushed northwards, indicated by the devastating magnitude 7.8 Nepal earthquake in 2015.

Lithosphere of the continental land masses

We are familiar with the rocks of the continental crust of the lithosphere because their sedimentary, metamorphic and igneous rocks are exposed at the surface. The continental crust is less dense and up to 90 km thick, according to the principles of isostatic balance, as indicated by a deeper Moho beneath the mountains. The collision compresses sedimentary rocks into folds cut by thrusts, changes them into metamorphic rocks and intrudes them by granitic igneous rocks.

Further into the continental interior, there may be stable, aseismic areas of ancient, deeply eroded fold mountains called *cratons* (or continental shields).

Aseismic areas

Earthquakes are mainly confined to active seismic zones around the world, where stresses are acting in the crust and upper mantle. Between these zones are *aseismic* areas with no stresses to cause dislocation and therefore no earthquakes. Aseismic areas are generally the continental shields (cratons) of old stable rocks and the ocean basins. Britain is, at the present time, an aseismic area (other than the minor tremors we occasionally feel).

Types of plates

Oceanic lithospheric plates have an oceanic crust of basalt. Continental lithospheric plates have a continental crust of sedimentary, metamorphic and igneous rocks but most of them also have some basaltic oceanic crust. The name given to a plate depends on the type of plate boundary in question.

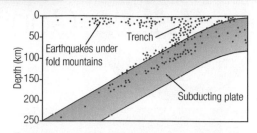

▲ **Figure 4** The Benioff zone

KEY DEFINITIONS

Island arc a curved line of andesitic volcanic islands. As the subducting plate heats up, partial melting occurs along the top surface of the descending oceanic crust, producing basaltic magma. As this wet, less dense, hot (1200°C) magma rises it may melt and differentiate into intermediate, andesite lavas.

Deep-ocean trench a long, narrow, linear submarine depression with relatively steep sides; the deepest example is over 10 000 m deep. Trenches occur alongside island arcs or fold mountain belts and indicate where a plate is beginning to descend along a subduction zone.

Benioff zone (named after the scientist who identified it) a zone of earthquake foci sloping down at an angle of about 45° from the deep-ocean trench. The zone descends beneath oceanic island arcs or continental fold mountains, marking the boundary where one plate subducts beneath another. At higher levels, where the subduction is beginning, the earthquakes are generated along the boundary itself partly due to friction. Further down, the foci occur within the plate as the interior part of the plate remains colder and more rigid, while the edges of the plate heat up.

KEY FACTS

Earthquakes are classified by their depth of focus, or the depth at which the stress acts:
- Shallow-focus 0–70 km deep
- Intermediate 70–300 km deep
- Deep-focus 300–700 km deep.

Earthquakes do not originate at depths greater than 720 km, because deeper rocks are not brittle enough to fracture.

QUESTIONS

1   Why are there no deep earthquakes under the Mid-Atlantic Ridge?

2   Why are there no earthquakes deeper than 720 km?

Figure 1 Andesite volcano, Mount Fuji, Japan

Figure 2 Eroded granite batholith, Andes fold mountains

CASE STUDY: JAPAN: A CONVERGENT OCEANIC–OCEANIC PLATE MARGIN

Along the length of the coast of Japan there is:
- A deep-ocean trench
- Shallow- to deep-focus earthquakes
- An arc of volcanic islands. The islands are made of volcanic, not continental, material.

The steep sides of Mount Fuji, Figure 1, tell us it is intermediate in composition.

CASE STUDY: ANDES: A CONVERGENT OCEANIC–CONTINENTAL PLATE MARGIN

The more dense Nazca oceanic Plate subducts beneath the continental South American Plate. The Andes are fold mountains over 5000 m high and 7000 km long, with some of the world's largest and most active volcanoes and strongest earthquakes. The earthquakes show a clear pattern, with the shallow-focus earthquakes close to the Peru-Chile trench and deeper-focus ones further inland.

As the Plate Tectonic theory was developed, it was quickly recognised that if new lithosphere is being created at divergent boundaries, then it must be destroyed somewhere else. What happens when plates converge?

Volcanic activity

Around the edges of ocean basins there are island arcs. Along the edge of some continents there are fold mountains with chains of andesite volcanoes. As the subducting plate heats up, flux melting occurs along the top surface of the descending oceanic crust, producing basaltic magma. As this less dense, hot (1200°C) magma rises, it may melt and mix with silicic continental crust to erupt intermediate, andesite lavas at the surface.

Batholiths

Where continental crust is melted by magma most of the material will remain separate because of the different viscosities. Melted silicic material forms large granite batholiths deep within fold mountains, later exposed by erosion.

Trenches

Deep-ocean trenches are long, narrow, linear, deepest parts of the Earth's surface. They are formed as the ocean floor is dragged down where two plates converge, indicating the boundary between the plates.

Fold mountain belts (orogenic belts)

Fold mountains form on the edges of the continents parallel to the subduction zone. They are compressional features made of folded and faulted sediments that have been scraped off the descending plate onto the non-subducting plate. High pressures and temperatures result in regional metamorphism.

Benioff zone

As the oceanic plate descends, the sloping plane of the boundary is marked by increasingly deep-focus earthquakes along the Benioff zone. At higher levels, where the subduction is beginning, the earthquakes are generated along the boundary itself partly due to friction. Further down, the foci occur within the plate as the interior part of the plate remains colder and more rigid, while the edges of the plate heat up.

Types of convergent plate boundaries

Oceanic–oceanic convergent plate boundary

Example: between the Pacific Plate and the Eurasian Plate in Japan

The slower or older/colder/denser oceanic plate is subducted at the trench into an area of higher temperature.

- Partial melting of the basaltic crust starts as the lower temperature minerals begin to melt.
- The melted minerals separate and, being less dense than the surrounding oceanic crust, rise as magma with an intermediate composition.
- The magma erupts at the surface as intermediate (andesite) and mafic (basalt) lava.
- The resulting island arc reflects the curving shape of the convergent plate boundary beneath.
- The rocks of the island arc, including sediment scraped off the descending plate, are metamorphosed by the increase in temperature and pressure.

Oceanic–continental convergent plate boundary

Example: between the Nazca Plate and South American Plate, forming the Andes.

The processes operating at oceanic–oceanic plate margins are the same here. However, the presence of continental crust makes it more complicated.

- The magma rising from the subducting oceanic plate is hot enough to cause partial melting of the continental crust that it passes through. Some silicic magma may reach the surface to form explosive rhyolitic volcanoes. Some will mix with the rising mafic magma to create large volumes of the intermediate lava that builds the strato-volcanoes of the Andes.
- Large batholiths are formed from melted continental crust and intruded deep in the core of the fold mountains.
- Compression of the continental crust and the sediments scraped off the subducting plate will form fold mountain chains. These are intensely folded and faulted, with major thrusts, nappes and overfolds. The continental crust shortens laterally forming high mountains with deep roots.
- Many of the rocks in the deeper part of the mountain chain are regionally metamorphosed by increased heat and pressure. Gneiss and **migmatite** in the core of the mountains grade outwards through schist and slate.
- Segments of the former oceanic crust may be broken off at the top of the subduction zone, trapping an ophiolite suite within the fold mountains.

Continental–continental convergent plate boundary

Example: the collision of India with Eurasia, forming the Himalayas.

Two continental plates of similar composition and density meet under the Himalayas. The Indian Plate had spent the last 100 million years as an island, drifting north away from Africa and Antarctica. Movement was initially slow but increased in the last 60 million years, with a maximum rate of 19 cm per year, until it collided with continental Asia.

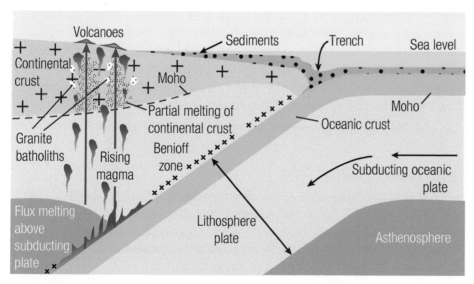

🔺 **Figure 3** Cross-section of convergent continental–oceanic plate boundary

Evidence for earlier convergent plate boundaries

There are many geological processes which occur only at plate boundaries. Looking at the distribution of these features allows us to identify not only modern plate boundaries but also ancient examples. Intermediate volcanic rocks, ophiolites, batholiths, folded and faulted rocks in an ancient fold mountain belt in Scotland tell us that this was an active plate boundary 400 million years ago.

So far we have only considered volcanoes that are found along active plate boundaries. A number of volcanoes are found in the middle of plates for which we must also find an explanation. The most notable example is the island chain of Hawaii in the middle of the large Pacific Plate.

Magma from the deep

The 'Big Island' of Hawaii is built of five shield volcanoes, three of which are active. Mauna Loa (4100 m) is the highest and Kilauea is the most active volcano on Earth, producing daily eruptions of lava. The actual height of the Hawaiian volcanoes, from the sea floor to their summits, is greater than the height of Everest, so the volume of lava that built them was immense. The volcanoes sit on oceanic crust that is only about 5 km thick, so where did all the lava come from?

Hawaii is located on a hot spot above a rising mantle plume. Mantle plumes are stationary, long-lived areas of high heat flow within the mantle. They consist of a long, thin conduit within the mantle, with a bulbous head that spreads out at the base of the lithosphere. As the plume head reaches the lithosphere, the reduction in pressure causes widespread partial melting of ultramafic mantle producing huge volumes of mafic magma. The mafic magma accumulates beneath the lithosphere until it is able to punch a hole through to form a shield volcano.

Evidence for mantle plumes

The idea of mantle plumes is just one of a number of hypotheses that have been suggested to explain hot spot volcanoes within plates:

- Seismic tomography and heat flow: using a network of seismometers, geophysicists use seismic wave velocities to construct three-dimensional computer-generated images of heat flow within the Earth. Seismic waves slow down when they travel through hotter, lower density material. There are 32 regions in the mantle, worldwide, where P waves travel slower than average and these have been interpreted as rising mantle plumes. Changes in S wave velocities indicate that the plumes extend to the core–mantle boundary.
- Geochemistry: chemical and isotopic analyses of basalts erupted at hot spots show them to be different from basalts erupted at mid-ocean ridges or island arcs.

The geochemistry suggests that the magma originates from a different source area of the mantle.

Moving island chains

All the islands of the Hawaiian-Emperor chain are made of basaltic shield volcanoes. Apart from those on Hawaii itself (the largest island), the others are extinct. They get progressively older and more eroded towards the north-west, continuing as a 6000 km line of submarine **guyots** and **seamounts,** which extend to the Aleutian Islands.

There are two possible explanations for this pattern:

1 The position of the mantle plume has moved towards the south-east.

2 The Pacific Plate has moved towards the north-west over a stationary plume.

The second suggestion is more likely. A volcano builds when a vent or series of vents is created through the lithosphere. Plate movement then carries the volcano off the mantle plume so it becomes extinct and a new vent and volcano form over the stationary plume. Eventually a chain of volcanoes forms, each one younger than the one before. The position and ages of these seamounts show the direction and rate of plate movement. In the case of Hawaii, the age of volcanic islands and seamounts show that the hot spot is 80 million years old. A new submarine volcano, Loihi, is growing 30 km south of Kilauea, showing the mantle plume is still active today. At the Hawaiian-Emperor Bend, the chain of seamounts turns north towards the Aleutian Trench, indicating that there was a change in the direction of plate movement 42 Ma.

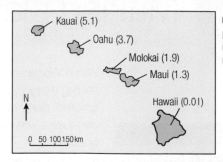

▲ **Figure 3** Map showing part of the Hawaiian island chain. Ages given in millions of years.

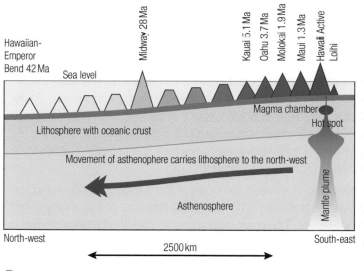

▲ **Figure 1** Cross-section through the Hawaiian-Emperor chain

▲ **Figure 2** Deeply eroded shield volcano of Kauai

MATHS BOX: THE RATE OF SEAFLOOR SPREADING

If the age of a volcanic island is known from radiometric dating and its distance from the active hot spot is measured, it is possible to calculate the rate of plate movement and hence the rate of seafloor spreading, using the formula:

$$\text{speed} = \frac{\text{distance}}{\text{time}}$$

The distance between Hawaii and Maui = 15 mm on map.

Using the scale: 15 mm = 225 km.

Age difference between Hawaii and Maui = 1.3 − 0.01 = 1.29 million years.

$$\text{Speed} = \frac{\text{distance}}{\text{time}} = \frac{225 \text{ km}}{1.29 \text{ million years}} = \frac{225 \times 1000 \times 100 \text{ cm}}{1\,290\,000 \text{ years}} = \frac{22\,500\,000}{1\,290\,000}$$
$$= 17.4 \text{ cm per year}$$

CASE STUDY: HOT SPOTS

There are numerous examples of oceanic hot spots around the world — Hawaii, the Canary Islands, the Azores, St Helena and Ascension, to name a few.

Due to the thickness of the crust, mantle plumes do not manifest themselves as obvious hot spots in continental areas, but they do exist. The Yellowstone 'super-volcano' in North America is a hot spot.

Mantle plumes have also been implicated in the break-up of large continents and in the formation of continental flood basalts which are huge volumes of basalt erupted in continental areas over a short time period. Flood basalts include the 66 Ma Deccan Traps in India, the 252 Ma Siberian Traps in Russia and the Karoo basalts and Ferrar dolerites of South Africa and Antarctica.

QUESTIONS

1 Describe the difference between a hot spot and a mantle plume.

2 Explain why the Hawaiian-Emperor chain gets older towards the north-west.

3 By using the points on Oahu and Molokai, shown on the map in Figure 3 showing part of the Hawaiian island chain, calculate the rate of seafloor spreading. Give your answer in cm per year.

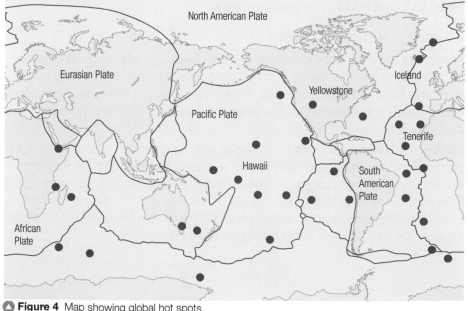

▲ **Figure 4** Map showing global hot spots

When Wegener proposed the theory of continental drift, his biggest problem was finding a mechanism by which the continents could move. His theory was largely ignored, partly because of limited knowledge about the interior of the Earth, but also because it was difficult to see how the continents could move over the top of the oceanic crust. To explain the theories of continental drift, seafloor spreading and plate tectonics, we need to find a mechanism by which the lithospheric plates can move.

Convection cells – an old theory

Convection cells in the mantle were thought to drive the motion of the plates and this was an accepted theory in the past, but is now thought to be incorrect. There is no correlation between plate area and the velocity of plate movement, which suggests that the process is not driven by movement in the mantle. In fact, the drag of the mantle itself may actually slow the movement of the plates, rather than participate in the mechanism for their movement.

There is also a good correlation between the edge of the subducting plate and the overall velocity of the plate as it moves. This suggests that slab pull is an important mechanism for plate movement. There are also some studies which have shown that the asthenosphere is too plastic (flexible) to generate enough friction to pull the tectonic plates along.

Ridge push and slab pull

Two processes, ridge push and slab pull, are thought to be responsible for the motion of lithospheric plates, and there is some uncertainty as to their relative contribution. The force applied by slab pull is around twice the force applied by ridge push.

Ridge push at the mid-ocean ridge

In this model, passively upwelling mantle at a divergent boundary is buoyant and raises the sea floor to form an MOR which is two to three kilometres higher than the abyssal plain. Gravity acts on the elevated lithosphere producing a pushing force down the tectonic plate and away from the elevated sea floor. As the sea-floor crust cools and thickens this directs an additional force away from the ridge. The net result of ridge push can be thought of as gravitational sliding off the elevated MOR which is a huge mountain range.

Slab pull at subduction zones

Attention has focused on the idea that negative buoyancy pulls subducting oceanic lithosphere down into the mantle at convergent plate boundaries. This may be the main driving force for lithospheric plate movement today. In this model, the mass of the cold, dense lithosphere sinks downwards at ocean trenches and pulls the rest of the oceanic lithosphere with it.

In the Archean, when the mantle was warmer and the geothermal gradient steeper, the lithosphere may have been too buoyant to subduct, and ridge push would have been the dominant process. However, some geologists argue that without the density contrast needed for subduction, plate tectonics would not be possible. If this was the case, then volcanic advection would have been the main heat transfer process at the time, as seen today on Venus, Mars and Io.

KEY DEFINITIONS

Ridge push a process where two sections of lithosphere are pushed apart by rising magma at the MOR.

Slab pull a process where a cold, dense section of lithosphere descends into the mantle at a subduction zone.

Pangaea (= all the Earth) a supercontinent that formed at the end of the Permian.

Recycling the Earth

We have already seen that no area of oceanic crust is older than 200 Ma. This shows that, in geological terms, oceans are temporary features that open and close due to the shifting balance between their formation at MORs and destruction at subduction zones. The complete cycle of opening and closing of an ocean is called a Wilson Cycle, after the Canadian geologist J Tuzo Wilson, who suggested the idea in 1967. If you travel along a 'moving walkway', for example in an airport, there is no danger of running out of floor because the belt is continuous and returns back to the beginning, out of sight, underneath the walkway. Similarly, the Earth will never run out of basalt because the material brought to the surface at a divergent plate margin is returned to the depths of the mantle along the subduction zone of the convergent margin. The melted material is then carried as the return current, gaining heat from the underlying core over which it passes, back to the 'start' of the convection cell where it rises once again under the MOR. Slab pull probably limits the maximum size of an ocean. One complete Wilson Cycle takes about 500 million years to complete!

🔺 **Figure 2** *Olenellus* from Newfoundland (coin for scale)

CASE STUDY: THE CALEDONIAN OROGENY

So far, we have considered the plate tectonic system which has prevailed for the last 250 million years. If each Wilson Cycle takes 500 million years, there must have been several other cycles to look for over geological history, such as the one which in Britain we call the Caledonian Orogeny.

In Britain, the 400 Ma Caledonian mountain belt marks an ancient convergent plate margin. Before 400 Ma, Scotland was part of North America (Laurentia), a long way from England and Wales which were part of Europe (Baltica), on the other side of a large ocean called Iapetus. In Greek mythology, Iapetus was the father of Atlantis. The collision of the plates brought the two parts of Britain together along the line of the Solway Firth to form a larger plate called Laurasia.

The evidence for an ancient plate margin includes:
- Extensive intermediate volcanic lavas and pyroclastic rocks in North Wales, the Lake District and southern Scotland;
- Folded and faulted rocks in an ancient fold mountain belt;
- Regionally metamorphosed rocks;
- Granite batholiths;
- Fossils of different genera of trilobites that must originally have been two separate faunas on either side of the Iapetus Ocean but are now geographically close together.

As a result of the collision, Europe and North America were joined together for the next 300 million years as part of the supercontinent **Pangaea**. Around 167 Ma, the present-day South Atlantic Ocean started to open as a result of rifting, followed by the North Atlantic. This divergence split the Caledonian orogenic belt into two parts, separated by the young Atlantic Ocean. The Appalachians and the Scottish Highlands have the same rock types, fossils and structures of the same age, showing they were once part of the same continuous mountain range.

CASE STUDY: TRILOBITES AND PLATE TECTONICS

Shallow water, benthonic trilobites are present on both continents either side of the Atlantic Ocean. They would not have been able to migrate from North-West Scotland across the Atlantic Ocean to Newfoundland. However, Ordovician trilobites (*Olenellus*) now found in these two areas are very similar. The Scottish species are more similar to those in Newfoundland than those in Wales. This shows that Scotland was close to Newfoundland and a long way from Wales during the Ordovician.

NW Moine Scottish Highlands   Midland Valley Southern Uplands   (Future line of closing) Iapetus Suture   Lake District North Wales   SE

Fold mountains with metamorphism

Iapetus Ocean

Thrust
Craton

Failed rift   OS

Granite batholiths

Subduction

Mid-ocean ridge (about to stop spreading)

Island arc andesite volcanoes

Partial melting

Continental lithosphere

Oceanic crust
Oceanic lithosphere

Asthenosphere

Asthenosphere

Key
OS   Ophiolite suite
→   Direction of plate movement
🔺   Volcanoes

🔺 **Figure 1** Plate movement 400 Ma. Length of section approximately 800 km.

QUESTIONS

1 Describe models of convection in the rheid mantle.

2 How different would the future development of the Earth be if the convection in the mantle stopped completely?

3 Explain the difference between ridge push and slab pull.

Thermal models predict a magma chamber beneath thin crust. Seismic surveys show a low-velocity zone thought to be a magma chamber system 5 to 10 km across. The chamber is fed from the centre, then spreading carries the two halves of the chamber, forming the wing shape identified by the survey.

A negative gravity anomaly which is only 2 km wide and just 1 km below the sea floor may be associated with a short-term upper magma reservoir.

KEY DEFINITIONS

Obduction a process by which the edge of oceanic crust is scraped onto the continental crust.

Multi-beam echo-sounder sends out up to 100 beams of sound covering an arc up to 10 km wide, allowing wide areas of the sea floor to be covered in one transect of the ship. The profiles are only 100 m apart.

Side-scan sonar towed at depth, sends out sound waves across the sea floor so that a side-view profile of submarine features is obtained. Strong echoes indicate a steep smooth surface facing the instrument; weaker echoes suggest a rough surface.

The mid-ocean ridge (MOR) is a 60 000 km long chain of mainly submarine volcanoes, ridges, rifts, fault zones, hot springs (and exotic animal communities) which encompasses the Earth 'like the seams on a tennis ball'. The system crosses through every ocean. The average depth to the MOR is 2500 m, ranging from 1800 m above sea level in Iceland, to 4000 m below sea level in the Cayman Trough in the Caribbean.

Basaltic volcanoes erupt more lava than the rest of the Earth's volcanoes put together. About 3 km^2 of new crust forms each year from 20 km^3 of magma. The zone of current volcanic activity is less than 10 km wide.

Every 50 to 500 km, the MOR is offset by transform faults which divide the MOR into many sections.

MORs are the principal means of transferring of mass and energy from the upper mantle to the surface by advection and partial melting.

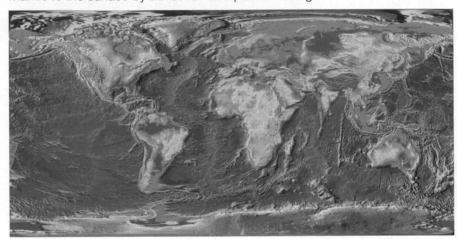

▲ **Figure 1** Topographic map of the world's oceans

Investigating the mid-oceanic ridge

The two main features of the MOR are volcanoes and features formed by extensional tectonics (rift valleys), suggesting that there is a source of magma at depth which is allowed to reach the surface by extension of the crust.

Direct evidence

The interior of the MOR cannot be studied directly because it is made of molten rock. We can investigate the lithosphere immediately adjacent to the MOR by sampling the ocean floor and sampling *ocean core complexes*. Deep ocean submersibles take a range of measurements and core samples.

No matter where you sample the oceanic crust, the overall structure is the same. All the igneous rocks form from magma rising at the MOR, cooling and then spreading apart symmetrically from the ridge.

▼ **Figure 2** Heat flow measurements across the MOR

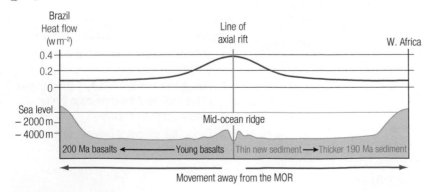

Remote surveys

The highest heat flow is directly above the axial high (rift) which indicates that passively upwelling mantle may be partially melting to form magma which migrates upwards, just beneath the seabed.

Gravity anomalies indicate the presence of a low-density zone in the upper mantle below the MOR. The broad, negative *Bouguer Anomaly* indicates a deficit of mass due to upwelling less dense mantle material.

Free air gravity profiles show no anomalies because the mass of the elevated sea bed is in isostatic equilibrium, supported by the low-density zone in the mantle.

Remote surveys sometimes give conflicting results. Seismic surveys identify a low velocity zone for seismic waves, suggesting that melting is taking place but electromagnetic surveys indicate that melting may not happen in the low velocity zone. Many other factors are involved such as hydration and composition of the magma, grain size and the presence of partial melt material.

Ophiolite suites

The photograph in Figure 3 shows a sample of rock from the Moho, which is at the base of the oceanic crust. How was it collected?

In a few locations around the world, segments of the oceanic crust and mantle have been thrust towards the surface during a plate collision, by a process called **obduction**. It is much easier to look at rocks of these *ophiolite suites* on land than to try to study them on the ocean floor. The structure of the oceanic crust has been worked out largely by using ophiolite suites on land: for example, Newfoundland and Cyprus.

▲ **Figure 3** The Moho in hand specimen

mantle peridotite   crustal gabbro
moho

▲ **Figure 4** The Moho on land. An inverted sequence of orange-weathered mantle peridotite separated from dark crustal gabbro by the Moho, Newfoundland.

▲ **Figure 5** Sheeted dykes, each about 1 m across, which acted as feeder dykes for the lavas. Cyprus

How do we find out what goes on at a mid-ocean ridge? Submersibles cover a very small area. Drill ships take a limited range of samples of sea-floor rocks. A detailed map showing all the topographical variations provides a virtual picture of the sea floor.

In 1972, a group of marine geologists decided to explore the MOR rift valley using the French bathyscaphe *Archimède* and the American submersible *Alvin* using the limited maps of the time. Marie Tharp and Bruce Heezen's classic 1957 'Floor of the Oceans' chart was very accurate for its time, but the survey transects were 10 to 100 km apart.

With the development of multi-beam deep water echo-sounders, a single transect by the towing ship could collect up to 100 profiles simultaneously, each transect just 100 m from the next. At the same time, GPS navigation became available at first for a few hours and then for 24 hours a day with precise fixes every 2 seconds. So, 10 km-wide bands of very accurate sea-floor data could be collected and located for each transect.

The proportion of the whole median ridge system which had been mapped increased from less than 1% to almost 50% within a period of ten years. Even so, away from the median rift, the coverage of information reduces to 1% again. The proportion of the MOR explored by submersibles is even less than that.

▲ **Figure 6** Reconstruction of the bathyscaphe *Archimède*.

CASE STUDY: THE REYKJANES RIDGE, ICELAND

BRIDGE (British Mid-Ocean Ridge Initiative) used gravity surveys, echo-sounding and seismic tomography to investigate the Reykjanes Ridge. In Iceland, the MOR reaches 1800 m above level. It has no median rift valley, more characteristic of a fast spreading ridge (see later), which suggests that the lithosphere is hotter than usual. Is there a mantle plume beneath Iceland?

A seismic tomography survey using 75 seismometers showed that velocities were 12% slower than average; the mantle is hotter than normal and less rigid. A narrow (200 km), hot plume was identified, extending at least 1600 km down into the mantle.

Images of the gravity field derived from satellite altimetry showed a series of offset ridges pointing down the ridge. BRIDGE mapped the sea floor using a **multi-beam echo-sounder** which had a resolution of 20 m. The offset ridges were aligned at right angles to the spreading direction, as expected for tension fractures. This type of mapping identified the offset ridges as lines of short-lived volcanoes.

The Towed Ocean Bottom Instrument (TOBI), an echo-sounder which works best about 300 metres above the sea floor, used a **side-scan sonar** to show that, after the volcanoes erupted sheet flows or hummocks tens of metres high, they were broken by faulting. The TOBI images suggested that the crust was stretched as the plates moved, causing faulting which destroyed fresh basalt mountains, sometimes as cleanly as if cut by a knife.

Magnetic surveys showed a weakened magnetic field indicating that the sea-floor basalt is hotter than usual.

BRIDGE concluded that Iceland is held up by a hot, less dense, buoyant mantle plume which produced extra magma. The thickened crust is more elevated due to isostatic adjustment.

QUESTIONS

1 Describe the changes in the sea floor as you move away from the MOR.

2 Explain how seafloor spreading causes continents to drift. Use the terms convection, asthenosphere, lithosphere and partial melting in your answer.

3 How can you prove that ophiolite suites actually originated in the ocean crust?

The MOR is the focus of seafloor spreading due to an extensional tectonic regime. The space that opens is filled with magma: lava flows, dykes and the remains of magma chambers. If there is not enough magma to fill the space, the crust fractures instead, stretching to form normal faults.

Spreading rates

The rate of spreading is found by radiometrically dating ocean floor basalts a known distance from the MOR, and by using pelagic microfossils to find the age of sediments at known distances from the MOR.

Each MOR has its own style of faulting, volcanic activity and spreading rate, depending on the strength of the ocean crust, and is dependent upon how cold and brittle the upper part of the lithospheric plate is, as this alters its properties.

The rate of spreading governs how much heat is carried upwards by magma and how quickly it is lost from the top of the magma chamber, mostly by hydrothermal activity.

Slow-spreading ridges

For example, the Mid-Atlantic Ridge:

South Atlantic – 4 cm per year   North Atlantic – 2 cm per year.

Slow-spreading MORs have median rift valleys up to 20 km wide and 3 km deep, with very rugged 'mountains' at the crest. Some parts of the ridge are covered by young lava flows and pillow eruptions. Other parts show faulting instead. Sonar images show vast areas of the ocean floors covered with rough abyssal hills on either side of the ridge.

At slow-spreading MORs there is often insufficient partial melting to maintain a magma chamber although there may be mush zones where small volumes of melt exist within softened mantle. Magma chambers are short lived and discontinuous in extent along the axial rift. Each eruption is a distinct event. Some sections of the Mid-Atlantic Ridge have numerous small volcanic cones formed in this way.

Fast-spreading ridges

For example, the East Pacific Rise:

- up to 16 cm per year between the Nazca and Pacific Plates.

The East Pacific Rise has a high elevation, dome-like topography resulting from many layers of lava built up from the greater supply of magma. Deep, hot rock rises quickly and more heat passes into the plate. Rock is a poor thermal conductor so heat cannot escape equally quickly and the lithosphere becomes hotter, weaker and more ductile. The ridge crest cannot subside due to rising magma. There is no rift valley, just a crack along the smooth crest.

The flanks are smooth and the ocean floors are covered with vast areas of smooth abyssal hills.

Effects on global sea level

Increased activity at a MOR makes it bigger, taking up more 'space' in the oceans. This causes a rise in sea level and the displacement of water onto the continents. Calculations of the volume of MORs correspond well with the Vail sea-level curve. Long-term high sea levels correspond with times of high plate velocities when lots of new material was being formed by seafloor spreading. The same happens if new MORs are created. A high rate of seafloor spreading leads to an increase in subduction causing subsidence and flooding of continental margins.

KEY DEFINITIONS

Adiabatic (decompression) melting rising magma experiences a decrease in pressure. Expansion causes a reduction in temperature with the loss of heat as molecules use energy to move further apart. Melting occurs because the melting point also decreases as the pressure decreases.

Mush a mixture of crystals and melt.

Continuous magma chamber continually produces lava.

Discontinuous magma chamber produces lava periodically, with long periods of inactivity in between when the chamber has solidified.

CASE STUDY: FINDING ROOTS

The Ramesses project combined two techniques to investigate the roots of the Mid-Atlantic Ridge down to a depth of more than 100 km:

- Seismic surveys, using small explosions to make images of the rocks below the ocean floor.
- Electromagnetic surveys, using receivers on the ocean floor to detect electromagnetic waves sent out from a deep-towed transmitter.

Ramesses found the first large magma chamber to be discovered beneath a slow-spreading MOR. It was very similar to those found beneath fast-spreading ridges although with a much shorter lifespan.

The amount of magma in the thin melt pool and the mush zone was much larger than for fast-spreading ridges, providing more evidence for the episodic nature of the magma chamber where the magma accumulated over time instead of erupting continuously.

Making magma

At depths of 100 km to 200 km, the mantle is at temperatures of about 1300°C, close to the melting point of peridotite. It is very hot and soft but still solid because the pressure is high.

As mantle material rises below the MOR, two changes take place:

1 The pressure drops and the mantle material expands. It experiences a cooling effect due solely to the increase in volume. This is an **adiabatic change**, well known to meteorologists.

2 As the pressure drops, the molecules in the rock move more freely and therefore the melting point falls.

If the rising mantle is hot enough and rises far enough, it starts to melt by **decompression melting**. Once melting starts, about 3% of the rock melts for each 10 km that the rock rises. More melt forms if the mantle beneath the MOR is hotter than normal, because it will start to melt deeper down, as happens with mantle plumes.

The amount of magma produced also depends on the latent heat of fusion (melting). It takes a lot of heat to change silicates from the solid to the liquid state, which helps to explain why melting of mantle material is usually partial.

Seismic tomography suggests that some magma chambers are made of a thin layer of molten rock lying on top of a partly-molten **mush zone**. The melted zone is only a few tens of metres thick and about a kilometre wide. **Magma chambers** which erupt magma all the time are said to be **continuous**, as at fast-spreading MORs.

The melt forms as thin films in between the crystals. The liquid moves upwards joining other films of melt, growing in size and accumulating in a shallow magma chamber where it is cooled by the ocean above, including by the infiltration of sea water. As more melt percolates upwards, the pressure inside the chamber rises until the top of the shallow chamber splits apart to give a narrow crack, about a metre wide. The pressurised melt flows up the cracks and erupts onto the sea floor as basalt lava. The melt in the crack solidifies to form a dolerite dyke. As the system cools, the magma chamber solidifies to form the gabbro of the lower ocean crust below the sheeted dykes and pillow lavas.

The thickness of the oceanic crust reflects the amount of melting. At normal mantle temperatures, enough melt forms to create ocean crust that is 7 km thick. Under Iceland, the mantle is much hotter due to a mantle plume. There is much more melting and the crust is up to 20 km thick, as determined by gravity surveys.

Slow-spreading MORs

At slow-spreading MORs there is not enough heat to maintain a molten magma chamber. The chamber is therefore short-lived or **discontinuous**, probably existing for no more than 10% of the time. Each eruption is a distinct event. Some sections of the Mid-Atlantic Ridge have numerous small volcanic cones formed in this way.

It is only possible to locate a magma chamber when the material inside is molten. If the chamber solidifies it is hidden from seismic surveys. This applies to 90% of the volcanoes. And by the time that new magma is formed, perhaps 20 000 years later, the old chamber will have been added to the lithosphere and spread away from the ridge.

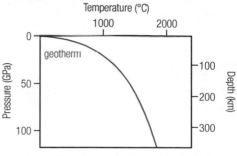

a   The geotherm: temperature increases with depth

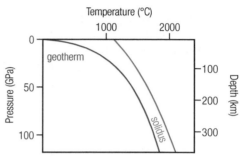

b   While the geotherm is at a lower temperature than the solidus, the mantle is solid. The melting point (solidus) increases with depth.

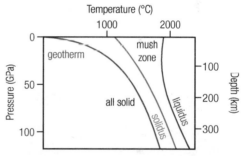

c   Between the solidus and liquidus, partial crystallisation occurs - the mush zone.

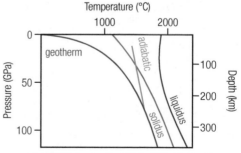

d   The adiabatic is a modified geotherm. If the rising mantle cools along the adiabatic, melting begins when it crosses the solidus.

▲ **Figure 1** Decompression melting

QUESTION

1 Explain what factors control the size and shape of a mid-ocean ridge.

Hydrothermal processes

Along fast-spreading mid-ocean ridges, hot rock rises from the mantle, giving rise to submarine fields of hot springs, each one a few hundred metres wide and some producing mineral deposits similar to those now mined on land.

This water becomes superheated to more than 400°C, and is now called a hydrothermal fluid. It is prevented from boiling by the high hydrostatic pressure. The hydrothermal fluid is much less dense than the cold sea water so that it gushes out of the sea floor at very high speeds. The jet of water looks like thick black smoke, hence the name of **black smoker** vent from which the fluids are expelled.

Basalt is made up of the minerals feldspar and augite, with some olivine. It also contains small amounts of accessory minerals with a range of metallic elements in their formulas. Due to the high temperature, the hot water dissolves metallic elements out of the basalt. The jets of hot solutions react with cold sea water. Fine particles of *chalcophile* minerals, sulfides of zinc, copper and iron, precipitate out, giving the appearance of black smoke. Pyrite is one of the minerals produced.

The precipitated minerals accumulate and, helped by chemosynthetic bacteria, rapidly build up chimneys and mounds of sulfide minerals around the hydrothermal vent, a few metres tall. Sometimes there are traces of platinum, gold and silver, but usually not enough to make these deposits economic. Such large quantities of minerals are ejected onto the sea floor that hydrothermal vents have formed rich ore deposits. Ancient deposits of hydrothermal minerals have been mined, for example from the ophiolite suites of Cyprus (which derives its name from 'copper').

Mounds of a potentially economic size accumulate by the growth of a chimney which then collapses, blocking the vent. Sea water percolates through the mound, precipitating sulfide and sea-water minerals such as anhydrite into the mound. After some time, hydrothermal water breaks though the collapsed mound, forming a new chimney. The increased temperature dissolves the anhydrite leaving gaps in which more sulfides can precipitate. This new chimney collapses in turn. Over a long period of time, an economic deposit accumulates.

The Broken Spur plume in the mid-Atlantic is a deposit of an estimated 10 000 tonnes of sulfides in mounds up to 10 m wide and 40 m high, scattered over an area 100 m across, rising from a surface of young lava flows. This and other examples have been drilled by JOIDES Resolution (ocean drilling research vessel) to give estimates of their size.

Typically, hot vents are identified in areas of active volcanism on the ocean floor. However, some vents have been discovered well away from any volcanic activity, located in areas of sediment rather than volcanic rocks. These smokers are located along fault lines on a slow-spreading MOR.

Metasomatism

Metasomatism means change of rock due to addition or subtraction of chemical constituents. It is essentially the chemical alteration of a rock by hydrothermal and other percolating fluids. The chemicals of the original minerals react with the hot water and its dissolved chemicals to form new minerals which replace the original crystals.

Peridotite contains the mineral olivine. Olivine reacts with hot water to form a new hydrated mineral called **serpentine**, so called because it looks like snake skin. This forms **serpentinite**. The water is incorporated into the atomic structure of olivine, so there is an increase in volume and a reduction of density, which may help to account for the size of the MOR. This exothermic reaction helps to raise the temperature of the hydrothermal water. In the reaction, other elements are given off including hydrogen.

KEY FACT: HOW SERPENTINE FORMS

Serpentine forms from mafic igneous rocks by the reaction:

olivine + pyroxene + water

$\rightarrow$ serpentine + magnetite + hydrogen

▲ **Figure 1** Serpentinite (coin for scale)

▲ **Figure 2** Chrysotile

HOW SCIENCE WORKS: THE DISCOVERY OF BLACK SMOKERS

Hydrothermal vents were discovered in 1977, at the Galapagos spreading centre on the East Pacific Rise, using the US deep-submersible *Alvin*. Since then, hydrothermal activity has been found on MORs in all oceans, although only about 100 are known worldwide.

Two years later, black smoker vents and their deposits were discovered. The high temperatures that *Alvin* recorded came as a complete surprise. Temperature probes were at first set to a maximum of 30°C, but were quickly recalibrated and temperatures of nearly 400°C were recorded. It was later realised that the melting temperature of *Alvin's* portholes was considerably less than 400°C!

Hydrothermal vents are rare. How can they be found in the vastness of the oceans? Modelling and deep-water studies show that hot water currents from vents spiral (due to the Earth's rotation) upwards before spreading out sideways as they cool. The resulting plume of mineral and 'smoke' laden water drifts down-current about 300 metres above the ocean floor. As with other mineral prospecting, all you have to do is to locate the plume – perhaps from as far away as a thousand kilometres and follow it as the concentration in the direction in which it becomes stronger. When the concentration suddenly reduces to zero, you have just passed over the vent.

An isotope of helium, ^{3}He, comes only from the mantle at MORs but it is difficult to measure. Manganese and methane are rare in the deep oceans but a million times more concentrated in water from hydrothermal vents. Deep sea water samples can be analysed easily on board ship.

You also need a bit of imagination. The plumes are cloudy, laden with mineral particles which scatter light passing through them in an instrument called a nephelometer (cloud meter) used in many other scientific disciplines. A nephelometer attached to a TOBI, designed to work 300 metres above the sea floor, gives continuous readings in real time which can be matched with sampling for manganese.

Heat flux from the sea floor

Hot water from the hydrothermal vents spirals upwards by convection. The height to which the plumes rise gives an estimate of the amount of heat they carry. Worldwide, this is estimated at about 12 terawatts (TW), compared with the 180 000 TW the Earth receives from the Sun. This is important for ocean geochemistry, geophysics, marine biology, and, some say, for the origin of life. Much of the heat goes into the ocean crust and then into the ocean water at MORs through the sea floor and from black smokers.

The whole of the ocean floor, not just the MORs, is involved in heat exchange. Cool sea water percolates into the crust along fractures and faults, cooling the crust as the water gets warmer.

HOW SCIENCE WORKS: CHEMOSYNTHESIS AND THE ORIGIN OF LIFE

Chemosynthetic (made by chemicals) microbes live around hydrothermal vents, extracting energy from chemical reactions. Hydrogen sulfide and methane in hydrothermal water react with oxygen and oxides in sea water. Microbes are adapted to speed up these reactions. The same amount of energy per square metre is available in a vent field as there is in a tropical rain forest from the Sun. Nutrients are abundant, so microbes flourish.

Energy may be abundant but the conditions are most hostile. Temperatures vary by 400 degrees in just a few centimetres and chemicals such as copper and hydrogen sulfide are highly toxic. Yet *hyperthermophilic* microbes, adapted to temperatures of well over 80°C, flourish around the vents.

Perhaps life on Earth began at hydrothermal vents. The most primitive known organisms are hyperthermophiles, and chemosynthesis must have developed before photosynthesis evolved. There was no molecular oxygen in the original atmosphere because there was no photosynthesis. However, simple reactions involving hydrogen can give enough energy to form organic molecules.

Metasomatism of mantle peridotite produces hydrogen. The young Earth had a high enough heat flow to form ultramafic lavas, which would have produced hydrogen in the same way.

Conditions for the origin of life include water with dissolved carbonate, a source of hydrogen, and temperatures around 100°C. These conditions are found around hydrothermal vents.

CASE STUDY: TROODOS OPHIOLITE, CYPRUS

The Mathiatis massive sulfide deposit was a low-grade (0.3% Cu), three million tonne ore body of pyrite and minor chalcopyrite occurring within basalt lavas. Mining stopped in 1987 leaving a deep open pit surrounded by large spoil heaps which are now undergoing oxidation and leaching, potentially causing problems for local water supplies.

Chrysotile is a fibrous variety of serpentine from which asbestos was manufactured. It was extensively mined, leaving vast quarries which are now being reclaimed.

CASE STUDY: BLACK SMOKERS

Hot vents (>300°C) are found close to volcanic activity. Here oxygenated sulfate-rich seawater penetrates the hot basaltic crust and is reduced to hydrogen sulfide-rich fluids. These fluids react with chalcophile metals and migrate upwards. The metals are deposited in zones of alteration or in hydrothermal chimneys. These are known as black smokers.

CASE STUDY: WHITE SMOKERS

Cooler vents (200°C to 300°C) have been discovered well away from any volcanic activity, located in areas of sediments rather than volcanic rocks. These white smokers are located along fault lines on slow-spreading MORs and ocean core complexes. Metal sulfides are precipitated below the sea bed in the zone of metasomatism. The precipitation of silica, anhydrite and calcite causes the smoke's colour.

QUESTIONS

1 Search the Internet to find out what the research ship JOIDES Resolution is doing today.

2 Is it possible to exploit metal deposits on the deep ocean floor?

3 Compare metasomatism and metamorphism.

4 Research how black smokers were discovered.

A **bed** is a layer of sedimentary rock that has been laid down under a specific set of environmental conditions. Usually the youngest beds will be found on top of older beds, though sometimes the strata have been inverted. Several beds are known as strata. A bed may be just a few millimetres thick or it could be several metres thick. Each bed is laid down parallel to the surface of deposition, which means it is roughly horizontal. A bed may be laid down by a single event over just a few hours, such as when a river floods and deposits a layer of mud over the flood plain. However, a bed can form very slowly over a period of hundreds of thousands of years, such as on the deep ocean floor where the sedimentation rate is extremely slow.

A **bedding plane** is a surface that separates each bed and is parallel to the surface of deposition. The bedding plane is usually an even line but where the surface of deposition is irregular it is uneven. Where erosion has occurred between beds being laid down, then the bedding plane may cross-cut older beds.

Beds will be different from each other in a variety of ways:

- *Colour*, for example, a bed may be red or green. More subtle changes may be noted, such as shades of grey or brown.
- *Grain size* which can vary between a fine clay and coarse pebbles. There could be minor changes between a fine sand and a medium sand.
- *Grain shape* will vary between angular and rounded.
- *Sorting* will vary between well sorted and poorly sorted.
- *Composition* of each bed will contain different amounts of minerals, such as quartz, mica or clay and the cement or the matrix will also vary.
- *Bed thickness* varies as some beds will be very thin, called *laminations*, and some can be very thick, known as *massive*.

Dipping beds

Beds are often tilted or folded by Earth movements, so that they are no longer horizontal. Understanding **dip** means that we can work out where a bed may occur under younger beds. Dip is measured as the maximum angle between a horizontal line and the dipping bedding plane.

We will always measure **true dip** if we can see a bedding plane, which allows us to determine the maximum angle of dip. If we can only see intersections between the beds or strata, then we will be measuring the **apparent dip**.

a
True dip, α

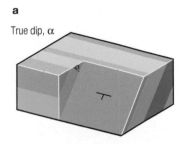

Apparent dip, α

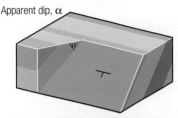

Apparent dip

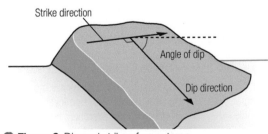
Strike direction

Angle of dip

Dip direction

⬆ Figure 2 Dip and strike of an outcrop

⬆ Figure 1 True and apparent dip **a** diagrams to show true and apparent dip **b** a photo showing apparent dip. True dip is pointing into the face away from the camera (large pebble at base of cliff is around 20 cm)

Using a compass-clinometer

The compass-clinometer is the basic tool for recording information about the inclination of beds.

The following checklist is one way of recording the information:

Measuring the strike

- Set the compass-clinometer into clinometer mode with the compass set West-East and the clinometer arrow able to move freely in the vertical plane against the protractor-like scale.
- On the bedding plane or other surface move the clinometer around until a position is found where there is zero dip – this is the **strike** of the rocks.
- *Mark this direction* on the bedding plane with a pencil or a piece of chalk.
- Hold the compass-clinometer flat so that the large red and white needle is able to move freely and align the side of the compass-clinometer along the chalk line.
- Move the dial so that the red arrow and the needle line up. Record the value indicated at the top, seen as a yellow triangle on this clinometer to the left of the photograph in Figure 3.
- You can use the *left hand rule* to help you decide which direction to record. Point the fingers of your left hand parallel to the strike direction with your thumb pointing along the direction of the dip, which will be at right angles to your fingers. Record the direction that your fingers point in as this is the strike direction.
- Strike is measured in degrees from North and recorded as three figures, for example 168.

Measuring the dip

- Set the compass-clinometer back into clinometer mode with the compass set West-East and the clinometer arrow able to move freely in the vertical plane against the protractor-like scale. The scale is usually only visible if you look at the back of the compass clinometer.
- Now measure the dip at right angles to the line you have drawn, this is the true dip of the surface in which you are interested.
- Dip is measured in degrees from the horizontal and recorded as two figures, for example 30.

The compass-clinometer can also be used for recording the orientation of other geological features such as joints, faults and dykes. It can also be used to record fossil orientations and information about some sedimentary structures, to determine palaeocurrent directions. These orientations are recorded as bearings from north.

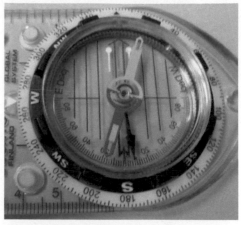

▲ **Figure 3** Compass-clinometer in clinometer mode, set West-East. The black arrow that indicates the inclination is seen pointing downwards.

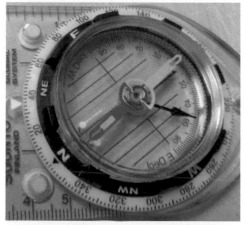

▲ **Figure 4** Compass-clinometer in compass mode. The compass should be flat so that the needle (red and white) can move freely.

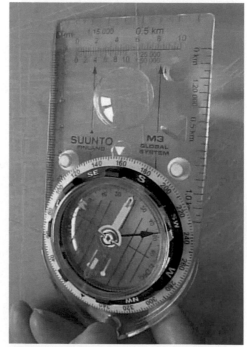

▲ **Figure 5** A compass-clinometer. Strikes are recorded by aligning the edge of the instrument along the horizontal.

HOW SCIENCE WORKS: THE COMPASS-CLINOMETER

This instrument can measure the strike and the dip of a structure, as below:

1. The compass measures a trend known as a strike, which is a direction between 0 and 360°, or degrees of a circle. Readings are recorded in three figures, for example 065.
2. The clinometer measures tilt on rocks from 0 to 90°, or a right angle. 0° is horizontal and 90° is vertical. Readings are recorded in two figures, for example 85.

So, a strike/dip reading in the field could be 160/25.

QUESTIONS

1. Explain how you would take measurements in the field and describe how you could record these in a notebook.
2. Explain why measuring the orientation of a mafic dyke using a compass might not be accurate.

Faults are brittle fractures where there is displacement, which can be horizontal, vertical or a combination. Faults are very common structures in the Earth's crust. They are of great importance as areas of weakness where earthquakes take place when stored stress is released. They range from a few centimetres to hundreds of kilometres in length, and are formed by tension, compression or shear forces.

Fault planes can be, but are not always simple fractures. There is often a relatively narrow zone of shattering along the fault. The fault may be brecciated or may contain finely ground rock, called fault gouge. This area of the fault zone may become a route for mineralising fluids or water.

Fault terminology

- Fault movement of **upthrow** and **downthrow** sides is usually shown on diagrams of faults by means of a half arrow symbol.
- On geological maps, the downthrow side of a fault is shown by means of a tick on the downthrow side of the outcrop of the fault plane.
- **Throw** is measured vertically between the top of the same bed seen on both sides of the fault.
- The **footwall** is the block below the fault and the **hanging wall** is the block above the fault.

Fault types

Faults are divided into those with largely vertical movement, dip-slip faults, and those with largely horizontal movement, strike-slip faults.

Dip-slip faults

Dip-slip faults are where the movement along the fault plane is parallel to the dip of the fault plane. There are two main types: normal faults (the result of tension) and reverse faults (the result of compression). These faults often have a steep dip and their outcrop pattern is usually straight.

Normal faults

The Earth's crust is being stretched and close to the surface, where the rocks are cold and brittle, they fracture to form normal faults. In a normal fault the hanging wall is the downthrow side. Extension of the crust is the result of the tensional (pulling apart) forces. This causes a lengthening of the crust and in cross-section there is clearly a gap created in the formerly continuous beds of rock. The downthrow side of the fault sinks under the influence of gravity. The principal stress direction forming these faults is vertical forces due to the weight of overlying rocks, with the minimum stress being horizontal tensional forces.

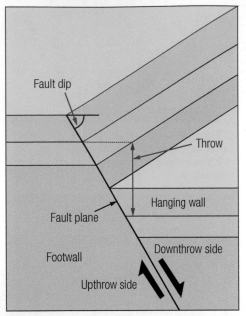

🔺 **Figure 1** Block diagram showing fault terminology. The green is the original land surface before faulting occurred.

KEY DEFINITIONS

Fault a fracture in a rock along which there has been an observable amount of displacement.

Fault plane a plane of fracture, along which the rocks have been displaced.

Upthrow the side of the fault where the movement is upwards, in relation to the other side.

Downthrow the side of the fault where the movement is downward, in relation to the other side.

Throw the vertical displacement of rocks along the fault plane.

Footwall the side of the fault that lies below the fault plane, if the fault is not vertical.

Hanging wall the side of the fault that lies above the the maximum inclination of the fault plane as measured from the horizontal.

KEY FACT

Throw is measured vertically between the top of the same bed seen on both sides of a fault.

🔺 **Figure 2** A normal fault

Graben and horst

Where two normal faults face each other, that is dip towards each other, then a graben or rift valley is formed. The graben is the area that forms the downthrow between the two faults. Rifts occur at the centre of mid-ocean ridges and form the East African Rift Valley system.

Where two normal faults face away from each other, that is dip away from each other, a horst structure is formed. The horst is the elevated block, as the areas either side of the horst have dropped downwards due to crustal extension.

Reverse faults

These faults are formed by compressional forces, which cause a shortening of the Earth's crust. In a reverse fault the hanging wall is the upthrow side. Reverse faults can be recognised because there is an overlap of the strata created by the fault movement, causing a repetition of the formerly continuous bed. The principal stress direction forming these faults is horizontal compressional forces, with the minimum stress being vertical. This suggests that these faults form relatively close to the Earth's surface where there is less overlying rock.

Thrust faults

Thrust faults are a type of reverse fault, where the **fault dip** is less than 45 degrees, but typically of a low angle between 10 and 40 degrees. They are formed by compressional forces. These structures are often associated with major fold mountain systems and have displacements which can be measured in kilometres (in some cases hundreds of kilometres). In north-west Scotland, the Moine Thrust brings Precambrian rocks over 560 Ma to rest above Cambrian rocks less than 560 Ma. This fault was formed in the Caledonian orogeny. Thrusts can result in inverted strata, particularly when they form on the limb of a fold.

△ **Figure 4** A thrust fault. The throw shown is approximately 50 cm.

In reality, many faults will have a dip-slip component as well as a strike-slip component.

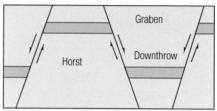

Dip-slip normal fault in cross-section

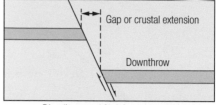

Horsts and grabens in cross-section

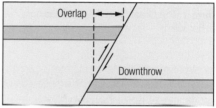

Dip-slip reverse fault in cross-section

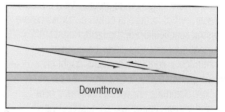

Dip-slip thrust fault in cross-section

△ **Figure 3** Cross-sections of different fault types

STUDY TIP

Fault movement of upthrow and downthrow is usually shown on diagrams of faults by means of a half arrow symbol.

On geological maps, the downthrow side of a fault is shown by means of a tick on the downthrow side of the outcrop of the fault plane.

QUESTIONS

1 Study the photo of the normal fault in Figure 2 and state the downthrow side, the dip of the fault plane and the throw. Draw a labelled field sketch of the fault and ensure that you measure the angle of dip shown. Include at least one other measurement on your sketch.

2 What is the difference between a fault and a joint?

3 Draw labelled cross-section diagrams to show the difference between normal and reverse faults.

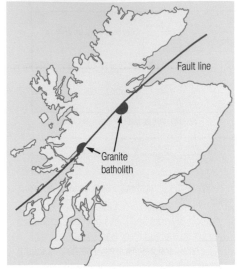

Figure 1 The Great Glen Fault displacing a granite batholith

Figure 3 A dyke displaced by a tear fault (red line)

Strike-slip faults

These are the faults where the fault plane is vertical (or near vertical) and the movement along the fault is horizontal, parallel to the strike of the fault plane. These faults are often large-scale structures, with large displacements. There are two main types: tear fault and transform fault.

Tear fault

A tear fault is the result of shearing forces applied to the rocks. A large-scale example is the Great Glen Fault in the Highlands of Scotland. This fault was thought to have a displacement of around 100 km and be a product of the formation of the Caledonian mountains. One now disputed piece of evidence used to recognise this fault is the existence of two granite bodies on either side, which are similar in composition, age and structure. These granites are now located at either end of the Great Glen on opposite sides of the fault.

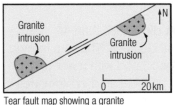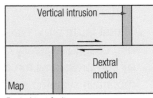

Tear fault map showing a granite displaced sinistrally

Sinistral tear fault

Dextral tear fault

Figure 2 Tear faults seen on maps

Transform faults

Transform faults are associated with plate margins and often described as conservative plate margins or transform plate boundaries. They are common at right angles to a mid-ocean ridge (MOR). They can be thousands of kilometres long. They are the result of different rates of movement within a plate and allow the rigid plates to adjust for these differences in the rate of movement. Superficially they resemble tear faults but when the movements along the fault plane are considered, they are clearly a different type of fault. The diagram shows how a transform fault can be distinguished from a tear fault.

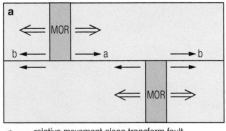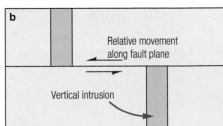

→ relative movement along transform fault
⇐ movement due to seafloor spreading

Earthquakes at **a** are caused by the tectonic plates moving in opposite directions due to seafloor spreading. There is little or no relative movement at **b**, so few or no earthquakes are produced.

Figure 4 Distinguishing **a** a transform fault from **b** a tear fault

Features associated with fault planes

Slickensides

Slickensides are scratches found on fault planes. They are left by the grinding of pieces of rock along the fault plane as the fault moves. They are best seen where the fault plane is coated in a mineral precipitated from a fluid that moved along the fault plane. The two surfaces are polished with linear grooves and ridges parallel to the direction of movement. Therefore the striations show the direction of the last movement along the fault plane.

Fault breccia

Fault breccia is found along fault planes and consists of fragments broken from the rocks on either side of the fault plane (which is often a zone rather than a simple plane). The fragments are large and angular and made of hard, competent rock. They may be cemented by minerals precipitated in the fault zone at a later stage.

Fault gouge

In some faults, such as thrust faults, the grinding of rock along the fault plane can be extreme, producing **fault gouge** consisting of very fine particles of rock. They have formed with high pressure and frictional heating and the particles can be fused together along the fault itself to form **mylonite**.

🔺 **Figure 5** Photos of slickensides and fault breccia

◀ **Figure 6** Fault zones

KEY DEFINITIONS

Slickensides the striations and polishing found on a fault plane indicating the direction of relative movement.

Fault breccia composed of fragments produced by rocks fracturing during faulting. It is found along the fault planes.

Fault gouge composed of very finely ground particles, produced by grinding of rock during faulting. These are often fused together due to frictional heat and found along fault planes.

Mylonite a rock produced by dynamic recrystallisation of minerals on a fault plane.

QUESTIONS

1 Figure 6 shows a number of faults close together. What forces produced these structures? Draw a fully labelled sketch with appropriate measurements to show structural features. The field of view is approximately 4m high and 6m wide.

2 Explain why small earthquakes regularly occur along transform faults at MORs.

3 Research information on two large-scale faults: the Great Glen Fault in Britain and the San Andreas Fault in California. What do they have in common and how do they differ?

KEY DEFINITIONS

Anticline an upright fold with the oldest rocks in the core.

Antiform an upward closing fold.

Syncline a downward pointing fold with the youngest rocks in the core.

Synform a downward closing fold.

Fold limb the section of a fold between one hinge and the next, forming the side of a fold.

Hinge the line along which there is a change in the amount and/or direction of dip, forming the most sharply curved part of the fold.

Crest the highest point of a folded bed.

Trough the lowest point of a folded bed.

Axial plane a plane that joins the hinges of all the beds. It bisects the fold.

Axial plane trace the outcrop of the axial plane at the Earth's surface.

Plunge the angle of dip of the axial plane from the horizontal.

Fold characteristics

A fold is a flexure in rocks, where there is a change in the amount of dip of a bed. All folds are produced by *compressive* forces acting horizontally, so are usually formed at destructive plate margins. Hard, brittle rock may fold and not break if the stress is applied slowly and continuously over a long period of time. They can be small-scale structures found over a few metres in a single outcrop or on an enormous scale over tens of kilometres. There are two broad categories:

- An upfold is **anticlinal** or an **antiform**, with dips pointing outwards. It is an anticline if the ages of the rocks are known, and an antiform if describing the shape only.
- A downfold is **synclinal** or a **synform**, with dips pointing inwards. It is a syncline if the ages of the rocks are known, and a synform if describing the shape only.

Describing folds

gentle	the inter-limb angle is between 180 and 120 degrees
open	the inter-limb angle is between 120 and 70 degrees
closed	the inter-limb angle is between 70 and 30 degrees
tight	the inter-limb angle is less than 30 degrees
symmetrical or *upright*	the axial plane is vertical, the limbs have the same dips
asymmetrical or *inclined*	the axial plane dips, limbs have different dips

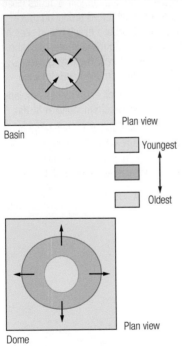

Basin — Plan view

Youngest ↕ Oldest

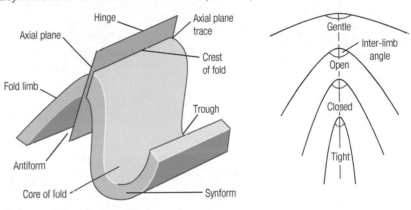

○ Figure 1 Fold characteristics

The axial plane bisects a fold but in a cliff face, quarry or cross-section diagram can only be seen as a single line. The **axial plane trace** can usually only be seen on a map in the centre of a fold. In some cases folds can be seen in 3D in the field, and you can walk along the axial plane trace.

Recognition of folds

Anticlines and synclines

An anticline is an upward closing fold, where the oldest rocks are in the core of the fold.

Conversely, a syncline is a downward closing fold, where the youngest rocks are in the core of the fold.

Symmetrical fold features:

- The angle of dip on each of the limbs is the same and the length of the limbs is the same.
- The **axial plane** will be vertical.

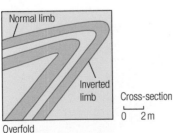

Dome — Plan view

Normal limb / Inverted limb / Overfold — Cross-section 0 2m

○ Figure 2 Dome and basin on map. Overfold in cross-section

Asymmetrical fold features:

- One limb has a higher angle of dip than the other.
- The limb with a low angle of dip will have a wider outcrop than the limb with a higher angle of dip.
- The axial plane will be inclined.

An anticline has the dip arrows pointing out from the axial plane trace and will have the oldest rocks in the centre of the fold.

A syncline has the dip arrows pointing in to the axial plane trace and will have the youngest rocks in the centre of the fold.

CASE STUDY: MODELLING FOLDING

Take a piece of modelling clay and form it into a bar. On one surface mark a series of circles – these will show the strain in the rocks. Bend the bar into a fold. The circles towards the outside of the fold will be elongated parallel to the shape of the fold, while those near the inside of the fold will be elongated at right angles to the pressure. These are described as concentric folds

An alternative process can be demonstrated using a pack of cards. Draw two bedding planes across the pack and then produce a fold – the fold is formed by a series of small movements as each card moves slightly with respect to those adjacent to it. This process is called shear folding and is common in incompetent rocks producing parallel folds. In practice, both of these processes operate together.

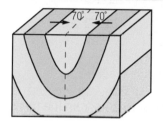
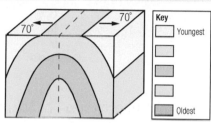

Key
- Youngest
-
-
-
- Oldest

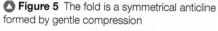

🔺 **Figure 4** Symmetrical syncline and anticline

🔺 **Figure 5** The fold is a symmetrical anticline formed by gentle compression

🔺 **Figure 6** An asymmetrical anticline formed by large compressive forces. The greatest force is from the right (south) side.

KEY FACT: PLUNGING FOLDS

Plunging anticlines and synclines do not have horizontal axes, instead they are tilted away from the horizontal. The **plunge** can be measured as a dip angle from the horizontal, along the axis of the fold.

Plunging folds can be recognised on maps as they form U or V shaped outcrop patterns.

a plunging anticline   **b** plunging syncline

🔺 **Figure 7** Block diagrams showing a plunging anticline and a plunging syncline respectively. Note the outcrop patterns at the surface.

KEY DEFINITIONS

Dome an anticline which dips away from the centre in all directions.

Basin a syncline which dips towards the centre from all directions.

Overfolds folds which have both fold limbs dipping in the same direction but by different amounts.

Recumbent folds have axial planes and fold limbs which are close to horizontal and always less than 30 degrees.

Nappes recumbent folds that are broken along thrust planes.

Isoclinal folds have parallel limbs that are nearly vertical and in very tight folds.

Recumbent fold

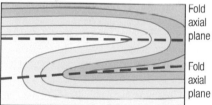

Fold axial plane

Fold axial plane

Nappe and thrust fault

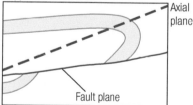

Axial plane

Fault plane

Isoclinal

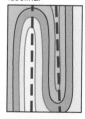

🔺 **Figure 3** Fold types

QUESTIONS

1. Describe how you would tell the difference between a symmetrical anticline and an asymmetrical syncline.

2. Make a detailed labelled sketch of Figure 5 using technical terms and measurements. The height of the outcrop is 5 m.

The geological map is a common way of recording the geology of an area in 2 dimensions. Information about the inclination of the rocks is also recorded so that a 3-dimensional interpretation can be made. Making such an interpretation is one of the key skills that a geologist needs to develop.

One way in which this interpretation can be achieved is by drawing sections across the map to show how the geology varies underground across an area.

- Symmetrical syncline
- Limbs same width
- Displaced by dip-slip fault
- Downthrows to the west – outcrops further apart on western side

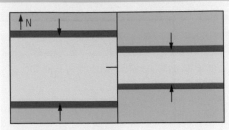

- Asymmetrical anticline
- Limbs of different width
- Displaced by dip-slip fault
- Downthrows to west – outcrops closer together on western side

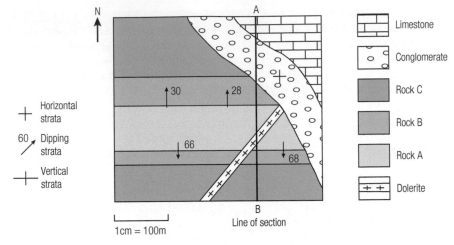

🔺 **Figure 1** Geology map

Figure 1 above shows the geology of an area. The key points to note from the map are:

- There is an unconformity shown on the map.
- The rocks below the unconformity are folded.
- There is a dyke intrusion in the rocks below the unconformity but it does not affect the rocks above the unconformity.

There is a line drawn on the map from **A** to **B**. This is the line of section which will be drawn across the map to illustrate the variation of geology with depth underground.

Transferring the information

The first step in the process of drawing the section is to place a piece of blank paper across the map along the line of section. Mark the positions of A and B on the piece of paper. Now the intersections that the bedding planes or intrusions make with the line of section are marked and labelled on the piece of paper. The finished version is shown in Figure 2.

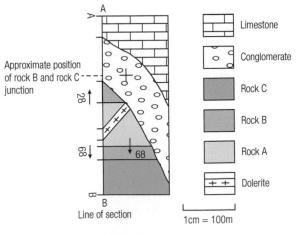

🔺 **Figure 2** Transferring the information from the map

The next stage of the process is to transfer the information to a topographic profile which shows the shape of the landscape between A and B. The topographic profile is shown in Figure 3.

You should note that the horizontal scale of the profile will be the same as the scale for the geological map. The vertical scale is not the same as the map scale and is usually exaggerated. This helps to illustrate the geology but the dip amounts will be slightly exaggerated by this process.

The piece of paper you recorded the information on is now placed onto the topographic profile so that the information can be transferred onto it. This process is shown in Figure 4.

Drawing the section

You are now in a position to complete the section. Firstly draw in the beds above the unconformity. These are horizontal, as shown by the cross symbol on the map and will be parallel to the base of the profile. The next structure that can be drawn in is the dyke intrusion. This is a vertical feature and can be drawn in with the edges of the intrusion parallel to the edges of the profile. Finally Bed B can be drawn in to illustrate the structure of the rocks below the unconformity. The southern limb of the fold dips towards the south at 68 degrees and the top and bottom of the bed are both visible at the surface. The northern limb of the fold dips towards the north at 28 degrees and the base of the bed is visible at the surface but the top of the bed is at depth, beneath the unconformity. The dotted line on the piece of paper shows its approximate position and this can be used to construct the bed on the northern side of the fold.

You can assume that the thickness of bed B does not change significantly across the area so you can complete the construction of the section. The dip amounts will not be accurate as the vertical scale has been exaggerated but they are close to the actual values. Figure 5 shows the completed section.

The section can then be labelled and the unconformity could be projected across the area.

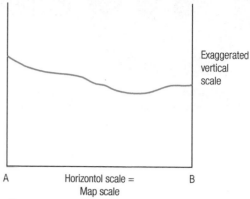

Figure 3 Topographic profile

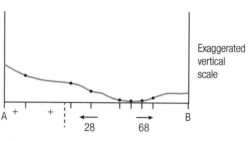

Figure 4 Transferring the information from the map to the topographic profile

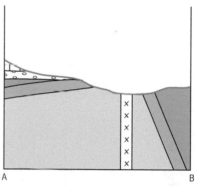

Figure 5 Completed section across the map to illustrate the geology

QUESTIONS

1 Draw a section from A to B across the map. Assume that the topography is flat and the fault is vertical.

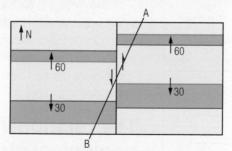

Figure 6 Geological map for section

2 Draw a section from A to B across the map. Assume that the topography is flat and the fault is vertical.

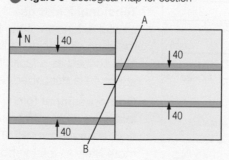

Figure 7 Geological map for section

3 List the map symbols for
   (a) Dipping beds
   (b) Horizontal beds
   (c) Downthrow side of a fault.

Stress is the term used to describe the forces applied to a rock by Earth movements or the mass of overlying rocks. **Strain** is the term used to describe the deformation caused by the applied forces. The strain is a change in volume or a change in shape, or both. Rocks behave in different ways to stress, depending on their physical properties:

- **Competent** rocks stay the same thickness when they are deformed and react in a brittle way. Sandstones, limestones and most igneous rocks are competent.
- **Incompetent** rocks will vary in thickness when they are deformed, as they behave in a plastic way. Mudstones and shales are typical incompetent rocks.

KEY DEFINITIONS

Stress the force applied to rocks.

Strain a change in the shape or volume of a body as a result of applied stress.

In cases where vertical stress is applied to perfectly elastic materials

$$\text{Strain} = \frac{\text{the change of length of line}}{\text{the original length of line}}$$

Competent rocks are strong and brittle and tend to joint and fault.

Incompetent rocks are weak and plastic and tend to fold and develop cleavage.

Tension the force trying to pull rocks apart, the Earth's crust will be lengthened.

Compression the force trying to push rocks together, the Earth's crust will be shortened.

Shear forces are those which act along a plane in the rock and promote sliding along that plane.

CASE STUDY: MODELLING DEFORMATION

If you put a bar of chocolate in a fridge for a while and then take it out and try to break it into pieces, you will find that it is difficult to break and when it breaks it will do so suddenly with a sharp noise. The chocolate is behaving as a brittle material like a competent rock.

By contrast, if you put a bar of chocolate into your pocket on a warm day, you may be surprised to find that the bar has deformed to the shape of your pocket. In this case, the bar has been deformed because it has become more plastic or ductile.

Rocks will behave similarly in the right circumstances.

The factors affecting stress and strain

There are three main factors which affect the stresses applied to a rock and the resulting strains. The important thing to remember is that all of these factors operate together:

1. The higher the temperature, the more plastic the rocks will be. At high temperatures, rocks will fold, not fault, even if the original rock is competent. Cold rocks will behave as brittle materials and therefore fracture.
2. The strength of rocks increases with confining pressure and therefore rocks at depth are more difficult to deform. The mass of the overlying rocks causes the confining pressure.
3. Time is crucial in determining the type of deformation that occurs. If the pressure is applied for a short period of time, then the rocks may well behave in a brittle manner, whilst pressure applied over an extended period of time can result in plastic deformation.

The relationship between forces and geological structures

Tensional forces usually result in the fracturing of rocks and crustal extension. If the forces result in a displacement of the rocks on either side of the fracture plane, then the structure is described as a fault. If there is no displacement, then the structure is described as a joint.

Compressional forces can result in either the fracturing or the folding of rocks. When rocks are cold they will tend to fracture and faults are formed. If the rocks are heated then they are more likely to fold. In both cases, the deformation will result in a shortening of the Earth's crust.

Shear forces result in the deformation of rocks in one plane – usually horizontally. These may result in faults or folds.

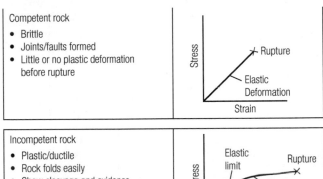

Competent rock
- Brittle
- Joints/faults formed
- Little or no plastic deformation before rupture

Incompetent rock
- Plastic/ductile
- Rock folds easily
- Show cleavage and evidence of plastic flow

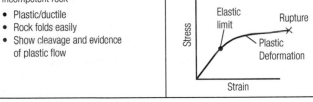

Effect of increased temperature
- Increased temperature decreases the strength of the rock

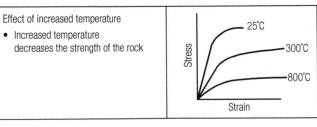

Effect of increased confining pressure
- Increased confining pressure raises the strength of the rock

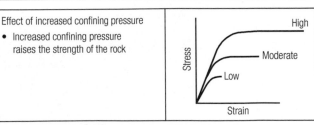

🔺 **Figure 1** The relationship between stress and strain

Using fossils and ooliths to measure strain

In order to measure the amount of strain that a rock has undergone, we can use bilaterally symmetrical fossils or ooliths. Undeformed fossils are used to establish the original shape and then the amount and direction of deformation can be analysed. Ooliths were originally spherical in shape and after deformation they become ellipsoid.

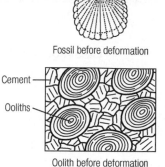
Fossil before deformation

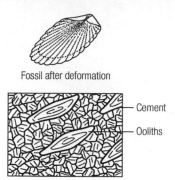
Fossil after deformation

Cement
Ooliths
Oolith before deformation

Cement
Ooliths
Oolith after deformation

🔺 **Figure 2** Determining the strain from deformed fossils and ooliths

Joints

Joints are produced by tensional forces. These forces may be the result of folding, cooling or unloading of rocks. They give rocks a distinctively 'blocky' appearance when they crop out at the Earth's surface. Joints will only form in competent rocks, which are brittle and break when put under tension. Joints in sedimentary rocks are usually found perpendicular to the beds.

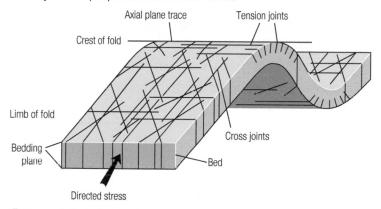

Axial plane trace
Tension joints
Crest of fold
Limb of fold
Bedding plane
Directed stress
Cross joints
Bed

🔺 **Figure 3** Tectonic joints in folded rocks

Joints produced by folding

Tectonic joints are produced by the tension created in rocks when they are folded by Earth movements. Compressive forces fold the rocks but cause some of the rocks around a fold to be under tension. The rocks fracture because the outer surface of the bed is stretched more than the inner surface. The rock is competent so it does not change in thickness as it is bent, so brittle fractures form.

Two types of tectonic joints are common:

- Tension joints, parallel to the axial plane trace of the fold.
- Cross joints, at an angle to the axial plane trace of the fold.

Where there are a number of joints with a similar orientation, they are described as a joint set.

Joints produced by cooling

Cooling joints are caused by tensional forces set up in an igneous rock as it cools steadily and contracts. The joints will be perpendicular to the cooling surfaces of thick basalt lava flows, dykes and sills, so will often be vertical. The igneous rock is insulated and then cooled with evenly spaced cooling centres. As the rock cools and crystallises, it contracts slightly and this results in a series of column-like structures, which are polygonal in cross-section.

Joints produced by unloading of rocks

Unloading joints often form in igneous rock which cooled deep below the surface where the pressure is high. The mass of the overlying rocks 'compressed' the rock – this is called the *load pressure*. When the rock is exposed at the surface, as a result of uplift and erosion, the lack of load pressure from overlying rocks allows them to expand. Joints form roughly parallel to the Earth's surface.

KEY DEFINITIONS

Joint a fracture in competent rocks along which no observable movement has occurred.

Tectonic joints form as a result of folding and cause tension joints parallel to the axial plane and cross joints on the limbs.

Cooling joints form as a result of contraction on cooling of igneous rocks.

Unloading joints are often horizontal as well as vertical and form as a result of lower pressure near to the surface.

HOW SCIENCE WORKS: DIFFERENTIAL STRESS

Differential stress occurs when forces applied are not equal in all directions. There are three principal stresses, σ max, σ inter and σ min operating at right angles. Load pressure is always one of the principal stresses, so the other two are horizontal. They may be compressional or tensional. Different orientations of σ max, σ inter and σ min produce different types of fault. In normal faults σ max is vertical and σ inter is parallel with the fault plane. In reverse and thrust faults σ max is horizontal in the direction of movement and σ min is vertical. Tear faults have σ inter as vertical.

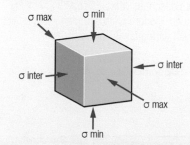

σ max
σ min
σ inter
σ inter
σ max
σ min

🔺 **Figure 4** Differential stress directions

QUESTIONS

1 Why do joints form mainly in competent rocks and not in incompetent rocks?

2 Using Figure 2, determine the maximum, minimum and intermediate stress directions in the deformed ooliths.

8.7 Slaty cleavage

Cleavage is a feature of rocks containing platy minerals. Platy minerals include clay minerals and micas, with a long thin shape. When these align, they form a series of planes along which the rock tends to split because they act as lines of weakness.

The origin of cleavage

Incompetent sedimentary rocks such as mudstone contain many clay minerals. When these rocks are heated and compressed, the minerals begin to recrystallise. They usually recrystallise as micas or mica-like minerals with a distinctive platy form. Recrystallisation causes a change in the rock so it is no longer a mudstone but altered by low grade metamorphism to a slate.

These platy minerals are recrystallising and growing in an environment where there is a strong stress field, so they will tend to grow in the direction of least resistance. This is at right angles to the stress field which means that all of these platy minerals form approximately parallel to each other. They are also parallel to the axial planes of the folds, which result from the applied stress.

On the normal limb of the fold, the bedding dips at a lower angle than the cleavage.

On the inverted limb, the opposite is true, that is the dip of the bedding is steeper than the dip of the cleavage.

> **KEY DEFINITION**
>
> **Cleavage** planes of weakness in incompetent sedimentary rocks (e.g. shale, made from clay minerals) and low-grade metamorphic rocks (e.g. slate) along which these rocks will split. It forms perpendicular to σ max in folds, so is at an angle to bedding planes.

> **STUDY TIP**
>
> Cleavage is only found in incompetent rocks.
>
> Competent rocks, such as igneous rocks or limestones often have a crystalline structure. This does not allow particles to change direction under pressure. Similarly, sandstones have rounded, equidimensional grains that simply rotate under pressure.
>
> Few materials share such high levels of resistance to weather, pollutants and freeze-thaw weathering as slate. It is impermeable and splits into thin sheets along cleavage planes, which makes it ideal as a roofing and paving material as well as for floors and work surfaces.

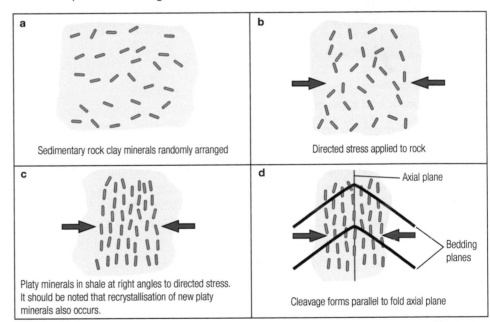

▲ **Figure 1** The formation of slaty cleavage from **a** to **d**.

Key for Figure 3:
- Sandstone
- Shale
- Cleavage planes

Cross-section
0   2
    m

▲ **Figure 3** Development of cleavage in a fold. Cleavage develops only in incompetent rocks.

> **CASE STUDY: SPAGHETTI AND CLEAVAGE**
>
> Strands of spaghetti make a useful simulation for cleavage. Clay minerals are so small that you need a very high-powered microscope to be able to see them. The spaghetti strands have a broadly similar shape, so react in a similar way to clay minerals. In the photos, the particles are clearly aligned at right angles to the pressure.

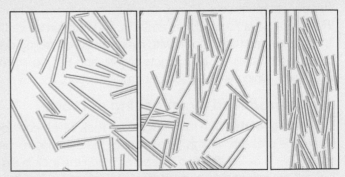

▲ **Figure 2** Spaghetti and cleavage

The photo shows part of a fold in which the competent beds have been folded and slaty cleavage has developed in the rocks. The rocks in this case are muddy sandstones. The beds are dipping towards the left-hand side and the cleavage is nearly vertical. The cleavage is formed by platy minerals in the rock, which formed when the clay minerals were altered by heat and pressure and recrystallised as micas. The mica minerals are aligned perpendicular to the pressure forming the fold and parallel to the fold axial plane.

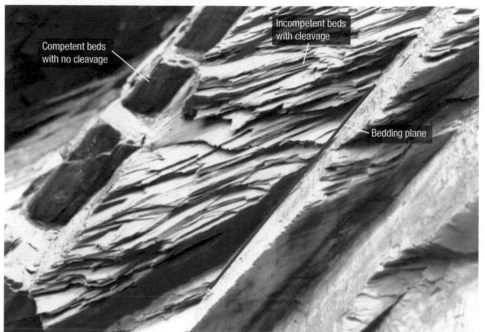

Competent beds with no cleavage

Incompetent beds with cleavage

Bedding plane

🔺 **Figure 4** Cleavage planes at an angle to the beds

0   2
cm

STUDY TIP

The structures described are very different, in theory. When you are out in the field or look in detail at photos, it can be tricky to distinguish between a bed and a cleavage plane.

At the crest of an anticline it is common to find both tension joints and cleavage planes, though not usually in the same beds.

The table summarises the main differences.

A comparison of bedding, jointing and cleavage

	Bedding	Jointing	Cleavage
How to recognise structures on photographs and in diagrams	Differences in colour or composition between beds	Fractures along which there is no displacement	Rock splits easily along parallel planes, making slates
	Differences in grain size between beds	Often perpendicular to bedding	Often parallel to the axial plane of folds
	Structures within the beds and along the bedding planes	Possible mineralisation along the joints	Planes very close together – a few mm
Type of rock in which they form	All types	Only develops in competent rocks such as sandstones and limestones	Only develops in incompetent rocks such as shales, which contain platy clay minerals and micas

QUESTIONS

1 Draw an overfold and a recumbent fold consisting of alternating beds of sandstone and shale and add cleavage planes in the correct places.

2 Explain why sandstone composed only of quartz will not develop cleavage.

3 Create a summary table of structures produced by compression and those formed under tension.

4 Identify and interpret the structural features shown in Figures 5 and 6.

🔺 **Figures 5 and 6** For use with question 4.

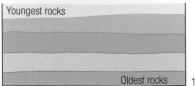

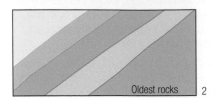

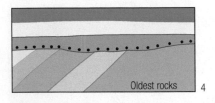

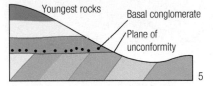

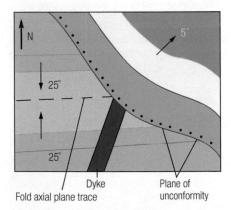

Youngest rocks
Oldest rocks 1
Oldest rocks 2
Oldest rocks 3
Oldest rocks 4
Youngest rocks
Basal conglomerate
Plane of unconformity 5
N
25°
5°
25°
Dyke
Fold axial plane trace
Plane of unconformity

Figure 1 Development of an angular unconformity is shown in cross-sections 1–5. The map shows a fold, a dyke and an unconformity.

All unconformities represent a break in time or a period when no sediment was deposited.

This may be the result of a change in the environment or the result of earth movements and erosion. An angular unconformity occurs where the rocks above the unconformity have a different dip, and possibly strike, from the rocks beneath the plane of the unconformity.

How an unconformity forms

Stage 1: Deposition of beds of rock, with the oldest at the bottom and the youngest at the top.

Stage 2: Beds of rock are tilted by earth movements, maybe as a result of compressive forces folding the rocks.

Stage 3: Erosion of the beds of rock. This could be a time gap of thousands of years or many millions of years.

Stage 4: Deposition of younger overlying beds of rock. They are deposited unconformably on the underlying series. They have a different dip. The eroded surface is the plane of unconformity. Fragments of the underlying beds may form conglomerate, which is a sedimentary rock formed of pebbles. This will be the first bed of the younger rocks called a basal conglomerate.

Stage 5: Uplift and erosion expose the unconformity. Often all the rocks will be folded again, so that the younger rocks are tilted as well.

How to recognise an unconformity

Angular unconformities can be recognised on a map in several ways:

- There is a difference in dip of the beds below and above the plane of unconformity.
- The presence of a basal conglomerate, rounded fragments left on the erosion surface are incorporated into the younger sediment.
- Older beds and structures (e.g. faults, fold axial plane traces and igneous intrusions) are cut off by the younger rocks above the plane of unconformity. The structures do not stop at the plane of unconformity, but continue beneath it.
- The rocks above the unconformity are younger than those beneath the unconformity.

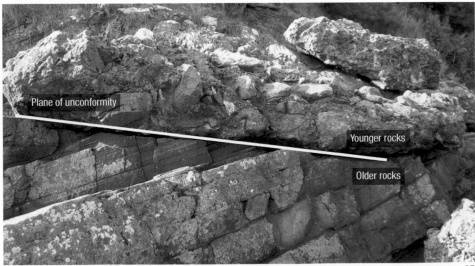

Plane of unconformity
Younger rocks
Older rocks
0 25 cm

Figure 2 Photo of an angular unconformity

Using cross-cutting relationships to deduce age relationships

In the block diagram (Figure 3) there is a clear sequence of events shown:

1 The first event is the lower sequence of rocks, which were laid down with the pebbly sandstone first, followed by the limestone and then the shale.
2 These beds were folded into a synclinal fold.
3 The area was eroded, leaving an uneven surface.
4 The conglomerate was laid down unconformably horizontally on top of the eroded limestone and shale. The sandstone, which is the youngest rock, was laid down.
5 The area was tilted to the south-east, followed by more erosion.

In the top map view of the block diagram, only the unconformable beds can be seen to crop out, while in the cross-section the older beds are visible.

On the map of a different area (Figure 4), the same ideas can be seen, though they look very different:

1 The beds formed first, with C being the oldest, then B and A.
2 A symmetrical synclinal fold formed with limbs dipping north and south at 30 degrees and the axial plane trace trending west–east.
3 The fold was cut by a strike-slip fault that has displaced the axial plane trace of the fold, without the outcrop width of the fold changing. The fault is sinistral.
4 Erosion followed and bed F was laid down as an unconformity. This cuts across all other features, so is the younger structure.
5 The area was tilted towards the north-west.

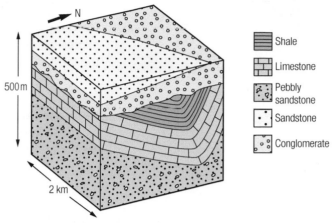

△ **Figure 3** Cross-cutting relationships in a block diagram

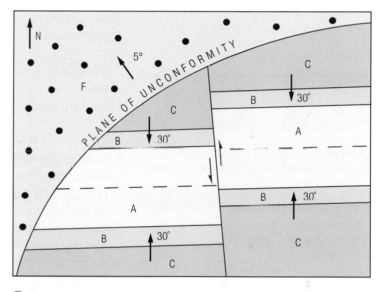

△ **Figure 4** Cross-cutting relationships in a map

QUESTIONS

1 Draw your own set of fully labelled diagrams to show the development of an angular unconformity.
2 Use the cross-cutting relationships to deduce age relationships for the map below.

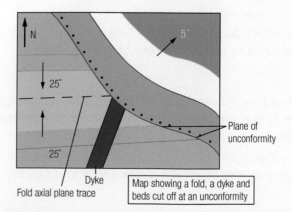

△ **Figure 5** Geological map to be used with Question 2

Fluid storage and movement in rocks

Porosity is the amount (volume) of pore space in a rock or sediment, usually expressed as a percentage of total rock volume. You can calculate porosity using the formula:

$$\% \text{ Porosity} = \frac{\text{total volume of pore space}}{\text{total volume of rock or sediment}} \times 100$$

Permeability is the ability of a rock or sediment to transmit fluids such as water, oil or gas, and can be expressed as a rate of flow of the fluid through the rock or sediment. You can calculate permeability using the formula:

$$\text{Permeability} = \frac{\text{distance fluid has travelled}}{\text{time taken}}$$

Factors affecting porosity

The most obvious factor affecting porosity is rock type. Crystalline igneous and metamorphic rocks have virtually zero porosity. Factors affecting porosity are shown below.

The degree of sorting

A well-sorted rock has a high porosity. A poorly sorted rock has a low porosity because the finer grains fill in the spaces between the coarser grains.

Amount of diagenesis

A loose unconsolidated rock has a much higher porosity than a rock that has undergone compaction and cementation. There is therefore a general decrease in porosity with depth, with some sedimentary rocks showing a decrease from 40% porosity close to the surface, to as little as 10% at depths of 10 km. As temperature also increases with depth there is a broad relationship between increasing temperature and decreasing porosity. Increasing temperature may also produce hydrothermal solutions by dissolution at grain boundaries, which may precipitate out as cement and further reduces porosity.

Grain shape

Rocks containing rounded grains have a higher porosity than rocks containing angular grains that fit together.

The packing of the grains

This is the way the grains fit together. Figure 1 shows how the packing of particles can reduce porosity. Again, depth is important here as compaction from overburden pressure will tend to cause particles to pack closer together.

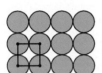
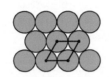

△ **Figure 1** The difference in pore space between cubic and rhombic packing for spheres

---

KEY FACTS: TYPES OF POROSITY

The porosity of a rock is a measure of its capacity to store or contain fluids. It is calculated as follows:

$$\text{Total porosity} = \frac{\text{Pore space volume}}{\text{Bulk volume}}$$

Primary porosity is the pore spaces at the time the sediments were deposited.

Secondary porosity develops after in the rock, and includes cavities or fractures that have formed.

Effective porosity is the total porosity minus any cement and this contributes to fluid flow through the rock. Clean sands have an effective porosity that is almost equal to the total porosity.

Other factors

Grain size is not a factor in porosity. Coarse grained rocks tend to have large pore spaces but fewer of them. Fine grained rocks have small pore spaces but many of them do not connect with each other. Clay is a good example of a rock with high porosity but the pores are not interconnecting and so it is more or less impermeable.

Rocks may also develop secondary porosity through the presence of geological structures, such as joints and faults, which are particularly important in increasing the porosity of limestones. Porosity may also develop in a rock after it forms by the dissolving or alteration of mineral grains and cements.

The porosity is only effective if the pore spaces are interconnected. Sandstones may typically have a porosity of 15% with most of the pores interconnecting. Although clay may have a porosity of 50%, its *effective porosity* is virtually zero.

Factors affecting permeability

Porosity

A rock that has a high effective porosity with good interconnections between the pore spaces will also have a high permeability. Although grain size is not a factor in porosity, it is important in determining permeability. Coarse grained rocks have a higher permeability than fine grained rocks because there is less resistance to flow around the coarse grains than around finer grains.

Temperature

This is an important control on permeability as it affects the viscosity of a fluid. This is especially true of oil where the viscosity decreases with increasing temperature so the oil is more fluid.

Secondary permeability

This results from the presence of fractures, such as joints and faults, and voids, such as caves produced by solution. The presence of these secondary structures is important in increasing the porosity of limestones, and also in increasing its permeability. Rocks, such as clay, mudstone and shale, with zero or very low permeability, are termed *impermeable*.

Capillary pressure

Capillary pressure is the force which exists across the interface which separates two immiscible fluids (such as oil and water). In a rock, there is a balance between the attractive force applied by each fluid on surface molecules (surface tension) and the adhesive forces between each fluid and the rock (wettability). Before a non-wetting fluid can displace a wetting fluid, the pressure must exceed this threshold.

As hydrocarbons begin to migrate into a rock, displacing the pore water because they are less dense, they first enter the pores with the largest pore entrances, leaving the formation or **connate water** in the pores with smaller entrances or in small nooks and crannies. As the hydrocarbon level increases, the capillary pressure increases, forcing hydrocarbons to enter smaller and smaller pores and displace the water in those pores. This process continues until either generation or migration ends or the trap reaches its spill point.

Typical values for porosity and permeability (for water)

Rock type	% Porosity	Permeability (m per day)
Sandstone	15	5
Sand	35	20
Chalk	20	20
Clay	50	0.0002
Shale	3	0.0001
Granite	1	0.0001
		Values <0.01 = impermeable rock Values >1 = potentially exploitable

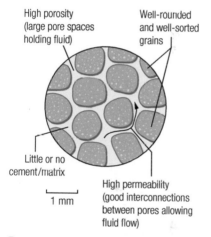

High porosity (large pore spaces holding fluid)

Well-rounded and well-sorted grains

Little or no cement/matrix

High permeability (good interconnections between pores allowing fluid flow)

1 mm

▲ **Figure 2** Thin section diagram of desert sandstone with high porosity and permeability

KEY DEFINITIONS

Porosity the volume of pore space.

Permeability the rate at which a fluid (usually water, oil or gas) flows through a rock.

Capillary pressure the pressure between two immiscible fluids in narrow pore spaces, resulting from interactions of forces between the fluids and the solid grains.

Connate water trapped in the pores of a rock as the rock formed. It includes water trapped in the original sediment and water released during diagenesis.

QUESTIONS

1 Explain how compaction and cementation reduce the porosity and permeability of a rock.

2 Draw labelled thin section diagrams to compare the porosity and permeability of a well-sorted oolitic limestone and a greywacke.

Groundwater and the water table

Water is vital to all life on Earth and the supply of water is a fundamental issue at all scales: local, regional and global. The two main sources of useable fresh water are surface water and groundwater. Surface water is found in lakes, rivers and reservoirs.

Groundwater is water held in the pore space of rocks beneath the ground, below the **water table**. Groundwater is the world's largest accessible store of fresh water and its significance cannot be underestimated, particularly in arid regions with little surface water. Today, hydrogeologists are playing an increasingly important role in locating suitable groundwater resources.

Most groundwater originates from rainwater that has infiltrated into soil and then percolated downwards through the pore space of rocks to reach the water table. The water table is the level at which water sits within the ground. Rocks above the water table are unsaturated and have air and water in their pore space. Rocks below the water table are saturated and only contain water in their pore space. The shape of the water table generally follows the surface topography, but with less relief, and intersects the ground surface at lakes and most rivers. The level of the water table may change depending on the season and how much rainfall the area receives.

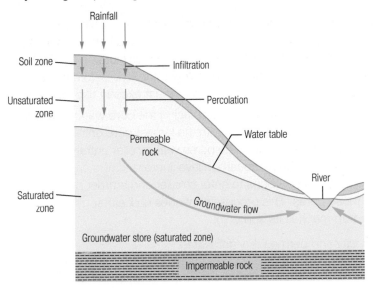

● **Figure 1** Cross-section showing groundwater zones and the water table

Hydrostatic pressure and hydraulic gradient

Hydrostatic pressure results from the mass of the overlying column of water and increases with depth. The height of the overlying column of water is known as the hydrostatic head.

Pressure differences cause groundwater to flow. Water always flows down the **hydraulic gradient** from areas of high pressure to areas of low pressure. The rate at which the water flows is proportional to the drop in height of the water table. These concepts are critical for accurately predicting the movement of groundwater. You can calculate the hydraulic gradient between two points using the formula:

$$\text{Hydraulic gradient} = \frac{\text{difference in hydrostatic pressure or hydrostatic head}}{\text{distance between two points}}$$

(quoted as a ratio)

MATHS BOX: CALCULATING THE HYDRAULIC GRADIENT

The diagram shown in Figure 2 is a cross-section through part of the South Downs in Sussex. Calculate the hydraulic gradient of the water table between points A and B.

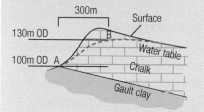

Key
OD – Ordnance datum (above sea level)
Vertical scale exaggerated

Figure 2 Cross-section through part of the South Downs in Sussex

$$\text{Hydraulic gradient} = \frac{\text{difference in hydrostatic pressure or hydrostatic head}}{\text{distance between two points}}$$

$$= \frac{130 - 100}{300} = \frac{30}{300} = 0.1 \text{ or } 1{:}10$$

The application of Darcy's law to model the flow of fluids in rocks

Darcy's law is a method of calculating the permeability of a rock or sediment. The value of permeability that is obtained also depends on the characteristics of the fluid being considered. The density and viscosity of the fluid will affect the results obtained. In a simplified version of the equation applied to the flow of water we can ignore these two characteristics. You can calculate the rate of flow using the formula:

$$Q = -KA((h_2-h_1)/L)$$

Where:

Q is the rate of flow: commonly measured in cubic metres per second.

K is the coefficient of permeability: a measure of the ability for flow through porous sediments or rocks.

A is the cross-sectional area in square metres.

$((h_2-h_1)/L)$ is the hydraulic gradient.

It is possible to calculate values for permeability for simple geological structures.

MATHS BOX: DARCY'S LAW

Let us assume that there are two wells in a sand **aquifer** 1000 m apart. The height of the water in the two wells is 55 m above sea level and 50 m above sea level respectively. The depth of the aquifer is 20 m and width of the aquifer is 500 m. The value for Q has been measured as 1375 m^3 per day. Determine a value for K.

$$Q = -KA((h_2 - h_1)/L)$$

$$1375 = -K(20 \times 500)((55 - 50)/1000)$$

$$-K = 1375 / (20 \times 500)((55 - 50)/1000)$$

$$= 1375 / 50$$

$$= 27.5 \text{ m per day}$$

Values for K for sediments and rocks are often determined in a laboratory using specialist apparatus, though there are methods which enable local permeability to be determined within wells.

The calculation above assumes that water passes equally through the cross-sectional area but this is not the case as the fluid will pass through the available pore spaces and the pores are interconnected.

Darcy's law allows an estimate of the following:

• The flow rate within the aquifer.
• The average time of travel from the area of recharge of the aquifer to a point located down the hydraulic gradient.
• The prediction of pollution plume arrival times.

QUESTIONS

1 A sand aquifer has a permeability of 30 metres per day. A pollution event occurs in the aquifer 2 km upstream of a well from which water is supplied for human consumption. How long before the water supply is potentially contaminated? How much sooner would the pollution plume arrive if the aquifer was composed of gravel with a permeability of 200 metres per day?

2 Explain why values of permeability for a sandstone will be less than those for sand.

3 Explain why permeability rates for water in a reservoir rock will be greater than those for oil.

4 Describe how the rate of flow varies with hydraulic gradient and horizontal distance travelled

CASE STUDY: LONDON BASIN

The London Basin is the best example of an artesian basin in the British Isles. The main aquifer is Cretaceous chalk, 180 to 245 m thick, sandwiched between the Gault Clay below and the London Clay above. The main recharge zones for the London Basin are the Chiltern Hills and the North Downs.

Groundwater was first *abstracted* from beneath London in the eighteenth century and made a significant contribution to the economic and industrial development of the city. Many businesses, including the Savoy Hotel and the Bank of England, had their own boreholes into the aquifer. At the peak of **abstraction** in the 1960s the water table under London had fallen by at least 50 m across an area of 200 km². This created problems of reduced water yield, poor water quality and surface subsidence.

Most of London's water supply now comes from surface water supplies, and this has led to a recovery of groundwater levels below the city to such an extent that the rising groundwater is now causing concern! Increased hydrostatic pressure within the chalk has caused saturation of the overlying London Clay. This is affecting the stability of foundations of some buildings and threatens flooding of tunnels, including the London Underground.

The Environment Agency currently has a network of 200 observation boreholes in London to monitor groundwater levels. Management of water levels is mainly done through control of water abstraction licences.

Types of aquifers

There are two main types of aquifer, *unconfined* and *confined*.

An *unconfined* aquifer is open to the atmosphere, under atmospheric pressure, and is recharged by rainwater from directly above. Water will need to be pumped to the surface from a well or borehole sunk into an unconfined aquifer.

A *confined* aquifer is overlain by impermeable rocks and the groundwater held within it is under hydrostatic pressure. Groundwater can only be replenished in a confined aquifer if it has **recharge zones** that are open to the atmosphere.

A *perched* aquifer sits above the regional water table and is underlain by a lens of impermeable rock which prevents the water from percolating further downwards.

Aquifers can also be described as 'live' or 'fossil':

- A *live aquifer* is one that is currently being replenished by rainwater via a recharge zone on the surface.
- A *fossil aquifer* is no longer being replenished and represents a relic of a past wetter climate.

Artesian basins and artesian wells

Large, synclinal, confined aquifers are termed **artesian basins**, after Artois in France. If a borehole is sunk into an artesian basin the water may flow up to the surface. We call this an **artesian well**. The level to which the water rises is known as the **piezometric surface**. The fountains installed in Trafalgar Square in 1843 initially flowed naturally under hydrostatic pressure, as they are in the centre of the London Basin. Once the hydrostatic pressure falls, the water has to be pumped to the surface.

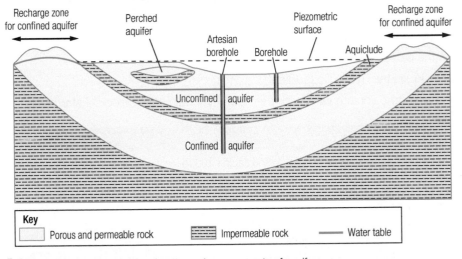

Figure 1 Cross-section showing the main components of aquifers

Figure 2 Vegetation at Skoura, a desert oasis in Morocco

What are springs?

Springs occur where the water table intersects the land surface and groundwater flows out onto the surface. This often occurs at the junction between permeable and impermeable rocks. Often a line of springs occurs along the boundary. We call this a *spring line*.

A large spring can yield more than 1000 litres of water per second. Springs can be very important for water supply and for sustaining river flow. Springs are subject to variations in the height of the water table. As a result some may only flow intermittently, depending on the amount of rainfall and the season.

In deserts, oases of lush vegetation occur where aquifers intersect the land surface and water seeps to the surface. Such oases have been extremely important for the location of transport and trade routes in arid areas.

Types of springs

Lithological springs

Lithological springs are the result of changes in rock type. They occur where porous and permeable rock overlies impermeable rock. The water table will intersect the land surface at the junction between the two rock types. There will be a spring line along the base of the permeable rock.

Springs resulting from lithology also occur where impermeable igneous intrusions, such as dykes, cut through porous and permeable sedimentary rocks. Springs occur where the contact between the two rock types crops out at the surface.

Springs at faults

Faults can produce springs if they have moved porous and permeable rock into contact with impermeable rock. A spring line will occur where the fault plane intersects the land surface.

Faults allow the formation of pressurised springs to form if they intersect confined aquifers. Water under hydrostatic pressure from the confined aquifer rises up the fault plane and flows out onto the surface.

Springs at unconformities

Unconformities can also result in the formation of springs. If porous and permeable rock lies unconformably on top of impermeable older rock, the water table will intersect the land surface at the junction between the two rock types. A spring line will occur where the plane of the unconformity intersects the land surface.

Seeps

In some parts of the world hydrocarbons rise naturally to the surface. These are known as seeps. One good example can be found in Kimmeridge Bay in Dorset where oil from the Kimmeridge Clay emerges at the surface and an oily film can be seen on the water nearby.

Lithological springs in a valley

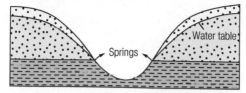

Lithological spring next to an igneous intrusion

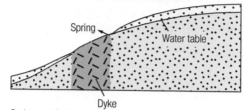

Spring at a fault

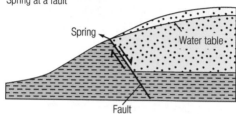

Spring at an unconformity

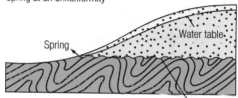

Key	
	Porous and permeable sediment (unsaturated)
	Porous and permeable sedimentary rock (saturated)
	Impermeable sedimentary rock
	Impermeable igneous rock
	Impermeable rock

🔺 **Figure 3** The location of springs

KEY FACTS

Aquiclude an impermeable rock that does not transmit water.

Aquitard a rock which has very low permeability and only allows the transmission of water at very low rates.

QUESTIONS

1 Describe the characteristics of a rock which make it a good aquifer.
2 Explain the difference between an unconfined and a confined aquifer.
3 Copy Figure 4. On your diagram:
   (a) Draw a line to show the position of the water table.
   (b) Shade the area of the aquifer that is confined.
   (c) Label the position of a spring.
   (d) Explain why there will be a spring at the position you have chosen.
4 Describe the geological conditions leading to the formation of springs as a result of lithology, faults and unconformities. Draw labelled diagrams to illustrate your answer.

Key

Porous and permeable sedimentary rock

Impermeable sedimentary rock

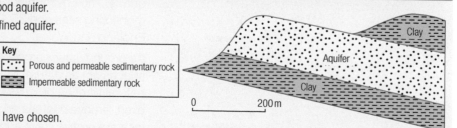

🔺 **Figure 4** Cross-section through part of an aquifer

KEY DEFINITION

Cones of depression occur when there is a lowering of the water table in the vicinity of a well, usually due to abstraction of water.

HOW SCIENCE WORKS: EVOLUTION OF GROUNDWATER

In the *soil zone*, there is free O_2 which allows aerobic respiration and decomposition to occur. This increases the CO_2 in this zone, increasing the overall pH. Leaching of soluble minerals is then more likely.

In the *recharge zone*, there is *young water* characterised by the formation of hydrogen carbonate ions HCO_3^-.

As *groundwater* becomes older (downflow), sulfate ions (SO_4^-) become more dominant. There are a number of chemical reactions which alter the chemistry and free chloride ions (Cl^-) mean that the groundwater may resemble seawater in composition.

Connate water

Connate water is the water included in a sediment when it is deposited, trapped in the pores of the sediment. The water is usually saline in nature with sodium and calcium chlorides present ($NaCl$ and $CaCl_2$). These brines may also be richer in silica (SiO_2) and bicarbonate (HCO_3^-) than normal sea-water. During diagenesis of the sediment the connate waters are often expelled but some of the solutes in the waters may be precipitated in the pores. The main effect of this is to reduce porosity and permeability and thus make the rock less suitable as a reservoir rock or aquifer.

Problems caused by groundwater abstraction

The main problems of groundwater abstraction are as follows.

Lowering of the water table

Shallow wells become dry and have to be sunk deeper. This is a particular problem if wells are situated too close together and their **cones of depression** overlap. This leads to a lowering of the whole water table.

Subsidence

This can occur at the surface, resulting from the removal of water from the pore spaces of rocks. Rocks overlying the aquifer collapse downwards creating depressions at the surface which can be several metres in diameter. Subsidence results in compaction of the aquifer and a permanently reduced water storage capacity.

Saltwater encroachment in coastal areas

Where aquifers occur at coastlines, the less dense fresh groundwater forms a lens floating on top of more dense sea water. Over-pumping disturbs the freshwater–saltwater interface and allows sea water to enter the aquifer. The groundwater becomes saline (brackish) and unfit for drinking. This has occurred at many locations around the coast of the British Isles.

Threats to groundwater supply

The two main threats to our groundwater supplies are over-pumping and pollution:

- Over-pumping occurs if too much groundwater is extracted so that there may not be enough left to provide a reliable public water supply.
- Pollution may enter groundwater as it is vulnerable to contamination from a variety of sources and once polluted is difficult to clean. Unconfined aquifers are more at risk from pollution than confined aquifers because the porous rock is exposed at the surface.

It is important that groundwater is tested to monitor its level and quality. The Environment Agency currently has 7300 groundwater level monitoring sites and 3500 groundwater quality monitoring sites in England and Wales.

As groundwater percolates through the pore spaces of the rocks, pore spaces act as natural filters, removing impurities from the water. This natural filtration process not only removes chemical impurities, but can even remove bacteria and viruses. This process is the result of contact with absorptive clays and also depends on the time that the water is resident in the aquifer. As a result, groundwater does not always require

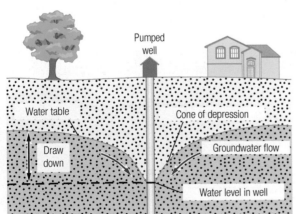

Figure 1 The effect of pumping water from a well

treatment such as chlorination to make it fit for drinking. For this reason, many people may prefer to drink bottled groundwater or spring water rather than tap water. Between 1995 and 2015, the UK bottled water market has grown from just 800 million litres to almost 2.8 billion litres per year. The UK bottled water market was worth over £2.1 billion in 2015.

Soluble minerals dissolved from the rocks are taken into solution as the groundwater percolates downwards. These dissolved salts or ions can make the water taste good or bad and can be beneficial or harmful. Some people believe drinking water containing certain dissolved ions is good for their health. The ions present will depend on the rock types the water has passed through.

'Hard' groundwater contains dissolved calcium (Ca^{2+}) and magnesium (Mg^{2+}) ions. Hard water is harmless to drink but it leaves limescale in kettles and can be difficult to form a lather when you add soap.

Some groundwater contains naturally dissolved fluoride (F^-) ions. Fluoride in drinking water has been proved to reduce tooth decay, but too much fluoride (greater than 4 parts per million) has been linked to dental fluorosis, where teeth become discoloured and pitted, and more serious health problems, such as joint pain, liver and kidney disease.

Unfortunately, some places, including parts of India and Bangladesh, have naturally high levels of toxic arsenic in their groundwater, making it unfit for human consumption.

Groundwater pollution

Groundwater is often under threat from pollution. This can be from a point source, such as a factory, which is relatively easy to pinpoint. It can be a more widespread, diffuse source, which is much harder to track down and prevent. Sources of groundwater pollution include:

- Nitrates, pesticides and microbes from agricultural run-off and sewage
- Hydrocarbons and solvents from petrol stations and factories
- Toxic fluids from landfill waste disposal sites
- **Acid mine drainage** (AMD) water containing toxic metals such as lead and cadmium from abandoned coal and metal mines.

Unfortunately, once groundwater in an aquifer does become polluted, the pollutants have a long residence time of thousands of years and it is virtually impossible to remove them.

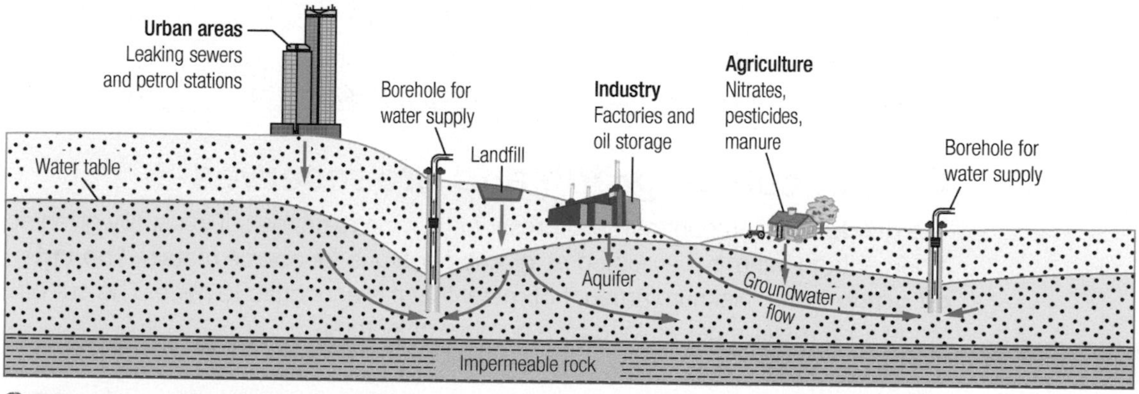

Typical mineral analysis	mg/l
Calcium	25.6
Magnesium	6.4
Sodium	6.4
Potassium	<1.0
Hydrogen carbonate	98.3
Sulfate	10.1
Nitrate	<2.5
Chloride	6.8
Fluoride	<0.1
Silicate	7.6
pH	7.4
Dry residue at 180°C	109.1

▲ **Figure 3** A label from a bottle of mineral water

KEY FACTS

Borehole is drilled down into an aquifer, with an average diameter of 200 mm.

Wells are boreholes which produce water. They are lined with a casing of steel, plastic, concrete or brick.

KEY DEFINITION

Acid mine drainage caused when water flows over or through sulfur-bearing materials, for example rocks containing iron pyrites. This causes acidification of the water by iron pyrites reacting with air and water to form sulfuric acid and dissolved iron ions.

▲ **Figure 2** Sources of groundwater pollution

QUESTIONS

1 Describe one problem caused by groundwater abstraction.
2 Give reasons why some people prefer to drink bottled water rather than tap water.
3 Explain why unconfined aquifers are more at risk from pollution than confined aquifers.

HOW SCIENCE WORKS: WHY ARE METALLIC MINERAL RESOURCES IMPORTANT?

Metals are the foundation of our civilised world. The first metals used were gold, along with other metals found in their native state in the ground. Then around 3500 BCE, humans learned how to get copper and tin by heating their **ores** and the Bronze Age began. It was tin that drew the Romans to Britain, silver that lured the Spanish Conquistadores to South America and gold that led settlers to California during the 1848 gold rush.

Imagine a world without metals. From cars to televisions, bridges to mobile phones, virtually every manufactured item you can think of contains metals. Our modern lifestyle relies on the use of metals and we are using up these non-renewable **resources** at an alarming rate. It is not just familiar metals like iron and copper which are being depleted. New technologies demand the use of other, often scarce, metals. For example, LCDs for flat screen TVs need indium, as do touch screen tablets and mobile phones. Estimates suggest viable global **reserves** of indium are only 6000 tonnes, which at current rates of extraction will last just 10 years.

KEY DEFINITIONS

Ore the rock containing valuable metal(s) that is economic to mine.

Resource a useful and valuable natural material.

Mineral resources can be metallic and non-metallic or industrial minerals.

Reserves the amount of the resource that can be extracted at a profit using existing technology.

Ore deposit an accumulation of metal that may be economic to mine.

Average crustal abundance describes the amount of metal in 'average' continental crust.

Concentration factor the amount by which the metal is concentrated to make an ore deposit.

Ore mineral a mineral containing valuable metal(s).

Gangue mineral a low-value waste mineral.

Grade refers to the concentration of valuable mineral within an ore.

Cut-off grade the grade below which it is uneconomical to mine.

Crustal abundance and concentration factor

Metals are scattered unevenly throughout the Earth's crust, often in amounts that are too low to be economic to mine. The amount of metal in a rock can be given as a percentage or as parts per million (ppm) or parts per billion (ppb) if the amount is very small. In order to form an **ore deposit**, the metal must be concentrated above its **average crustal abundance** by geological processes such as igneous activity or weathering and erosion. The **concentration factor** is the amount by which the metal has been concentrated above its average crustal abundance to form an ore deposit.

Ore deposits consist of rocks called ores. Ores are a mixture of valuable **ore minerals** and low-value **gangue minerals**. Quartz, calcite and pyrite (fool's gold) are common gangue minerals. Compare the photographs of gold and pyrite; the differences really are very obvious! The table shows some common ore minerals and their properties. Many ore minerals are compounds of either oxygen (oxides) or sulfur (sulfides).

Metal	Ore mineral	Formula	Colour	Form	Lustre	Hardness	Streak	Density g cm^{-3}	Cleavage
copper	chalcopyrite	$CuFeS_2$	brassy	tetragonal	metallic	3.5–4	green-black	4.2	none
gold	gold	Au	yellow	cubic	metallic	3	–	19.3	none
iron	magnetite	Fe_3O_4	black	cubic	metallic	6	black	5.2	poor
lead	galena	PbS	grey	cubic	metallic	2.5	grey	7.5	3 at 90°
tin	cassiterite	SnO_2	brown	tetragonal	adamantine	6–7	brown	7.0	poor
zinc	sphalerite	ZnS	brown	cubic	adamantine	3.5–4	brown	4.1	6 at 60°

Grade, cut-off grade and ore reserves

The **grade** is the amount of metal present in a mineral deposit. It is usually given as a percentage but may be quoted as grams per tonne for scarce metals. **The cut-off grade** is the minimum amount of metal that it is economic to mine. The cut-off grade for a metal is determined by a number of factors:

- The value of the metal – the more valuable the metal, the lower its cut-off grade.
- Demand – the higher the demand for the metal, the more valuable and lower its cut-off grade will be.
- The abundance of the metal – if they are useful, scarce metals will have a lower cut-off grade.
- The size of the ore deposit – large deposits will be economic to mine at lower cut-off grades than small deposits.

STUDY TIP

Make sure you learn all the key definitions in this spread.

You also need to learn the formula for calculating concentration factors and be able to rearrange the subject if necessary. Practise converting units – make sure you are confident about doing it.

- Cost of mining and extraction – if it is costly to mine and extract a metal then the cut-off grade will be higher.

The table below shows the relationship between average crustal abundance, cut-off grade and minimum concentration factor for six metals.

Metal	Average crustal abundance (%)	Cut-off grade (%)	Minimum concentration factor
copper	0.007	0.4	57.14
gold	0.0000004	0.001	2500
iron	5.0	40	8
lead	0.0015	4	2666.67
tin	0.002	1	500
zinc	0.007	4	571.43

There is a relationship between cut-off grade and reserves of a metal. If the cut-off grade goes up then the reserves will go down, because it will no longer be economic to mine lower grade deposits. If the cut-off grade goes down, then the reserves will go up.

In the last hundred years, the cut-off grade for copper has reduced from 3% to 0.4%. This is because higher grade deposits have been worked out and technology has improved allowing profitable extraction of lower concentrations of copper.

▲ **Figure 1** A gold nugget

▲ **Figure 2** Pyrite (fool's gold).

MATHS BOX: CALCULATING CONCENTRATION FACTORS

You can calculate the concentration factor using the formula:

$$\text{Concentration factor} = \frac{\text{concentration of metal in ore (grade)}}{\text{average crustal abundance}}$$

For example, if an ore deposit contains 3.5% copper and the average crustal abundance of copper is 0.007%, the concentration factor is $3.5 \div 0.007 = 500$. So, the copper has been concentrated 500 times above its average crustal abundance in the ore deposit.

You can calculate the cut-off grade by rearranging the formula to change the subject:

Cut-off grade = average crustal abundance × minimum concentration factor

It is important that you can convert between units and use numbers in both decimal and standard form. Standard form is a way of writing down very large or very small numbers easily. $10^3 = 1000$, so $4 \times 10^3 = 4000$. In the case of numbers less than 1 the index will be negative.

For example, the average crustal abundance of gold at 0.0000004% can be written in standard form as 4×10^{-7} because the decimal place has been moved 7 places to the right. The minimum concentration factor for gold can be written as 2.5×10^3.

1% is equal to 10 000 ppm or 1×10^4 ppm and 10 000 000 ppb or 1×10^7 ppb.

1 gram per tonne is equal to 0.0001 %, 1 ppm or 1000 ppb.

QUESTIONS

1   Ore = ore minerals + gangue minerals. Define these terms.
2   Define the term concentration factor and explain how it is calculated.
3   The table shows data for three ore deposits. Calculate the missing values.

metal in ore deposit	average crustal abundance (%)	grade (%)	concentration factor
iron		50	10
gold	0.0000004	0.02	
tin	0.002		1000

4   Explain why gold is more valuable than pyrite (fool's gold).

Secondary enrichment occurs when metals are leached from within surface rocks and precipitated just below the water table.

Chemical weathering the in situ breakdown of rocks at the Earth's surface due to chemical reactions.

Leaching where ions are dissolved from rocks and carried downwards in solution.

Gossan an insoluble cap of iron oxides at the surface of a mineral vein.

Oxidising describes oxygen-rich conditions, allowing elements to combine with oxygen to form oxides.

Reducing describes oxygen-poor, anoxic conditions.

Enriched deposit a zone of high-grade ore just below the water table, formed by secondary enrichment.

Porphyry a large igneous intrusion with a porphyritic texture.

HOW SCIENCE WORKS: WHY IS SECONDARY ENRICHMENT IMPORTANT?

Mining companies often mine the area of secondary enrichment first to offset the capital costs of exploration and putting the mine into production. This is an important factor in the exploitation of a specific type of low grade, high tonnage copper deposit called a porphyry copper deposit.

▲ **Figure 2** Copper minerals: **a** Chalcopyrite; **b** Azurite

Secondary enrichment of chalcopyrite in copper deposits

Secondary enrichment is not an ore-forming process but an ore-concentrating process. It is a very important natural process for increasing the grade of otherwise uneconomic copper deposits.

Chalcopyrite, a copper iron sulfide with the formula $CuFeS_2$, is the main ore mineral of copper. Large, low-grade ore deposits of chalcopyrite are formed by igneous processes. When rocks containing chalcopyrite are exposed at the Earth's surface they may undergo **chemical weathering** that can cause secondary enrichment.

Above the water table

Rainwater infiltrates into the exposed copper deposit and percolates downwards through the pore spaces. In the zone of oxidation above the water table, chemical reactions change insoluble copper sulfides such as chalcopyrite into soluble copper sulfates. The copper sulfates are dissolved and carried downwards by the groundwater. The simplified chemical reaction is as follows:

chalcopyrite + water + oxygen → copper sulfate + iron (III) hydroxide

A barren, **leached** zone is left near the surface covered by an insoluble iron oxide capping called a **gossan**. Although of no economic value, gossans are useful exploration targets as their presence suggests there may be ore deposits underneath.

Below the water table

The copper is carried downwards in solution through permeable rocks to the water table. At the water table the conditions change from **oxidising** above to **reducing** conditions below. Chemical reactions change the copper sulfates back into insoluble copper sulfides, resulting in immediate reprecipitation of the insoluble 'secondary' copper sulfides, just below the water table. In addition to chalcopyrite, new copper sulfide minerals such as bornite (Cu_5FeS_4) and chalcocite (Cu_2S) with a higher copper content also form. All the copper that was spread out in the rock above the water table is now concentrated in a much smaller volume so the grade is higher. This high-grade **enriched deposit** overlies the unweathered lower grade 'primary' copper ore at depth.

Just above the water table, brightly coloured blue and green copper oxides and carbonates such as green malachite ($Cu_2CO_3(OH)_2$) and turquoise/blue azurite ($Cu_3(CO_3)_2(OH)_2$) are precipitated but these are usually of minor importance.

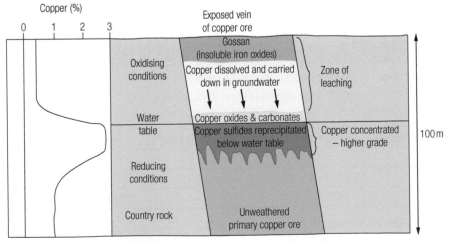

▲ **Figure 1** Cross-section showing how secondary enrichment occurs

CASE STUDY: BINGHAM CANYON, UTAH, USA

Bingham Canyon porphyry copper mine is the largest man-made hole in the ground. The mine is owned by Rio Tinto plc and has been in production since 1906. The open pit is 4 km wide and 1.2 km deep – it takes half an hour to drive to the bottom and is visible to the naked eye from the space shuttle. The mine employs 1400 people and mineral processing, smelting and refining are carried out on site. In its 100 years of operation, Bingham Canyon has been one of the world's most productive mines. To date, the mine has produced 15.4 million tonnes of copper, 715 tonnes of gold, 5900 tonnes of silver and 386 000 tonnes of molybdenum.

The ore deposit is associated with an igneous intrusion (35 Ma) that intruded into sandstones and limestones. Metallic ore minerals carried in solution by late stage hydrothermal fluids from the intrusion were precipitated in fractures in the surrounding country rocks. More recently, the top of the ore deposit was exposed by erosion and underwent secondary enrichment. The zones of secondary enrichment were the first parts of the ore deposit to be mined. When the high grade enriched ore was worked out in 1980, innovations in mineral processing technology allowed profitable extraction of the lower grade primary ore.

The mine experienced a massive landslide in April 2013 – the largest non-volcanic landslide in the recorded history of North America. Around 65–70 million cubic metres of dirt and rock thundered down the side of the open pit. Fortunately monitoring systems were in place and workers were evacuated to safety beforehand. The current mine plan will expire in 2019 but Rio Tinto has been studying a plan to extend the open pit 305 m southward, which will extend the life of the mine into the mid-2030s, provided that the required environmental permits are approved.

▲ **Figure 3** An opencast porphyry copper mine

Formation of porphyry copper deposits

Porphyry copper deposits form as a result of hydrothermal processes associated with granite intrusions at convergent plate margins. They acquired this name because the igneous rocks with which they are associated often have a porphyritic texture. A large number of porphyry copper deposits, including Bingham Canyon in Utah, USA, and Chuquicamata in Chile, are located along the western side of North and South America along the plate margin with the Pacific Plate.

Partial melting of the subducted plate and base of the continental crust produces wet silicic magmas. These are the source of water, heat and metals in porphyry copper deposits. Late in the cooling history of these silicic intrusions, water and 'incompatible' metals, including copper, that do not fit into silicate minerals collect at the top of the intrusion to form a hydrothermal fluid. The hydrothermal fluid moves out into the surrounding country rock, exploiting any weaknesses such as faults, joints and bedding planes. Groundwater may be drawn into the intrusion and a hydrothermal circulation system is set up. As the hydrothermal fluid moves away from the intrusion it cools and may encounter chemically reactive country rocks such as limestone. This results in the precipitation of a mixture of ore and gangue minerals from the fluid. *Veins* are formed when minerals precipitate within fractures. *Disseminations* are formed when minerals precipitate within the pore spaces of rocks.

The upper parts of these porphyry copper deposits are then exposed by erosion and undergo chemical weathering, resulting in the formation of high grade secondary enrichment deposits. In the case of Chuquicamata, the primary copper ore has a grade between 0.5 and 1% copper and the zone of secondary enrichment reaches grades of up to 15% copper – a concentration factor of 300.

QUESTIONS

1 The data in the table shows the changes in copper concentration with depth in an ore deposit.

Depth (metres)	Copper (%)
0	0.2
10	0.3
20	2.9
30	1.5
40	1.4

Plot a line graph of the data, with depth downwards on the Y axis and copper concentration on the X axis, as indicated.

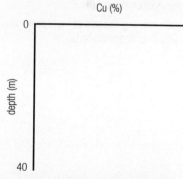

Draw a horizontal line on your graph to indicate the likely position of the water table. Label the position of the zone of leaching, enriched ore and primary copper ore.

2 Draw a labelled diagram to explain how the grade of a copper ore deposit can be increased by the process of secondary enrichment.

3 Explain why the process of secondary enrichment is important to the economics of a copper mining operation.

What are placer deposits?

Probably the best known and often romanticised type of mineral exploitation is that of prospectors panning for gold. The prospectors fill their shallow pans with river sediment and swirl it around with water in the hope of finding a gold nugget at the bottom of the pan when less dense sediment has been washed out of the pan! These prospectors are exploiting placer deposits. The word placer comes from Spanish and means 'reef' or 'sand bank'.

Placer deposits are surface deposits formed by the sedimentary processes of weathering, erosion, transport and deposition. Dense, physically and chemically resistant minerals including cassiterite (tin oxide, SnO_2), gold and diamonds can be concentrated by these processes into usually small, but high grade ore deposits.

How placer deposits form

Just like any rock, mineral veins exposed at the Earth's surface will be weathered. During weathering ore is broken up by mechanical weathering or left as insoluble material by chemical weathering. The ore and gangue minerals are separated into individual grains.

The weathered material is then transported. Most placer deposits are concentrated by the action of moving water in rivers and the sea. During transport the sediment is sorted by grain size, hardness and density. Less resistant minerals are worn away by the erosion processes of abrasion and attrition or dissolved by water, leaving the more resistant insoluble minerals to form placer deposits.

Minerals found in placer deposits have similar properties:

- Hard, with little or no cleavage, so they survive abrasion and attrition during transport. Gold is an exception. Gold is soft but because it is malleable it rolls into nuggets rather than being broken up.
- Chemically unreactive, so they are not dissolved and taken into solution.
- Dense, so they are deposited first when the current velocity slackens.

You can see the properties of some placer minerals in the table.

Mineral	Density (g cm^{-3})	Hardness	Cleavage	Solubility
cassiterite	7.0	6–7	poor	insoluble
diamond	3.5	10	4 perfect	insoluble
gold	19.3	3	none	low

When the current velocity slows, these minerals are preferentially deposited in one place, though mixed with unconsolidated sand and gravel.

CASE STUDY: WELSH GOLD

The Queen's wedding ring was made from a nugget of Welsh gold and the Prince of Wales' crown jewels were made from Welsh gold. In 2011, this tradition of the British Royal family using Welsh gold was carried into its 88th year when Catherine, the Duchess of Cambridge, commissioned her wedding ring to be made from pure Welsh gold.

Most Welsh gold comes from the Dolgellau Gold Belt. Placer gold is found in rivers sourced from small hydrothermal lodes. Total recorded production is 4 tonnes and because the gold is rare it fetches a very high price. Although the small Clogau Gold Mine closed in 1998, a 2012 report suggested that as much as 14 175 kilograms of gold could be lying untouched in the surrounding area, giving hope that once again gold may be extracted from the mine.

CASE STUDY: GOLD FEVER

The discovery of placer deposits sparked many of the famous gold rushes – the 1848 Californian, 1893 Kalgoorlie in Western Australia and 1897 Klondike in the Yukon, Canada, to name a few. Unfortunately, most of the prospectors did not achieve fame and fortune, but endured hardship and heartbreak!

▲ **Figure 1** A gold pan

KEY DEFINITIONS

Dredging where material is scraped or sucked from the river or sea bed.

Hydraulic mining the use of high-pressure water jets to dislodge material.

Sites of deposition of placer minerals

Meander bends

When a river flows around a meander bend the current swings around outside of the bend. The current velocity is fastest on the outside of the bend and slowest on the inside. This results in erosion on the outside of the bend and deposition on the inside, to form a point bar. Placer deposits are found on the inside of meander bends.

Plunge pools

When a river flows from hard rock to less resistant rock, it erodes downwards producing a waterfall. Turbulent water and boulders at the bottom of the waterfall scour out a deep hollow called a plunge pool. Dense placer minerals become trapped in the sediments of the plunge pool.

Upstream of projections

Projections from the riverbed will trap dense placer minerals on the upstream side. This may be where a hard rock such as a dyke projects upwards and/or, on a small scale, on the upstream side of ripples.

Downstream of confluences

Where a fast-flowing tributary joins a slower flowing river the current velocity will drop. This results in dense placer minerals being deposited to form a mid-channel sandbar.

On beaches

Rivers transport sediment into the sea. The sediment may then move along the coast by longshore drift. Waves throw sediment up the beach on the swash and as the energy of the water reduces on the backwash, dense placer minerals can be left behind forming beach placer deposits.

Advantages and disadvantages of placer mining

Recent placer deposits are loose, unconsolidated sands and gravels that are easily accessible and cheap to mine. Common mining methods include **dredging** and **hydraulic mining**. To some extent the ore minerals are already separated from the gangue minerals so less waste rock is produced. As a result, it could be argued that placer mining has less environmental impact than underground mining.

However, the immediate environmental impacts of placer mining are very obvious – scars on the landscape, noise and dust during mining and dredging and hydraulic mining which stir up silt causing surface water pollution.

In addition, because placer deposits tend to be small they are quickly exhausted.

CASE STUDY: OLD GOLD

The Witswatersrand Goldfield in South Africa is a fossil placer deposit that accounts for 50% of the world's gold production. The Witswatersrand gold mines are up to 6 km deep.

The geological setting is a 400 km-wide Precambrian sedimentary basin. Gold occurs in the matrix of quartz pebble conglomerates deposited in a large lake. The source of the gold is thought to be from weathered and eroded hydrothermal veins that cropped out in hills surrounding the lake.

QUESTIONS

1 Use labelled diagrams to explain how placer deposits can form at different locations along rivers.

2 Explain why cassiterite is found in placer deposits but galena is not.

3 Research where placer mining is taking place today and write a case study, which should include the minerals being mined and the environmental consequences of the mining.

Meander bend (plan)

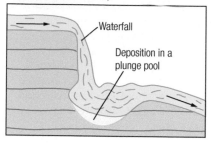

Plunge pool (cross-section)

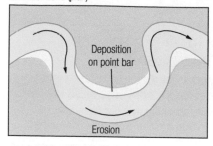

Projections in riverbed (cross-section)

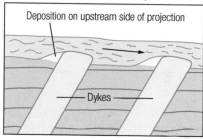

Confluence of tributary (plan)

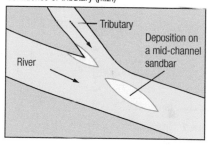

Beach (cross-section)

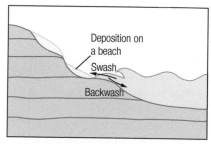

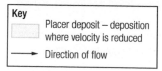

Figure 2 Sites of deposition of placer minerals (not to scale)

CASE STUDY: HANS MERENSKY

In 1924, Hans Merensky, a geologist and mining engineer, used prospecting techniques to discover platinum deposits in the Bushveld Igneous Complex, South Africa. The Merensky Reef varies from 30 to 90 cm in thickness and contains an average of 10 ppm platinum group metals. It contains 80% of the world's known platinum reserves.

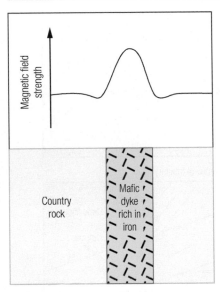

▲ **Figure 1** Cross-section showing a transect magnetic survey

KEY DEFINITIONS

Geochemical anomaly a concentration of a metal above its normal background value.

Dispersion occurs when small amounts of metals are spread out around the ore deposit by the surface processes of weathering, erosion and transport.

Catastrophic dilution occurs where tributaries meet and water and sediment from other sources are added.

Geophysical exploration techniques used to find metals

Exploration geophysics is used to measure the physical properties of the rocks underlying an area. It can be used to detect the presence and position of metallic mineral deposits. Geophysical instruments can be mounted in helicopters or planes to allow rapid coverage of large areas. Hand-held, land-based instruments can be used for detailed follow-up surveys in target areas. Geophysical surveys are carried out along a line called a transect or on a grid pattern to allow plotting of data on maps. Survey points are located using GPS. Geophysical surveys detect anomalies which may indicate the presence of metallic ore minerals or rocks that may contain them. A **geophysical anomaly** may be more positive (higher) or negative (lower) than normal.

Gravity surveys

An instrument called a gravimeter is used to measure small variations in the Earth's gravitational field strength. The units of measurement are milligals (mGal). Gravity data are corrected for the effects of latitude, altitude and topography leaving the variations resulting from the underlying rocks and minerals. Gravity maps are plotted with lines joining points of equal gravitational field strength allowing any anomalies to be identified.

A positive gravity anomaly could be due to:
- A high density mafic or ultramafic intrusion, which may contain cumulate ore deposits formed by magmatic differentiation and gravity settling
- The presence of dense, metallic ore minerals.

A negative gravity anomaly could be due to:
- A low density silicic intrusion, which may have hydrothermal mineral veins around it.

Magnetic surveys

An instrument called a magnetometer is used to measure small variations in the Earth's magnetic field strength. Measurements of the Earth's magnetic field strength are often quoted in units of nanotesla (nT).

Minerals rich in iron produce positive magnetic anomalies. This could be due to:
- A mafic or ultramafic intrusion, which may contain cumulate ore deposits
- The presence of the mineral magnetite.

Electromagnetic (EM) surveys

EM surveys measure ground conductivity by the process of electromagnetic induction. During an EM survey, electromagnetic waves produced by a transmitter induce currents in conducting materials such as metallic ore minerals. EM surveys often make use of electromagnetic waves from very low frequency (VLF) signals that are transmitted around the world by governments for navigation and communication purposes. There are about 42 VLF transmitters worldwide, including one at Rugby, UK. The values are usually given as a percentage of the induced secondary field relative to the primary field. EM surveys are particularly useful for identifying the presence of metal sulfide minerals such as chalcopyrite.

Electrical resistivity surveys

In an electrical resistivity survey, two electrodes are placed in the ground and an electric current passed between them. If the underlying rock is a good conductor it will have a low resistance. Metals are good conductors so rocks containing metallic ore minerals will have a lower resistance than other rocks. This method has been successfully used for gold exploration.

Geochemical exploration

Geochemical exploration involves collecting samples and analysing them to find **geochemical anomalies**. Samples of rock, soil, stream sediment, water, drill core and even vegetation or atmospheric gases can be taken for analysis. Techniques used to chemically analyse the samples include X-ray fluorescence (XRF) and inductively coupled plasma atomic emission spectroscopy (ICP-AES).

Geochemical exploration relies on the fact that metals from an ore deposit will undergo **dispersion** by weathering, erosion and transport processes producing a 'halo' around the deposit. So, interpretation of geochemical data depends on a good understanding of how particular metals travel at the surface and near-surface parts of the rock cycle. This is determined by the stability of the metals and the climate and terrain of the area.

Stream sediment sampling

Stream sediment surveys are used for regional exploration in temperate climates where there are rivers. Stream sediment samples of 1 to 2 kilograms are collected from riverbeds and analysed for the metals of interest. Interpretation is based on:

- Stream sediments downstream of the source will have anomalous metal values.
- Stream sediments upstream of the source will have normal metal values.
- The size of the anomaly decreases downstream due to dilution by surface water run-off and sediment entering the river.
- **Catastrophic dilution** occurs where tributaries meet.

The best sampling strategy is to sample each tributary immediately upstream of each confluence. The anomaly can be traced back upstream to its source area, which is then the target for more detailed exploration.

Soil sampling

Soil surveys are used for local follow-up surveys in target areas where there is a lack of outcrop. Hand-held tools such as shovels or soil augers are used to collect samples of the soil and subsoil. Soil samples are usually collected on a systematic grid pattern. Interpretation depends on:

- Soil samples on top and downslope of the source will have anomalous metal values.
- Soil samples upslope of the source will have normal metal values.

Water sampling

Either surface water or groundwater can be collected and tested in *water surveys*. In the case of surface waters, samples are taken at regular intervals along a river course. The background metal content of most natural waters is only a few parts per billion so sensitive analytical techniques are required and seasonal variations have to be taken into account. An increase in the metal content of the water to a few parts per million may indicate the presence of metallic mineral deposits nearby.

Vegetation sampling

Metal elements can be taken up by plants through their roots. *Biogeochemical surveys* involve analysing the trace element content of a selected plant species. Samples of leaves, twigs or seeds are collected from the chosen plant. Elevated metal content may indicate the presence of underlying metallic mineral deposits.

Another type of survey called a *geobotanical* survey involves mapping the distribution of an indicator plant species that can tolerate high concentrations of a particular metal in the soil. Identification of plants with a high metal content is another method that can be used to locate mineral deposits.

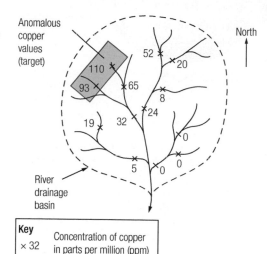

Key: ×32 Concentration of copper in parts per million (ppm)

Figure 2 Map of a stream sediment survey

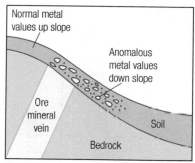

Figure 3 Cross-section showing how metals are dispersed in soils

QUESTIONS

1. The table below shows the relative magnetic strengths of some geological materials.

Material	Relative strength of magnetism
chalcopyrite	0.4
gold	0.0
hematite	7.0
river sediment	0.0
granite	3.0
limestone	0.3

How easy would it be to detect the following minerals deposits using a magnetic survey?

   (a) A hematite vein in limestone.

   (b) A placer deposit rich in gold.

   (c) A chalcopyrite vein in granite.

2. Explain why sampling each tributary immediately upstream of each confluence is the best sampling strategy to use for a stream sediment survey.

After regional geophysical surveys, the next stages include:

- Target selection, followed up by detailed geological mapping, further geophysical and geochemical surveys, trench digging and exploration drilling to identify the extent of any ore deposits found.
- Calculation of ore reserves using computer modelling. The size and grade of an ore deposit must be accurately determined to decide if it is economic to mine.

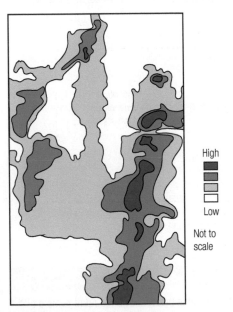

High

Low

Not to scale

▲ **Figure 1** Map of a generalised geophysical survey

KEY DEFINITION

Geographical Information System (GIS) a computer application used to store, view and analyse spatial information, especially maps.

Following the regional exploration described in the previous pages, we now move on to the more detailed stages of the search for mineral wealth, where to search and how much of the resource we are likely to find. Our surveys move from the air and down onto the ground.

Target selection

Regional geophysical surveys indicate areas where it may be worth spending time and money on further investigations. Equally importantly, they show areas which have no mineral resources. Promising areas are targeted for surface mapping and the collection of rock and mineral samples. Geophysics has shown where the ore body is; geochemical analysis, using chemical, metallurgical, X-ray, spectrographic and radiometric evaluation techniques, the composition of the ore body and the likely grade of its ores.

Surface mapping and structural analysis indicates the best sites for exploration drilling. Target selection is assisted by **Geographical Information System (GIS)** software. Mineral exploration geoscientists enter diverse types of datasets into the GIS program so that they can view and analyse qualitative spatial relationships quickly and easily using multiple geophysical maps, charts and tables.

Figure 1 shows a hypothetical map for a generalised geophysical survey. The obvious target areas are those shown in red where the anomaly is strongest. However, siting the borehole away from the strongest anomaly not only shows the minerals but begins to delineate the extent of the ore body.

Other boreholes are needed between the two 'ridges' of high anomalies in Figure 1 to show if the pattern is repeated by folding, faulting or the presence of two separate deposits.

Exploration drilling

Exploration drilling provides the most accurate information, but it is also the most expensive method. Rotary, percussion, or diamond drills can be used; the latter allows core samples to be taken from the borehole. Downhole logging provides additional geophysical data.

Exploration drilling aims to determine:

- What useful ores, if any, are present at that location
- How much ore might be present in terms of tonnage and grade
- How much gangue and other barren material there is in the ore body
- The depth and extent of the ore body and how it may be accessed when mining starts.

Estimating reserves

Reserve estimation is a valuation of the deposit. The crucial question: is it a metal ore body, or just another mineral deposit? While there is no right answer for a reserve estimate, a wrong answer may cause the company to fail.

Reserves make up the part of the resource that can be economically extracted. Reserves can be divided into proven and probable mineral reserves. The assessment of a mineral reserve makes allowances for waste rock which may 'dilute' the ore mineral when it reaches the processor and for unavoidable losses that may occur during mining.

Reserve estimation is a complex geostatistical operation which uses sophisticated computer software. The only way to find out how much of a resource can be mined profitably is to mine it, by which time it is no longer a reserve! In view of some well-documented frauds giving inflated values for reserves, stringent guidelines are now set out by the mining industry.

Grade

A key factor in profitability is the grade of the ore. This may vary considerably across the ore body. How much variation in the grade is there between the exploratory boreholes? Statistical probability techniques are used to suggest an answer, such as *kriging* (named after Daniel Krige who pioneered the technique to estimate probable distributions of gold based on samples from few boreholes at the Witwatersrand reef complex in South Africa). It is probable that the rank of coal will not change much between two boreholes; a copper vein is much more variable.

The shape of some ore bodies is far from uniform. Three-dimensional computer modelling is used in which the ore body is divided mathematically into 'blocks'. Each block is assessed for grade and other factors, based on borehole and other data. The dimensions for the blocks depend on the spacing of the sampling boreholes and the scale of the mining machinery. Each block has the same size but different characteristics of mineral grade and rock type. The blocks are assembled by the computer program into a block model of the ore body, and all the data analysed.

Geological considerations

The type of computer modelling selected depends on:

- The geological setting
- Whether there are areas of very high grade (the 'nugget' effect)
- The styles and zonation of mineralisation
- Whether the mineralisation has sharp or gradational boundaries
- How easy the rock is to work
- The presence of toxic elements, such as arsenic.

Operational considerations

Whether ore can be extracted at a profit depends on:

- The type of mine and its stability, open pit or underground mining
- The methods and rates of extraction and processing
- Pollution management and acid mine drainage
- Dilution by waste rock mixed with ore
- Water control, drainage and pumping.

Economic considerations

By definition, a mineral reserve must have 'reasonable prospects of economic extraction' and must account for:

- Set-up costs (including assembling 'plant')
- Level at which the cut-off grade is set
- Long-term projections for the metal prices for the ore
- Closure and reclamation plans, as laid out in the planning consent.

When the reserves have been estimated, the project may be developed as a fully operational mine, traded to another company or abandoned altogether.

Accuracy of estimates of reserves

A number of factors can affect the accuracy of the estimation of reserves:

- The detail of the survey
- Variation in grade of the mineral and the size of the ore body
- Unexpected geological conditions, such as faulting
- Unrecorded earlier workings, especially underground
- Variations in the economic climate and in demand
- Improvements in extraction technology.

A change in any one of these will change the reserve estimates.

MATHS BOX: MAKING ESTIMATES

You may have taken part in the game 'Guess the number of sweets in a jar'. The rules do not allow you to tip the sweets out and actually count them. You have to make an estimate!

The equivalent to exploratory drilling is to look at the jar from all sides. This will show that all the sweets are the same type. The probability is that the sweets in the middle are also the same size. If the jar is very big, the probability is lower; indeed, the organiser of the competition may have padded out the middle with bubble wrap!

You can then count the number of layers of the sweets and, by looking at the bottom, count how many sweets there are in each layer and complete your estimate by multiplying the two numbers.

It is important to make estimates for your geological calculations to check that your final answer is realistic.

HOW SCIENCE WORKS: RESERVES

Metallic minerals are a non-renewable resource and metals' mining is unsustainable on a global scale. The geological processes that create economic ore deposits take millions of years. Dwindling ore reserves should attract some concern. There are no synthetic alternatives for metals and recycling is not always feasible due to technical difficulties and high energy requirements. Demand for scarce metals for new technologies is outstripping their reserves to such an extent that it is not worth developing the technology any further.

It is very difficult to estimate global reserves of metals. Some reserves have not yet been discovered. For example, in 2015, based on the tectonic history of the Tethys Sea, USGS estimated that 180 million metric tonnes of copper may lie under the Middle East. New technologies may in future make unprofitable deposits economical to mine. To give a few examples, 2007 statistics from the United Nations and the USGS suggest that at current rates of extraction, global copper reserves will last 61 years, uranium 59 years, gold 45 years, and tin 40 years. Future generations will be faced with the prospect of having to exploit lower grade, less accessible ore deposits with ever increasing energy costs, unless we make a determined effort to find ways of conserving metallic mineral resources.

QUESTIONS

1. When does a resource become a reserve?
2. Why can the estimate for a reserve change?
3. Research the kriging technique.

Sustainability using a resource so that it is not depleted or harmful to the environment, supporting a long-term balance in ecological and other systems.

Non-renewable resources do not renew themselves at a sufficient rate for sustainable economic extraction within a human lifetime.

Open cast extraction is mining from surface quarries.

Shafts vertical openings to an underground mine.

Stope the void left when the ore has been extracted from an underground mine.

CASE STUDY: OPEN CAST COAL MINING

Open cast coal mining takes place from quarries or 'open pits' at the surface. All the rock above the coal seam has to be removed and the depth of this overburden will be critical to the economics of the mining operation. The sides of the open pit are dug at an angle and stepped to increase stability and prevent collapse. The flat parts of the steps are called benches.

The angle of the sides will depend on:
- Rock type – weak, incompetent rocks like clays will require shallower slope angles.
- Weathering – heavily weathered rock will be weaker and need shallower sides.
- Structures such as faults and joints weaken the rock and may need rock bolts, wire netting, shotcrete, rock drains or other ground improvement strategies.

The overburden is removed and piled up as spoil heaps near the edge of the open pit. Blasting is used to break the coal up and large machines called dragline excavators are used to extract it. A large bucket at the end of the dragline is dragged along the surface to scoop up the coal. The excavator can remove as much as 450 tonnes of coal in one pass.

In Britain the thickest seams being worked are 2.5 m thick, while in the USA the thickest seams being worked are nearly 30 m thick. After mining is complete the land can be restored by backfilling the pit with the overburden.

KEY FACT

Health and Safety Executive (HSE) in the UK is a national independent watchdog for work-related health, safety and illness.

In 1987 the Brundtland Commission defined sustainable development (**sustainability**) as 'meeting the needs of the present, without compromising the ability of future generations to meet their own needs'. How does this apply to mining, which extracts irreplaceable ores from holes in the ground, especially as it uses fuels and produces greenhouse gases? It could be argued that minerals are not being lost through mining, just converted into new forms, which can be recycled later. However, they are considered **non-renewable resources**.

Mining can be sustainable if the positive economic, social and environmental benefits outweigh any negative impacts it brings, such as dust, noise, heavy traffic, visual pollution, landscape degradation and the clearing of vegetation and loss of habitat. Impacts may extend beyond the mine site, affecting groundwater, producing silt from dredging or hydraulic mining and needing the disposal of waste.

There needs to be compensation for the loss of agricultural land with funding left over to pay for the provision of better services and facilities in the community, along with robust planning for closure, after the quarry or mine has finished.

The life cycle of a mine

Stage 1 – Design and planning

If surveys are promising, the design and planning stage assesses if the project will be economically viable, environmentally sound and socially responsible.

In a phased development plan, the project grows from small beginnings, avoiding economic risks, starting with limited open pit mining of the highest grade ore, perhaps from secondary enrichment, and low cost processing. Profits are reinvested in underground operations, based on updated information during the mining. More profit leads to further expansion, for example with a new milling or processing plant. The plan is more likely to be successful if the needs of local communities are met so that the benefits of mining are shared by all and so local employment and community projects are important.

It should be noted that plans to deal with the site after extraction has ceased are also considered at the planning stage; rehabilitation and reclamation is usually a condition of planning consent.

Stage 2 – Construction

Construction includes extraction and processing facilities (plant), roads, environmental management systems, housing and other facilities.

Stage 3 – Production

Mining may be by open pit or by underground methods depending on the depth, shape and size of the target body and the limits imposed by health and safety (HSE), technology, environmental and economic concerns.

Stage 4 – Processing

The mined materials are sent through huge crushers or mills to separate commercially valuable ores from the gangue and waste rock, so that transport to the smelter is more economical.

Stage 5 – Rehabilitation and reclamation

Once a mine has been exhausted, the facilities on the site are dismantled and the land returned to its original state, although with a large open pit this may be impossible. The aim is to ensure public health and safety, minimise environmental effects, remove waste and hazardous material, preserve water quality, stabilise land against erosion and to establish new landforms and vegetation.

Surface mining as open pit or open cast mining

If the ore is relatively close to the surface, it is often more economical to strip away the overburden by quarrying to reach the ore than it is to construct an underground mine. **Open cast** mining is very efficient and high rates of production can be achieved. Setup and working costs are lower and only a small workforce is required. Although the machinery is expensive, it is cheaper than the high-tech machinery needed for underground mining. Ventilation equipment is not required and thinner seams and low-grade ore can be mined at a profit. Open pit mining is also safer. There are environmental and slope stability implications regarding any spoil heaps or earth bunds created by the excavated overburden.

▲ **Figure 1** Copper mine, Bingham Canyon, Utah

Underground mining

Stope mining

It is most productive to extract the ore from the ore body itself without the need to construct additional **shafts**. This leaves a cavity or **stope** which may not need support if the rocks are strong. In practice it is common to drill shafts vertically downwards and then drive horizontal levels through to the ore body. Supports are usually provided. The stope may be backfilled with tailings or allowed to collapse in a controlled way.

The specific method of stoping depends on the dip and width of the deposit, the grade of the ore, the hardness and strength of the surrounding rock and the cost of materials for supports. In horizontal seams, such as coal seams, mining may be carried out by pillar and stall stoping or by *longwall retreat mining*. In steeply-dipping ore bodies, such as lodes of tin, the stopes become long narrow and almost vertical. In earlier times, the ore was won by working down into the ore body, an advantage when using hand tools. When using explosives and power tools to break up the rock and gravity to help transport it, mining proceeds from the bottom upwards, in slices with the broken ore being left in place for miners to work from. Forty percent of the broken ore must be removed to provide working space for blasting the next ore slice. Once the top of the stope is reached, all of the remaining broken ore is removed, often using remote controlled dump trucks.

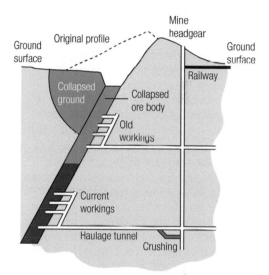

▲ **Figure 2** Dump truck in an open cast mine

▲ **Figure 3** Stope mining

Longwall retreat mining

Longwall underground mining is highly mechanised and in good geological conditions can achieve production rates comparable to open cast mining. It was the main method used for underground coal mining in Britain until the last deep mine (Kellingley, Yorkshire) closed in December 2015.

A main shaft is dug from the surface and tunnels or roadways are driven out from the shaft. There are also ventilation shafts. Two horizontal roadways are driven out to the furthest point to be extracted. This allows the geological conditions to be assessed prior to mining. The 'longwall' or coal face, up to 400 m long, is established between the two roadways. An extracting machine such as a *shearer* moves to and fro along the face, removing slices of coal which falls onto a conveyer belt and is transported to the shaft and up to the surface.

The roof is held up by closely spaced, mobile hydraulic steel chocks. Once a slice of coal is removed, the chocks are moved forward and the mined-out area is allowed to collapse. Mining takes place backwards, retreating towards the shaft. This system of deliberate collapse can cause subsidence on the surface but in a uniform manner. It also means collapse of the mine will not happen in the future, as it does with so many old pillar and stall mines of the past.

QUESTIONS

1 Compare and contrast how coal is mined using longwall retreat and open cast methods.

2 Is the 1987 Brundtland Commission definition of sustainability a good one? Give reasons for your answer.

3 Suggest three uses for a large open cast pit, once mining has stopped.

4 When Kellingley colliery closed in December 2015, reserves of coal were estimated at 30 million tonnes. What is the volume of this amount of coal? If the seams are all 2 m thick, what is the total surface area of the seams?

CASE STUDY: WHEAL JANE, CORNWALL

Wheal Jane, one of the last operating Cornish tin mines, closed in 1991 and began to fill with water when the drainage pumps were switched off.

In January 1992, a unique and very visual pollution incident caught the attention of the world's media when over 50 million litres of heavily contaminated acid mine drainage water burst from the disused mine and flowed into the Fal Estuary. The water had a pH of 3.1 and was laden with toxic metals including cadmium, arsenic, lead, zinc, copper and iron. Precipitation of iron hydroxides produced a bright orange ochreous plume that swept downstream and out to sea. The incident led to widespread public concerns about health and the impact on the environment and tourism. The authorities immediately set up a temporary treatment system.

Water was pumped from the mine, collected in a tailings dam, and treated using lime to neutralise the acid and precipitate the dissolved metals. An experimental passive treatment system using limestone drains, reed beds and rock filters was also put into place. The passive treatment experiment has since ended and all of the mine water is now treated by active chemical treatment. To date, treating the acid mine water discharge from Wheal Jane has cost in excess of £20 million!

A 'hidden' cost is the fossil fuel energy needed to transport lime several hundred kilometres from a site in the Midlands to Cornwall.

In situ and heap leaching

Leaching uses chemical solutions to dissolve the ore from the rock. The solutions are treated to release the target minerals and the leaching solution is re-used.

In situ leaching, at depth

Boreholes are drilled into the ore deposit and opened up by explosive or hydraulic fracturing. The leaching solution is pumped down to the ore. The solution carrying the dissolved ore is pumped back to the surface through a second borehole and processed. Some examples:

- Evaporites are extracted using water.
- Uranium uses acids or sodium hydrogen carbonate.
- Copper, especially malachite and azurite, uses acids.

Alternatively, the ore is sometimes placed in a stope for leaching.

Heap leaching at the surface

The ore is crushed and heaped onto an impermeable clay or plastic liner. The leaching solution is applied either by a spray or by drip system and percolates through the crushed ore. The solution of dissolved minerals accumulates in a pond and is then taken for processing. This method is cheap but usually recovers only 60–70% of the ore. It takes between two months and two years.

Froth flotation

Froth flotation is the most important mineral recovery process. It separates 'water-repelling' (*hydrophobic*) minerals from 'water-loving' (*hydrophilic*) material. It yields significant amounts of metal from lower grade ore and can separate a wide range of sulfides, carbonates and oxides from each other.

The ore is crushed so that individual minerals occur as individual grains and are then mixed with water to form a slurry. The target mineral is made hydrophobic by adding a suitable chemical, a different one for each different mineral. The slurry is fed into a flotation cell which is aerated to produce bubbles. The hydrophobic particles attach to the air bubbles, which rise to the surface as a froth, a concentrate of the target mineral. The **flotation tailings** are left as they do not attach to bubbles in the froth flotation process. These tailings are removed. The process is repeated to maximise recovery.

Crushing and tailings disposal

Crushing produces fine grained waste called tailings or slimes, which not only contain toxic metals, but also harmful chemicals used in the extraction process. Radioactive uranium tailings are a particular hazard. Previously, the main options were storage in large surface ponds or disposal in underground mine workings, rivers or the sea. Unacceptable environmental disasters have been caused by the leakage of tailings. Problems may continue for many years after mines are closed and tailing dams are decommissioned.

Similarly, spoil heaps of waste rock may contain toxic metals and can be unstable. Metal-rich acids form in the same way as for acid mine drainage, but may be stronger because of the fine grained nature of the waste material.

Today, mining in developed countries is tightly regulated.

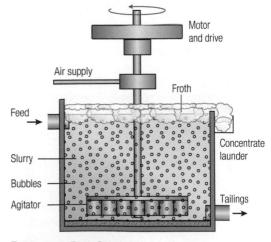

▲ **Figure 1** Froth flotation

Smelting

Smelting involves the extraction of the elemental metal from the ore. Iron is extracted from hematite in a blast furnace. Coal, as carbon, reacts with the oxygen in the hematite and coal, as a fuel (in the form of coke), combines with the oxygen in air. This releases carbon dioxide, which is a greenhouse gas. Sulfide ores produce sulfur dioxide when heated. Smelters produce 8% of global emissions of sulfur dioxide, the main cause of acid rain. Dead zones occur around some smelters where soils are contaminated and vegetation has died leaving a barren, rocky wasteland. One of the most notorious examples is the nickel smelter at Sudbury, Ontario, Canada.

Managing minewater

Mining extracts sulfide minerals such as chalcopyrite, sphalerite and galena, often accompanied by pyrite. Mining operations break up rocks, disturb groundwater, create voids for groundwater to fill and introduce oxygen. This combination of water and oxygen, together with iron-oxidising bacteria, leads to the formation of sulfur dioxide and then sulfuric acid. Toxic metals such as lead, mercury and especially arsenic are leached by the acid from the ores and from acidic salts on exposed surfaces of the chambers. Crucially, they are taken into solution as acid mine drainage (AMD), a major threat to surface and underground water supplies.

AMD forms during mining operations but amounts are kept low by pumping. It becomes a major problem when the mine is abandoned as the pumping stops and the mine fills with water. It is rarely possible to establish who is responsible for paying for the clean-up. AMD is not limited to metal mining, as coal contains a small amount of pyrite and so is an issue for abandoned coal mines.

Source control

AMD may be inhibited by preventing oxygen or water reaching the ores. Dewatering a mine is expensive but an effective method, although the water removed will need treating. Conversely, the abandoned mine could be flooded and sealed with concrete or clay, making use of any aquicludes present.

Migration control

A more common method is to treat AMD when it issues from the mine. The aim is to neutralise the acidity and precipitate the metals as non-toxic salts which can be disposed of correctly.

Active treatment

This involves adding bases such as lime, calcium carbonate, sodium carbonate or sodium hydroxide to neutralise the AMD. It requires constant resupply of chemicals and disposal of the salts, and so is expensive but more reliable and effective when dealing with high rates of flow of mine water.

Oxidising agents accelerate the precipitation of the metal ions as hydroxides and carbonates to form an iron-rich sludge.

Active treatment plants can either be permanently erected, with the mine water pumped directly to the plant, or a portable plant set up next to the affected lake or stream.

Passive treatment

This uses natural and constructed wetland ecosystems. They are expensive to set up and need more land but require only minimal maintenance and no electrical power or hazardous chemicals. They are suitable for low acidity and low flow AMD in remote areas.

The wetlands are usually shallow ponds nearly filled with limestone gravel. Natural oxidation precipitates iron, manganese and other metal salts. If the limestone is covered by organic compost, microorganisms create an anaerobic environment in which non-hazardous metal sulfides precipitate, aided by sulfate-reducing bacteria.

50 cm

▲ **Figure 2** Pool of acid mine water

QUESTIONS

1 Compile a list of possible hazards for an extraction company who is deciding whether to extract metalliferous deposits on the surface or underground. What might be the deciding factor?

2 Research one example of pollution caused by a metal mining or processing operation. Include the chemical reactions that cause the pollution.

Environmental consequences of waste disposal in the ground

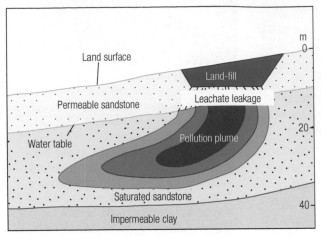

Figure 1 Plume of leachate pollution migrating laterally from a landfill site into permeable rocks below the water table

Leachate

As rainwater percolates through waste, for example in a landfill site, it dissolves soluble chemicals and collects microbial contaminants, producing a liquid called **leachate**. The exact composition of the leachate depends on the waste through which the water has percolated. Surrounding soils and groundwater in underlying aquifers are vulnerable to contamination by leachate, which forms a **pollution plume** spreading out laterally in response to groundwater flow.

Rock type

Fine grained, impermeable rocks such as clays prevent leakage of pollutants into groundwater. Thick, uniform, flat lying beds are best. Porous and permeable rocks such as limestone and sandstone allow flow of leachate. Limestone may be dissolved by acidic leachate leading to the formation of solution cavities. Cementation inhibits leachate flow, but weathering can increase the permeability and the cement has a finite life span. Crystalline igneous and metamorphic rocks may be suitable unless affected by jointing.

Geological structures

- Faults and joints increase the permeability of rocks.
- Tilted or folded beds allow down-dip and lateral movement of leachate.
- Anticlines may have tension joints at their crests, increasing permeability.

Groundwater

If the water table is high then there is less distance for leachate to travel before reaching underlying groundwater.

Nuclear waste disposal

Nuclear waste presents a special problem because it remains radioactive for thousands or in some cases millions of years, depending on the half-life of the isotopes present. The nuclear waste we produce today is a toxic legacy that will require very careful management well into the future.

Nuclear waste is classified as low, intermediate, or high level and **transuranic** waste.

Low level radioactive waste is usually disposed of in secure landfill sites. The high level and transuranic wastes present us with the biggest problems because they emit high levels of radiation, are often thermally hot, and have long half-lives. In the United Kingdom, high level waste is stored for at least 50 years prior to solidification and disposal. In terms of volume, the amount of high level waste is small, just 10 000 m³ a year world-wide.

Safe disposal of high level radioactive waste must meet the following criteria:
- Isolation for at least 250 000 years
- Secure from accidental or deliberate entry
- Safe from natural disasters such as floods, hurricanes and earthquakes
- No chance of leakage into the surrounding environment.

Suggestions for disposing of nuclear waste have included launching it into space, burying it in sea-floor sediments close to subduction zones and placing it in secure containers on the ice sheets of Greenland or Antarctica. Burying it in an underground geological repository is probably the least problematic of the choices.

An underground geological repository for nuclear waste would need to be:

- In a tectonically stable area
- Within dry, impermeable rocks with a low water table
- Free from the effects of potential natural hazards.

Evaporites have been suggested as a suitable rock type (investigated in Germany), as salt is dry and a good conductor of heat. Unfortunately, some hydrated evaporite minerals give out water when heated so pools of saline water could form. These would corrode storage containers and allow leakage of the radioactive waste.

The best option is storage in dry, competent rocks such as granite (used in Japan, Russia and France) or volcanic rocks (investigated in the USA). Granite contains naturally high levels of radioactive elements making it less attractive. Burial deep in crystalline basement rocks below younger sedimentary cover rocks is a good option geologically and has been suggested for disposal of nuclear waste from Sellafield in Cumbria but has proved unpopular amongst local people.

Storing carbon dioxide

Millions of tonnes of carbon dioxide are produced annually by burning and processing fossil fuels and in the manufacture of steel, cement and chemicals. The gas is released into the atmosphere where it acts as a greenhouse gas and may lead to ocean acidification. Carbon Capture and Storage (CCS), or **carbon sequestration**, aims to meet internationally agreed targets for the reduction of carbon dioxide in the atmosphere by permanently storing the gas underground.

Carbon dioxide is separated from the flue gases produced by combustion of coal or natural gas by chemical processes. Capturing and compressing the gas uses electricity produced by the power station. It is then usually transported by pipeline, the cheapest way; alternatively, road or ship tankers are used.

The aim is to store the carbon dioxide permanently underground with no leaks to the atmosphere.

Suggested sites include depleted oil and gas fields, saline formations, uneconomic coal seams and basalt lavas. Deep ocean storage is not currently permitted due to the risk of ocean acidification.

A ready-made, tried and tested storage facility is the porous reservoir rock of a depleted oil or gas trap. Enhanced oil recovery techniques are already used to pump gas under pressure into reservoir rocks to increase the flow from these rocks.

In saline formations, the carbon dioxide dissolves into the surrounding salty groundwater, making it denser so that it sinks to the bottom of the injection zone. Eventually the carbon dioxide may combine (by calcification) with the surrounding rock to form new stable minerals. Therefore the longer carbon dioxide is stored, the more secure it is.

Reacting carbon dioxide with metallic ions (carbonation) in rocks to form new carbonate minerals is a possible method. Rocks which contain magnesium-rich minerals such as olivine can form a new magnesium carbonate mineral (ankerite). The resultant carbonates are stable, but the reactions are slow under the expected temperatures and pressures.

In the UK, carbon dioxide will predominantly be stored in depleted North Sea oil and gas fields and underground saline formations. There is sufficient offshore capacity to store more than 100 years of emissions. Onshore storage is not currently being considered.

KEY DEFINITIONS

Leachate the liquid that drains or leaches from a landfill site.

Pollution plumes are zones of contamination, usually into an aquifer or water source. This can be from various sources, including leachate from landfill sites.

Transuranic elements have atomic numbers greater than 92, the atomic number of uranium. These are unstable and decay radioactively into daughter elements.

Carbon sequestration a process where CO_2 is removed from the atmosphere and held in an alternative form, either solid or liquid.

**CASE STUDY:
DISPOSAL OF NUCLEAR WASTE IN THE UK**

In 2014, the UK government published a White Paper setting out the approach to making geological disposal happen. In the UK, the *Radioactive Waste Management* (RWM) works with the Nuclear Decommissioning Authority (NDA) to develop a geological disposal facility (GDF) which will permanently dispose of the UK's most radioactive waste in sealed chambers deep underground, generally agreed to be the best method. RWM works with waste producers to ensure that radioactive waste is safely and effectively managed.

Safe disposal of low-level waste has been on-going since 1959. High-level waste from nuclear fuel reprocessing is vitrified and currently stored at Sellafield. In 2016, RWM initiated a screening exercise to inform the public about the geology of the UK. It was also set up to examine how the RWM might work with communities to identify a suitable GDF site and how to assess planning permission. This will help communities decide in due course if they would be interested in hosting a GDF in their area.

QUESTIONS

1 Describe the key requirements for a geological repository for nuclear waste.

2 Research the details of the disposal of nuclear waste on World Nuclear Association website; http://www.world-nuclear.org, looking for radioactive waste management in the alphabetical index.

Geological factors affecting the construction of tunnels

Tunnelling through hard rock, soft rock or unconsolidated material all require different approaches.

Crystalline igneous and metamorphic 'hard rocks'

These are very strong so tunnels can sometimes be left unsupported. Tunnelling is usually by drilling and blasting, which is slow and expensive. The amount of explosives used is carefully calculated, otherwise **overbreak** or **underbreak** can occur. At depth in hard rock tunnels, the high confining pressure can cause dangerous rock bursts.

Soft rocks

These are cheap and relatively easy to tunnel through but require lining with concrete or steel ribs. A specially designed tunnel boring machine (TBM) can achieve tunnelling rates of 30 metres per day in soft rocks. Sandstone, limestone and chalk are all fairly strong and make ideal tunnelling materials.

Weak rock

Clay and shale, and unconsolidated materials such as sands and gravels, are prone to collapse and leakage so require support and dewatering techniques.

Lateral variation and changes in rocks types

This makes tunnelling difficult, as the rocks may have different strengths. Weathering weakens rocks, and variations in compaction and cementation also cause problems. Porous and permeable rocks allow water seepage into tunnels and the possibility of flooding.

Attitude of the strata

Flat-lying, competent, uniform strata are best for tunnelling. In dipping beds, different rock types may be encountered along the length of the tunnel. Slippage along dipping bedding planes may lead to rock falls into the tunnel and problems with water.

Geological structures

Geological structures create some of the biggest challenges:

- Faults are zones of weakness which may have breccia and fault gouge clay along them. They are zones of permeability that may allow the tunnel to flood. There may be different rock types on either side of the fault and, in the event of an earthquake, movement may cause the tunnel to collapse.
- Joints are also zones of weakness and permeability. They are often more closely spaced than faults. Loose blocks of rock between joints may fall out of the tunnel roof.
- Bedding planes in sedimentary rocks and foliation in metamorphic rocks are also planes of weakness and may allow slippage or leakage of water.
- Folded rock sequences have changing angles of dip, and slippage may occur on the fold limbs. However, if the fold is a gentle syncline, the tunnel can follow the dip of the fold and stay in one bed (Channel Tunnel).

KEY DEFINITIONS

Overbreak when too much rock is extracted.
Underbreak when too little rock is extracted.

🔺 **Figure 1** One of many tunnels in Japan showing slope stabilisation methods.

CASE STUDY: NEW AUSTRIAN TUNNELING METHOD (NATM)

NATM, developed in the 1960s for Alpine tunnels, incorporates the strength of the surrounding rock or soil formation into the overall ring-like supporting structure.

The optimal cross-section is computed so that only a thin layer of shotcrete need be applied, immediately behind the tunnel boring machine (TBM), to create a load-bearing ring. The tunnel is strengthened by rock bolts, wire mesh and steel ribs.

Monitoring uses instrumentation embedded in the lining, ground and boreholes. If unexpected fractures or groundwater are encountered, the tunnel design is modified appropriately.

Groundwater

If the tunnel is below the water table, then flooding may occur:

- Water may be free-flowing through unconsolidated sediments.
- Water flows may occur along joints in limestone.
- Sandstones can develop high pore fluid pressures.
- Saturation of clays can lead to mobilisation, movement and failure by slumping.

Spoil

Millions of tonnes of material are excavated from tunnels. This has sometimes been used to build embankments elsewhere on the route. Crossrail shipped spoil to the Essex coast to make a new salt marsh nature reserve (Wallasey Island). Chalk marl from the Channel Tunnel was landscaped at the foot of Shakespeare Cliff near Dover (Samphire Hoe).

Ground improvement methods to prevent collapse and flooding of tunnels

Strategies to prevent tunnel collapse include:

- Lining with concrete segments or steel ribs.
- Using rock bolts to secure loose blocks.

Strategies to prevent tunnel flooding include:

- Grouting the surrounding rocks.
- Using rock drains.

CASE STUDY: CROSSRAIL – THE ELIZABETH LINE

The Crossrail railway, due to begin operating in 2018, runs from Reading to Shenfield and Abbey Wood, east of London. A 21 km stretch of the route is in twin tunnels beneath London.

Before tunnelling started in 2012, more than 1000 boreholes – mostly about 50 m deep – were drilled, looking at rock types, groundwater and faulting. Records from 650 earlier boreholes added to the data.

Tunnel boring machines (TBM) created smooth-walled tunnels with a circular cross-section, mainly through Tertiary clays and sands. Porous sands of the Lambeth Group were deposited in sinuous channels crossing tidal mudflats between mangrove swamps and were therefore difficult to predict and locate when dealing with groundwater. In places, the lowest Tertiary and even the Cretaceous Chalk are brought nearer the surface by anticlines. Tunnels in the clays were given a sprayed concrete lining. In sands and chalk, 200 000 bolted concrete segment linings were used.

Nodules, irregular lumps of rock hardened by chemicals, played havoc with the teeth and cutters of machines designed for soft rock, and slowed down progress. They often occurred in particular horizons and could therefore be predicted, but not always. Pyrite reacted with oxygen and water to produce acidic groundwater which could attack cast iron and concrete structures in tunnelling environments.

Faults presented a problem because of sudden changes between clay and sand and because they provided a pathway for groundwater. Faults usually occurred in zones, breaking up the ground, resulting in increased secondary permeability and decreased stability. Differential settlement can occur across fault zones.

CASE STUDY: CHANNEL TUNNEL

At 50.45 km long, the Channel Tunnel is the third longest rail tunnel in the world. It consists of three tunnels; two for trains, with a smaller service tunnel in between that acts as an emergency escape route. The first attempt was in 1881, when engineers started digging trial tunnels from either side of the Channel, but progress came to a halt in 1883 due to British fears of an invasion from France.

It was not until 100 years later that geologists started to investigate the underlying geology, using geophysical surveys and boreholes. The structure was found to be a gentle syncline cut by a few faults. 85% of the tunnel was dug by following the Chalk Marl, chalk with mudstone mixed in, ideal for tunnelling; soft, but strong and with low permeability.

It took nearly 8 years for tunnel boring machines from France and England to cut their way through the Chalk Marl and meet in the middle. The total cost was over £10 billion; an 80% overspend!

HOW SCIENCE WORKS: GROUND INVESTIGATIONS

Currently, every major engineering project has to recreate a significant proportion of geodata, adding to the cost. It is only when all geo-information is collected together that a realistic model can be generated.

Geographical Information Systems (GIS) and 3D modelling technology can handle large datasets on just a desktop PC, leading to a revolution in the way geo-environmental information can be viewed, manipulated and interpreted.

QUESTIONS

1 Describe the likely problems that would be encountered during tunnelling through (a) granite, (b) limestone, and (c) sand and gravel.

2 Describe how the leakage of water into a tunnel could be stopped.

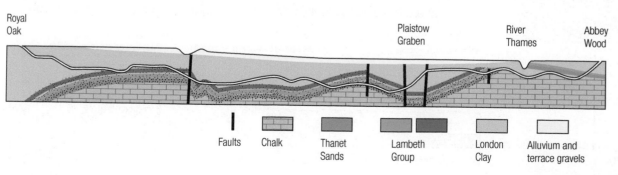

Royal Oak   Plaistow Graben   River Thames   Abbey Wood

Faults   Chalk   Thanet Sands   Lambeth Group   London Clay   Alluvium and terrace gravels

Figure 2 Simplified cross-section showing the geology of the Crossrail tunnel (not to scale)

Geological factors affecting the construction of dams and reservoirs

CASE STUDY: THE ST FRANCIS DAM COLLAPSE, SAN FRANCISQUITO, CALIFORNIA, 1928

The St Francis dam is one of many built on faults, which later failed. The underlying geology, sandstone/conglomerate separated from mica schist by a 2 m wide fault zone, was ignored.

During construction, several cracks appeared in the dam, some of which leaked as the reservoir filled. On March 12th, less than twelve hours after the project engineer declared the dam to be safe, it failed catastrophically. A wall of water, carrying huge pieces of the broken dam, rushed downstream destroying everything in its path; more than 600 people died.

General conditions

The area should be geologically stable, preferably away from earthquakes, faulting and volcanic eruptions.

Valley sides should be stable so that mass movement is unlikely. If the dip of the beds on the valley sides is towards the **reservoir**, then landslips may occur, especially if competent permeable rocks and incompetent impermeable rocks are interbedded.

The river catchment needs to have sufficient rainfall and be underlain by impermeable rocks to promote surface runoff. If the reservoir is for drinking water, there should be no exposed mineral veins containing toxic elements such as lead or arsenic that could get into the water supply. Rivers should have a low sediment load so the reservoir does not silt up too quickly.

Underlying rock types

Strong, competent rocks with a high load-bearing strength are needed to support the combined mass of the **dam** and water, such as crystalline igneous or metamorphic rock or a well-cemented sedimentary rock. Clay, mudstone and shale are weak and incompetent with a low load-bearing strength and will not support the dam.

Impermeable rocks prevent leakage of water from the reservoir. Competent rocks with joints cause leakage problems. Although strong, limestone is usually permeable and may contain caves and solution cavities, so is best avoided.

The rock type should be uniform. If sited across two rock types, differential subsidence may occur making the dam unstable. The depth of weathering, which weakens the rock and increases its permeability, is important. Old underground mine workings could cause collapse or leakage.

Attitude of the strata

Beds that are horizontal or dip upstream provide stable foundations. Beds dipping downstream are unstable. There may be leakage, and slippage along the bedding planes may lead to collapse of the dam.

Geological structures

Faults and joints are zones of permeability and weakness. Reactivation of old faults can lead to an increase in seismic activity. Joints are often more closely spaced than faults.

Anticlines make unstable foundations because slippage can occur along bedding planes on the fold limbs. Tension joints on the crest of the anticline may allow leakage of water from the reservoir.

Synclines are more stable, but water can pass under the dam, through permeable beds.

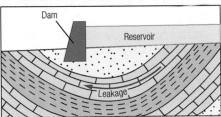

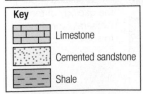

Key
Limestone
Cemented sandstone
Shale

▲ **Figure 1** How folding and dip direction affect dam and reservoir stability.

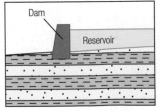

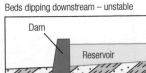

Types of dams

The type of dam chosen depends on the profile of the valley and the availability locally of construction materials, which are bulky and costly to transport.

Masonry/concrete dams

Masonry/concrete dams need a supply of cement and aggregate to make the concrete.

- An *arch dam* is curved upstream so that the hydrostatic pressure from the water compresses the structure against the sides of the valley, and strengthens it. The dam is therefore thinner than other types, using less construction material. Arch dams are suitable for narrow gorges with steep sides of strong rock, in remote areas.
- A *gravity dam* is held in place by the force of gravity due to the immense mass of the concrete or masonry. When built on an impermeable, high load-bearing foundation, this is often the best type of dam. Sometimes they are hollow, making them more economical to construct and may be supported on the downstream side by a series of inflexible buttresses.
- An *arch-gravity* dam combines the strength of the arch with the force of the gravity and does not have to be so massive, useful in areas with a high flow of water but limited material for dam building.

Embankment/earth dams

Embankment dams have an impermeable clay or concrete core held in place by piles of rock or earth, sand and clay, with an impervious covering. The material binds itself together by friction between the particles so that cement is unnecessary. These dams are built in broad, shallow valleys where the mass of the dam is spread over a wide area, so the foundations do not need to be as strong, but they do require very large quantities of fill material.

Ground improvement methods to prevent leakage from reservoirs

Most reservoirs leak to some extent. Ground improvement strategies reduce leakage.

Grouting

Holes are drilled into the rock and liquid cement pumped in. The cement fills pore spaces, joints and fissures, reducing permeability and increasing rock strength.

Clay or plastic lining/geomembrane

Prior to filling, the reservoir is lined with an impermeable material such as clay or plastic to prevent leakage of water into the underlying rock. Clay is a good choice because if there is a local supply it will be cheap.

Cut-off curtain

An impermeable barrier, usually made of concrete, is constructed as an extension below the dam, preventing leakage especially from dams situated on synclines. It also strengthens the foundations and prevents slippage of beds dipping downstream.

Hydroelectric power This is releasing water, stored behind a dam wall, which turns a turbine to generate electricity.

▲ **Figure 2** A concrete arch dam showing the curved wall and hydroelectric plant

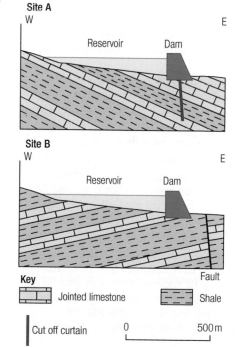

Key

⬛ Jointed limestone   ⬛ Shale

| Cut off curtain   0   500 m

▲ **Figure 3** Two possible sites for dam and reservoir construction

---

QUESTIONS

1 Describe the geological factors that are important in the siting of dams and reservoirs for water supply. Refer to the examples in Figure 3.
2 Compare and contrast the ideal geological setting for dams and for tunnels.

9.15 The effects of dams and reservoirs

CASE STUDY: EROSION OF EARTH DAMS

The most common type of erosion in an earth dam is called *internal erosion*. This is where water seeps through the dam and carries particles of soil away from the embankment and foundations. There is discharge of the particles downstream. Cavities or pipes form inside the dam itself, and these widen and enlarge back towards the reservoir. Once it reaches the reservoir, there is catastrophic failure. Signs of imminent danger include water discharging through the dam on the downstream side of the dam wall and subsidence anywhere on the embankment.

CASE STUDY: EXAMPLES OF POSSIBLE RESERVOIR-INDUCED SEISMICITY

The Sichuan earthquake (M 7.9, 2008) involved movement along a 300 km section of a fault and damaged more than 2000 dams. One example was the 156 m high Zipingpu Dam which was filled in 2004 with 300 million tonnes of water, causing much more stress than would result from plate movement. A week before the earthquake, the water level in the reservoir had fallen rapidly, suddenly releasing stress.

In Koyna, India, 35 000 tremors were recorded in the reservoir area after it was filled in 1962. The main earthquake (M 6.3, 1967) occurred after the reservoir filled rapidly due to monsoon rainfall. The dam and powerhouse were seriously damaged.

The Kariba Dam, on the Zimbabwe-Zambia border, is in the seismically active southern Rift Valley. As the reservoir filled with 180 billion tonnes of water, there were 20 earthquakes of M>5, the largest being M 6.2 (1963).

The Oroville Dam, California, the tallest earth dam in the US, was built on a fault previously thought to be inactive. Twenty years after filling, the area experienced a series of earthquakes (the biggest was M 5.7, 1975) after the reservoir had been drawn down to its lowest level since filling.

The Hoover Dam, on the Arizona-Nevada border, was built in 1963 on the Colorado River. More than 600 earthquakes were recorded in the 10 years following completion of the concrete arch dam. Most were minor tremors, but one was M5, and two others were M>4, attributed to an increase in load due to the water in Lake Mead, the reservoir behind the dam. This water also lubricated and reactivated old underlying faults.

Environmental and social consequences of dam and reservoir construction

Flooding of land for the reservoir may result in the loss of agricultural land or villages requiring people to relocate. Valuable archaeological sites may be lost or need to be relocated and reconstructed, where possible. Where forests are flooded, the decaying vegetation releases large quantities of carbon dioxide and methane, both potent greenhouse gases.

Damage to aquatic ecosystems is caused downstream of dams by changes in water depth, temperature and dissolved oxygen content. Dams may prevent fish such as salmon swimming upstream to their breeding grounds. Further afield, for example, the Mediterranean sardine fisheries collapsed due to lack of nutrients being carried down the River Nile after the Aswan Dam was built.

Over time, reservoirs gradually silt up with sediment carried from upstream. Water released from dam spillways therefore contains very little sediment which may lead to *clear water erosion* downstream.

Water engineers calculate how much water to release to ensure reservoirs do not overflow. More than average melting of snow in the spring or prolonged heavy rainfall may mean there is not enough time to draw down the reservoir before it overflows, causing flooding.

Figure 1 Reconstructed temples, Abu Simbel, Lake Nasser behind the Aswan Dam, Egypt

Dams, reservoirs and seismic activity

The occurrence of earthquakes in the area of newly constructed 'large dams' has led to a growing concern that this is *reservoir-induced seismicity* (RIS). While it is thought that, worldwide, some 100 earthquakes have been triggered by dam building, the link is difficult to prove. The M 7.9 Sichuan, China, earthquake (May 2008), which killed 80 000 people, highlighted the possible link to the construction of the Zipingpu Dam.

It is known that:

- Some reservoirs are constructed (and more are planned) in areas with known seismic activity or where there are known fault lines.
- Tremors occur near to some reservoirs while, or soon after, they are filled with water. Sometimes this is in areas with no known history of seismic activity, for example the Latur, India, earthquake (M 6.1, 1993) which killed 10 000 people.
- The dam structure places an immediate load on the rocks beneath.
- When filling a reservoir with billions of tonnes of water, extra loading is immediately put on the rocks, creating stress; the depth of water leads to an immediate increase in hydrostatic pressure.

- Over a period of time, the high hydrostatic pressure causes water to infiltrate into pores and fissures in the underlying rock, increasing pore pressure and providing lubrication for any fractures which would otherwise be prevented from slipping by friction.

Demonstrating the link is difficult because:

- Little research has so far been funded for what is a relatively new phenomenon.
- Historical seismic records may be inadequate for monitoring the site.
- Each dam site has different geological characteristics.
- Some earthquakes occur months or years after the reservoir was filled.

It is unlikely that reservoirs can by themselves generate 'new' earthquakes but they could bring forward the activity by hundreds or even thousands of years.

A number of recommendations have been put forward:

- More research and monitoring should be carried out into RIS in view of the potentially disastrous consequences, including the modelling of changes in pore pressure and rock mechanics underground.
- No further high dams should be constructed until this research has been carried out.
- When deep (100+ m) reservoirs are planned, the design should take into account the increased seismic risk due to RIS.
- Dam builders should monitor seismicity during the filling of the reservoir, ready to change or reverse the operation if RIS starts.
- The possibility that RIS may occur in areas previously thought to have no seismic activity should be considered. Such areas lack any sort of earthquake risk assessment. While dams may be built to withstand RIS, the surrounding cities and villages often are not.

CASE STUDY: THE VAJONT DAM DISASTER, ITALY, 1963

The deep, narrow Vajont River valley in northern Italy seemed like an excellent site for a dam and reservoir. Unfortunately, the engineers ignored the geological warning signs:
- The interbedded limestone and clay that dipped towards the reservoir.
- The scar of an ancient landslide above the reservoir.
- A small landslide that occurred in 1960 while the reservoir was filling.

The concrete dam was completed in 1961 and was the world's second highest dam (262 m). In October 1963, there was a period of prolonged and heavy rainfall. Rainwater percolated through the permeable limestone and collected on top of a layer of impermeable clay. A slide plane developed between the limestone and clay and a huge mass of 260 million m³ of limestone slid down the southern valley side into the reservoir. The dam withstood the force of the landslide but rock debris filled the reservoir, displacing the water over the top of the dam as a huge wave. The ensuing flood killed more than 2000 people. The Vajont dam still stands which is a testament to good engineering and construction, but poor choice of site!

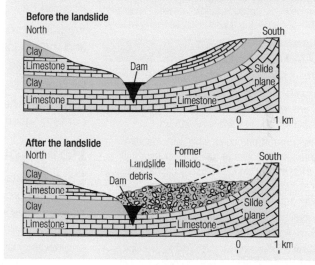

▲ **Figure 3** Cross-sections before and after the Vajont dam disaster

HOW SCIENCE WORKS: THE MECHANICS OF DAMS

As the reservoir fills with water, the hydrostatic pressure acting on the dam increases. The force of the water has two effects on the dam:
- It acts to push the dam downstream.
- It acts to tip the dam over about a fulcrum at the toe of the dam.

The hydrostatic pressure forces the water into the pores, joints and other fissures of the rocks underlying the reservoir and into the rocks below the dam itself. It also seeps into the junction between the dam and the bedrock. The water percolating through to the downstream side of the dam sets up a hydraulic gradient. The upward force which results from this movement of water is called *hydrostatic uplift*.

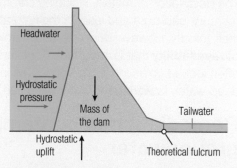

▲ **Figure 2** Forces acting on a gravity dam

In terms of levers, the moment of the hydrostatic uplift acts in the same direction as the moment of the tipping effect. Both act against the downwards mass of the dam. The effective mass of the dam is therefore reduced, making it unstable.

Possible solutions include:
- Keying the dam into the bedrock and sides of the valley to prevent slipping.
- Being aware of the porosity and permeability of the bedrock beneath the dam, including joints and fissures.
- Incorporating drainage holes into the lower part of the dam to collect the percolating water and channelling it to the tailwater; recording the flow helps to monitor the situation.

CASE STUDY: ULLEY, ROTHERHAM 2007

At the end of June 2007, cracks appeared in the Ulley dam, near Rotherham, leading to fears it might collapse. 700 people were evacuated from local villages and the M1 was closed. The emergency services pumped millions of gallons of water from the reservoir to ease the pressure on the damaged dam. Fortunately, the dam held and disaster was averted.

QUESTIONS

1 Describe some of the environmental and social consequences of dam and reservoir construction.

2 Explain how dam and reservoir construction can lead to an increase in seismic activity.

3 Choose one dam failure from around the world and research its causes and consequences.

Figure 1 Organic matter in domestic waste breaks down to produce acidic waters. These are capable of mobilising toxins such as heavy metals.

KEY DEFINITIONS

Inert substances are chemically inactive.

Bioavailability the proportion of total metals that are available for incorporation into biota (bioaccumulation). It will vary with pH.

Bioaccumulation occurs when an organism absorbs a substance at a rate faster than that at which the substance is lost by catabolism and excretion. The substance becomes concentrated in the organism.

Ion exchange an exchange of ions between two electrolytes or between an electrolyte solution and a complex. It is used to describe the process of purification or decontamination of ion-containing solutions.

Adsorption the process by which particles, such as heavy metal ions, become attached, semi-permanently, to the surface of a clay mineral.

Isomorphous substitution the replacement of one atom by another of similar size in a crystal lattice during crystal growth without changing the crystal structure. It takes place only between ions differing by less than about 10% to 15% in ionic radii.

HOW SCIENCE WORKS: HYDROCARBON AND SOLVENT CONTAMINATED SOIL

Hydrocarbon and solvent contamination can be found in brownfield sites or where industry is still operating. Contamination can be from many places including: filling stations, automotive, haulage yards, scrap metal industry, aerospace industry, waste processing, paint industry, metalworking, waste disposal sites and anyone who uses solvents. Some pollutants will be held in pore spaces in the soil, whilst others dissolve in water to pollute groundwater as *plumes*. Volatile components can remain as gases in the soil itself.

The UK has a high population density and a long history of heavy industry. The 75% of land area used for agriculture is valued as it stands, so there is an increasing pressure to re-use brownfield sites for purposes such as new housing. The treatment of derelict land to ensure that it is made safe requires an understanding of the interactions between geology and pollutants, in addition to a detailed knowledge of chemistry.

Sources of pollution

The extraction of geological resources is responsible for the greatest mass of land pollutants; fortunately, most of this is **inert** waste. There are problems with coal and metal-ore waste which contain sulfides. Once exposed to oxygen and water, these create acid waters with metals in solution. Oil and gas extraction and refining have resulted in contamination by organic pollutants; the current development of fracking has raised environmental concerns.

Agriculture contributes organic waste and hazardous chemicals (e.g. pesticides), and the disposal of domestic and commercial waste requires detailed consideration of the underlying geology.

Brownfield sites describe land that has been used for commercial or industrial purposes. These are often well known to contain pollutants, often due to soil contamination by hazardous waste.

Pollution problems

The pollutants must somehow be prevented from harming people, livestock and wildlife. To prevent the pollutant from being ingested or breathed in, it can be isolated. This is easier to achieve if the pollutant cannot be moved by water (not taken into solution or carried as an emulsion or suspended particles). Geological and engineering knowledge is required to predict how the water may move through the subsurface rocks. Disturbing the land during excavations will inevitably change the nature of the problem by increasing permeability and thus oxidation and surface area.

The destination of metals, including chromium, nickel, copper, manganese, mercury, cadmium and lead, in the natural environment, is of great concern due to their toxicity. However, a simple geochemical analysis may be misleading: it is the **bioavailability** that is important. For example, in reducing conditions metals are often locked up as insoluble sulfides, as in sediments below the water table or in deep water. Lowering the water table can lead to oxidation and mobilisation.

Remediation of contaminated land and groundwater

Superficial deposits can be treated in situ or dug up and treated elsewhere. The treatments are as varied as the pollutants, generally being based on chemical treatments, designed using laboratory samples. The in situ treatments use devices such as soil augers to distribute treatment into the soil profile.

Phytoremediation

This uses the ability of plants to deal with substances that do not benefit them. Heavy metals cannot damage plants, but are stored in the leaves as they are drawn through in solution. This beneficial use of **bioaccumulation** means that plants are effectively removing the metals from the soil. Removing and disposing of the leaves therefore removes the metal from the soil.

Stabilisation and solidification (S/S)

This aims to immobilise the toxic constituents of hazardous waste to prevent them leaching from the disposal site. Immobilisation is accomplished by reducing the solubility of the waste components and by physically isolating the waste. S/S treatments involve mixing a variety of binding reagents into the contaminated ground or waste. These commonly include Portland cement, lime, limestone, fly ash, slag, gypsum and phosphate mixtures. This has been a contentious subject in the UK. In the USA, S/S has been identified as the Best Demonstrated Available Technology (BDAT) for treating a wide range of hazardous wastes (including radioactive waste). It is increasingly used for treating contaminated soil at sites being redeveloped.

Groundwater treatments

Groundwater can be pumped up and treated. Clays and their minerals have routinely been cheaply and successfully used as an adsorbent for removing toxic heavy metals from aqueous solutions. Organic contaminants such as Perfluorinated Compound (PFCs) have recently been found to respond to **ion exchange** treatments. The resins can be regenerated for reuse, unlike the previous treatments, using activated carbon. In situ treatments involve chemicals introduced into the subsurface at the periphery of polluted sites, designed to intercept and immobilise water-borne contaminants.

Treatment for metals can be as simple as adding caustic soda to increase the pH and precipitate metal hydroxides. There is often a minimum solubility at a particular pH, so it is not just about raising the alkalinity indiscriminately. The hydroxides are removed by adding coagulants such as aluminium salts or polymers to cause flocculation.

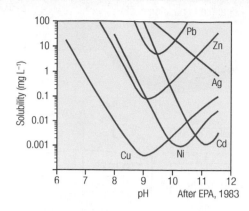

▲ **Figure 2** Solubility of metals changing with pH used in treatment of groundwater. If cadmium were the problem, adjusting the pH to approximately 11.3 would cause the maximum precipitation

HOW SCIENCE WORKS: CLAY MINERALS AND THE ADSORPTION OF HEAVY METALS

Geological theories and ideas have developed to provide an explanation of how clay minerals **adsorb** metal ions. It has long been known that clays are good at removing toxins; Native Americans call bentonite 'the mud that heals'. The reason for this was only understood with the development of X-ray diffraction. We now appreciate the sheet-silicate structure of the clay minerals and the subtle differences between the groups of this complex and useful class of minerals.

Clay minerals make up a large proportion of soils and because they are fine grained and platy, they have large reactive surface areas, making them the primary adsorbent. Clays exchange ions at their surface against an equivalent amount of ions of equal charge in solution (cation or anion exchange). These permanent charges mostly result from **isomorphous substitution**, e.g. Al^{3+} for Si^{4+}. In some clays, e.g. kaolinite, the main source of charge is pH-dependent charges, which can be positive or negative. Some are due to the ionisation of hydrogen in the exposed hydroxyl groups at the broken crystal edges. The net result of all these clay surface charges is the ability to adsorb ions.

KEY FACT

Biomagnification refers to the increases in concentration of substances, often toxins, in the tissues of organisms as they pass up the food chain.

CASE STUDY: THE FALLOUT FROM CHERNOBYL

The accident at Chernobyl in 1986 caused severe pollution of the surrounding area resulting in appalling illness and loss of life in Ukraine, estimated to be as many as 60 000 deaths in a 2006 report. The explosion and fire allowed radioactive particles to enter the atmosphere where winds spread them around the northern hemisphere. In the UK rainfall over upland areas washed out a proportion of the radionuclides and contaminated the soils. It took several weeks before a ban on vegetables, milk and meat grown in the Lakes, North Wales and south-west Scotland was enacted. Over a million sheep were affected by Ministry of Agriculture prohibitions. The final restrictions on sheep movement in England and Wales were lifted in 2012, 26 years after the disaster.

Dangerous materials such as caesium-134 (half-life of 2 years) and caesium-137 (half-life of 30 years) have the greatest long-term radiological impacts. Some of the caesium-137 washed out by the rainfall was found in river sediment (up to 1450 becquerels per gram) but most was taken up by the soil and adsorbed onto clay minerals. It is so effectively bound into the upper soil clays that it is not found at deeper levels. Caesium-137 can accumulate in fungi (including edible species), lichens, and mosses, as a result of their shallow root systems and the way they take up potassium. Supplying potassium reduces the uptake of caesium into plants.

▲ **Figure 3** Warning signs. The European Commission imposed a maximum limit of 1000 becquerels per kilogram of radiocaesium in sheep meat affected by the accident to protect consumers. Levels up to four times that were seen in Germany.

QUESTION

1 Acid mine drainage contains significant levels of metals in solution. Suggest an economical method of ensuring that these fluids do not get into the environment.

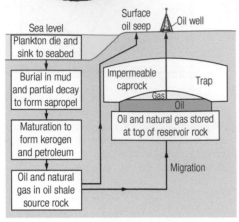

Figure 1 Geological requirements for the formation and accumulation of oil and natural gas

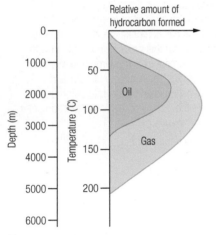

Figure 2 Oil and gas windows

Over the last 200 years, we have steadily increased the amount of the **fossil fuels** (coal, oil and natural gas) we take from the Earth. Our hydrocarbon-based society is dominated by the burning and use of **petroleum** and its products: fuels, plastics, paints, synthetic fibres, synthetic rubber, pharmaceuticals and fertilisers.

Formation of petroleum

HOW SCIENCE WORKS: BASIN ANALYSIS

Petroleum geologists use the theory of basin formation, including aspects of structural style, subsidence and sedimentation, to look for certain geological conditions which may lead to the development of petroleum. These include:

- A possible **source rock** which is fine grained, organic rich, deposited in a low energy anoxic marine environment.
- A possible **reservoir rock** which is porous and permeable, deposited in a high-energy environment or fractured after lithification.
- A possible **caprock** which is impermeable. This includes evaporites.
- The ages of these rocks relative to each other.
- Location of stratigraphic or structural traps.
- Identification of the stage reached in the maturation process using burial history curves.
- Tracing migration routes that the **hydrocarbons** might follow through permeable beds and along faults.

Onshore, this information is gathered by field work, mapping and by satellite imagery. Exploration boreholes build up a one-dimensional picture of the underlying geology. This is converted to a three-dimensional pattern by adding information from reflection seismology, electromagnetic and gravity surveys. Stratigraphic and structural features are extrapolated to offshore areas, guided by the results of remote surveys over the sea. Basin analysis is usually carried out before exploratory drilling for hydrocarbons, but valuable information is added after drilling has started and predictions can be checked and modified.

Environment of deposition of source rocks

Oil and natural gas form from microscopic, planktonic, marine algae. These floating organisms die and sink through the water column to settle and accumulate on the sea bed in a low-energy environment. Conditions must be anoxic (water lacking in oxygen), so the plankton will not decay or be scavenged and eaten by other organisms. Anaerobic bacteria cause partial decay of the plankton to an organic mud called a **sapropel,** the organic-rich sedimentary source rock. Examples include black oil shales and mudstones, dark in colour because of the high organic carbon content.

Maturation of petroleum

On burial, the source rock is subjected to compaction and an increase in temperature. Over time the organic matter breaks down to **kerogen**, a mixture of organic compounds of carbon, hydrogen, oxygen, nitrogen and sulfur, and then to petroleum. This process is called **maturation**.

Below 50°C, biogenic gas forms but is usually lost due to its shallow burial. Oil forms between 50 and 200°C, mostly between 50 and 100°C, referred to as the **oil window**. Natural gas, mainly methane, forms between 100 and 200°C, referred to as the **gas window**. Above 200°C, hydrocarbons denature and are destroyed.

Migration from source rock to reservoir rock

Compaction of the fine grained source rock squeezes out water carrying tiny droplets of any oil that have formed. The hydrocarbons then **migrate** into a reservoir rock.

Reservoir rock

A reservoir rock has a high porosity to store significant amounts of oil and natural gas and a high permeability to allow them to migrate into it and then be extracted

KEY FACTS: OIL AND GAS MIGRATION

- Permeable rocks provide a pathway between the source rock and the reservoir rock.
- Oil and natural gas migrate down the pressure gradient from high to low pressure, usually upwards towards the surface but in certain conditions this may be downwards.
- Oil and natural gas are less dense than the water in the pore spaces of rocks so they percolate upwards, above the water, until they either encounter an impermeable layer or reach the surface.
- At higher temperatures, the viscosity of the oil is lower so it flows more easily.

from it. The properties are the same for an aquifer for water. Reservoir rocks include poorly cemented sandstones, jointed and other limestones and fractured chalk.

Caprock

Further upward migration of hydrocarbons is prevented by an overlying impermeable caprock, otherwise they will continue to rise, forming seeps of gas and oil and tar pits on the surface. Caprocks include fine grained clay, mudstone, shale and evaporites.

Loss of oil and natural gas

Oil and natural gas are destroyed if the temperature increases above 200°C due to deep burial, regional metamorphism or heat from an igneous intrusion or volcanic activity. It may also be lost from a trap by erosion and removal of the overlying caprock or by migration along an unsealed fault plane.

CASE STUDY: OIL AND GAS TRAPS

Traps

Traps allow hydrocarbons to be concentrated in one place, making them economic to extract. Migration from the reservoir rock must be prevented by impermeable barriers to the side as well as above as, for example, in an anticline. Gas forms in a layer above the horizontal layer of oil, which in turn floats above the water. This all takes place in the pore space of the reservoir rock.

Anticline traps

Anticline traps hold 80% of the world's known reserves. The larger the fold, the more oil it may store. An open fold is likely to contain far more oil than a tight fold. Once filled to capacity, hydrocarbons leak out laterally, at spill points, and migrate into adjacent rocks.

Fault traps

A fault brings an impermeable rock next to a reservoir rock, preventing the oil and natural gas escaping laterally. Hydrocarbons migrate upwards and are trapped by the caprock. The fault itself must be sealed to prevent the hydrocarbons escaping.

Salt dome traps

Evaporites such as halite and gypsum have a lower density ($2.3\,g\,cm^{-3}$) than the surrounding rocks (2.5–$2.7\,g\,cm^{-3}$) and, much like magma, form diapirs that rise upwards. These uplift and pierce the surrounding rocks to form a salt dome. Evaporites are crystalline and impermeable so form good caprocks. Hydrocarbons do not accumulate in the salt itself but in the dipping reservoir rocks adjacent to the salt dome or in gentle anticline traps above it, if there is a caprock.

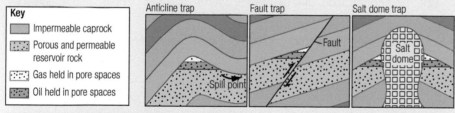

Key
▦ Impermeable caprock
▨ Porous and permeable reservoir rock
▨ Gas held in pore spaces
▨ Oil held in pore spaces

Figure 3 Structural traps for oil and natural gas

Unconformity traps

Hydrocarbons migrating upwards into a reservoir rock below an angular unconformity are trapped by caprocks above the unconformity. There must be impermeable rocks on either side.

Lithological traps

Fossilised limestone reefs make good lithological traps, if surrounded by impermeable rocks. Reefs have very high porosities and form in areas with abundant life for petroleum formation. River channel, point bar or deltaic sandstones often have a lens shape, making small but common traps, surrounded by impermeable river or delta clays.

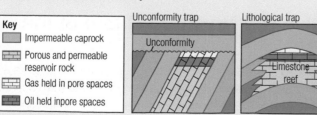

Key
▦ Impermeable caprock
▨ Porous and permeable reservoir rock
▨ Gas held in pore spaces
▨ Oil held in pore spaces

Figure 4 Stratigraphic traps for oil and natural gas

KEY DEFINITIONS

Fossil fuel made from the remains of once living organisms and are coal, oil and natural gas.

Petroleum (*petra* = rock and *oleum* = oil) includes natural gas (gas), crude oil (liquid) and asphalt (solid).

Source rock organic-rich mudstone or shale which contained abundant plankton that formed in low-energy, anoxic, marine conditions.

Reservoir rock highly porous and permeable rock capable of storing and yielding significant amounts of petroleum.

Caprock the impermeable rock above the reservoir rock preventing further upwards migration of petroleum.

Hydrocarbon includes any compound made just of hydrogen and carbon. It is often used as a general term to refer to oil and natural gas. The simplest hydrocarbon is methane (CH_4), also known as natural gas, marsh gas and fire damp.

Sapropel the unconsolidated, dark-coloured, organic-rich deposit which becomes the source rock.

Kerogen a solid bituminous material in source rocks which forms oil when distilled.

Maturation process converts plankton into petroleum by the effects of increasing temperature and pressure during burial.

Oil window the narrow temperature and pressure conditions where oil can form.

Gas window the narrow temperature and pressure conditions where gas can form.

Migration the movement of petroleum from a source rock to a reservoir rock.

Trap geological situations that concentrate petroleum in one place.

QUESTIONS

1 Explain how oil and natural gas form in a source rock.

2 Describe and explain two factors that control the migration of oil and natural gas.

3 Describe the properties of a good reservoir rock for oil.

Most British oil and natural gas fields are in Mesozoic and Cenozoic sedimentary basins. Older rocks are too deeply buried (over 200°C), have a low porosity due to compaction and are faulted allowing hydrocarbons to escape. North Sea oil was discovered in the early 1960s, coming on line in 1971. Extraction was costly and it was only when oil prices rose in the 1980s that it became economic to exploit.

Gas fields also occur in the Irish Sea. Of the small onshore oil and gas fields, Wytch Farm in Dorset is the largest.

The North Sea Basin during the Mesozoic

The area now known as the North Sea underwent rifting during the Mesozoic, with thick sedimentary deposits built up as the area subsided in a basin. The palaeogeography was varied, beginning with continental conditions followed by marine and deltaic environments.

The rift system, established in the Jurassic, was associated with the initial opening of the North Atlantic. The complex structure of linear horsts and grabens is summarised in the generalised section. Subsidence between the converging faults of the graben caused the faulting and folding within the central block which provides the structures needed for the accumulation of hydrocarbons. The major faults do not affect the Cretaceous, showing that the main tectonics had finished by then. Subsidence and deposition occurred on a large scale in the Tertiary, a vital component in the development of oil and gas fields, and continued in the Quaternary while mainland Britain was affected by glaciation.

The North Sea is separated into two sedimentary basins, Northern and Southern, by an east–west trending ridge or 'high'.

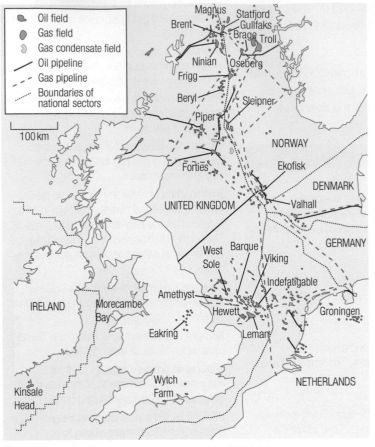

Figure 1 Oil and gas fields in and around the British Isles

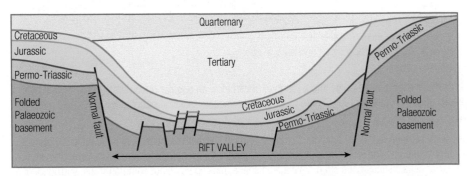

Figure 2 Generalised east-west cross-section across the North Sea Basin (not to scale).

Why is there oil in the Northern Basin of the North Sea?

Synsedimentary faults formed at the same time as the deposition of sediments and controlled the rate and type of sedimentation. In the Northern Basin the main source rock is the Late Jurassic Kimmeridge Clay. Reservoir rocks are marine sandstones and fractured chalk. Caprocks are mainly clays, with a great variety of traps.

MATHS BOX: INTERPRETATION OF A BURIAL HISTORY CURVE

Red burial history curve
Source rock deposited 200 Ma. Between 200 and 15 Ma there was slow accumulation sediment and slow burial. This was followed between 150 and 50 Ma with more rapid sedimentation and shows a steeper curve. At around 100 Ma there was 1700 m of overburden and a temperature of 50°C, and oil started to form. Between 50 and 10 Ma there was formation of oil, followed by gas. At 10 Ma the oil and gas are denaturing.

Blue burial history curve
Source rock deposited 20 Ma. Burial is rapid sedimentation up until present day. No hydrocarbons formed yet as the source rock has not reached the oil window.

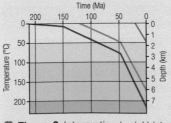

Figure 3 Interpreting burial history

Source rock

Jurassic rifting formed deep marine basins in which fine grained sediment with a high organic content was deposited, which became the *Kimmeridge Clay*. Deep anoxic waters prevented the decay of organic material, which instead changed into varieties of *kerogen* consisting of high molecular mass organic compounds.

Maturation

The kerogen matured by breaking down into compounds of smaller molecular masses. Assuming a geothermal gradient of 30°C per km for all these examples, the oil window is at a depth of 2–4 km and the gas window is at a depth of 3–6 km.

The source rock was buried by younger sediments deposited on top, implying subsidence of the sea floor, aided by rifting. The thicknesses, ages and rock types making up the stratigraphic succession of the overburden are determined from boreholes, remote surveying and correlation with outcrops on land. The events which led to the maturation of oil are recorded in a burial history chart.

Why is there gas in the Southern Basin of the North Sea?

During the Permian, the area of the North Sea was land. As the rifting of the North Atlantic began, synsedimentary faults developed which controlled the development of alluvial fans and fan deltas which derived their sediments from the fault footwalls. Other sandstones were deposited in aeolian dunes (the main reservoir rock), fluvial channels and sabkhas, separated by silty shales. Later, Permian evaporites were formed in enclosed shallow seas. These continental deposits were buried by Jurassic marine sediments.

The Southern North Sea Basin reservoir rocks contain only natural gas. The source rocks are coals and organic shales of the underlying Carboniferous Coal Measures which did not contain the marine plankton needed for oil formation. Gas escaped from the coal as it formed and is mainly held in the overlying Permian dune sandstones. The caprocks are often Permian evaporites and the traps are mainly salt domes and associated anticlines.

CASE STUDY: EXAMPLES OF NORTHERN BASIN OIL FIELDS

The source rock for all these oil fields is the Kimmeridge Clay

Name	Type of trap	Reservoir rock	Caprock
Piper	Folded, tilted fault blocks	Upper Jurassic deltaic sandstone	Lower Cretaceous clays
Forties	Lithological: sandstone lenses	Early Tertiary channel sandstones	Tertiary mudstones
Ekofisk	Anticline in central graben formed by Permian Zechstein salt movement	Early Tertiary and Upper Cretaceous chalk	Tertiary mudstones
Brent	Tilted Jurassic fault block	Middle Jurassic deltaic sandstones	Kimmeridge Clay and Cretaceous clays

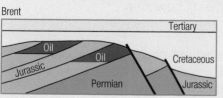
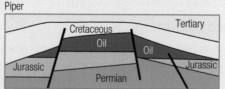

Figure 5 North Sea hydrocarbon traps **a** Brent oilfield and **b** Piper oilfield

HOW SCIENCE WORKS:
BURIAL HISTORY CHART

The burial history chart is a graph which shows the depth of burial by overburden against time since deposition of the source rock.

▼ **Figure 4** Burial history charts

Temperature
There is an increase in temperature as a bed is gradually buried by overburden. The line is not the bed itself, but the temperature increase with depth. Here the graph is a straight line, implying steady, continuous deposition.

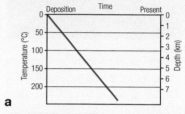

a

Oil and gas windows
The graph indicates that oil formation started when the overlying sediment was almost 2000 m thick.
Rapid subsidence, for example in a graben, takes the bed down to the level of the gas window.

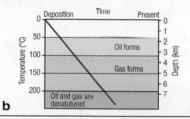

b

Uplift
The bed gets cooler and oil stops forming, reducing the amount which migrates to the trap, although the oil which has already formed remains in the bed. This is indicated by the peak in the middle of the chart.

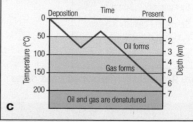

c

KEY DEFINITION

Synsedimentary faults show displacement over an extended time period, usually within sediments during deposition.

QUESTION

1 Describe the pattern of distribution of oil and gas fields in the North Sea Basin.

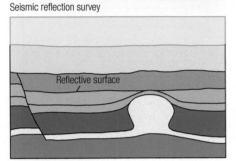

Seismic reflection survey

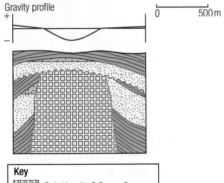

Interpreted seismic profile

🔺 **Figure 1** A seismic reflection survey and the resultant seismic profile

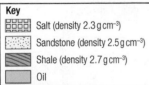

🔺 **Figure 2** Gravity profile across a salt dome

Key
- Salt (density 2.3 g cm⁻³)
- Sandstone (density 2.5 g cm⁻³)
- Shale (density 2.7 g cm⁻³)
- Oil

KEY DEFINITIONS

Travel time the time taken for a pulse to travel to a boundary and back again. To calculate the distance to the boundary, the travel time must first be divided by two.

Primary recovery oil initially gushes to the surface under natural pressure and is then pumped out.

Secondary recovery water is injected below or natural gas injected above the oil to maintain the pressure.

Geophysical exploration techniques

Seismic reflection surveys

On land artificial seismic waves are generated by explosions or by vibrations from dropping a heavy 'thumper' mounted on a truck. At sea, air guns are used.

Seismic waves travel into the Earth and are reflected at layer boundaries within sedimentary sequences.

Reflected waves travel back up to the surface where they are detected by an array of geophones or hydrophones. Their location is accurately pinpointed using GPS. Offshore surveys are particularly efficient as a large number of hydrophones can be towed behind a ship and its path is not restricted, unlike 'vibrotrucks' on land.

The time taken for the reflected waves to arrive back at the receivers, the **travel time**, is used to calculate the depth to the reflective layer. The data are used to plot a seismic profile showing the subsurface layering which shows the subsurface geometry and can be interpreted to identify potential traps.

Gravity surveys

Gravimeters can be mounted in road vehicles, helicopters or planes allowing rapid coverage of large areas. Survey points are located using GPS.

The gravity data is corrected for latitude, altitude and topography to give the Bouguer anomaly which indicates variations in the underlying rock types.

Anomalies are identified from maps showing isogals, lines joining points of equal gravitational field strength.

A positive gravity anomaly results from an excess of mass. This may be due to an anticline or an uplifted fault block which could be potential trap structures.

A negative gravity anomaly results from a deficit of mass. This may be due to a low-density salt dome. The exploration target would be around the edge of the salt dome at the zero milligal line.

Exploration drilling

When potential hydrocarbon traps have been located, exploration drilling is carried out using diamond-studded cylindrical drill bits. The rotating drill bit is cooled and lubricated by drilling mud containing the dense mineral barite to make it dense enough to reach the bottom of the hole and to flush up the millimetre-sized rock chips. Sometimes a continuous rock core is recovered from the borehole.

Mud logging

Rock chips are sieved from the drilling mud, washed and examined under a microscope by 'mud loggers' who identify the rock types and microfossils present at different depths down the hole. They build up a picture of the changing rock types and correlate the geology between different boreholes.

Core sampling

Using a hollow drilling bit, a length of rock is cut and brought to the surface. Changing the drilling bit takes time so this is only done at critical depths. It is also more expensive than using a drill bit and examining chippings.

Down-hole logging

When the drill string has been removed from the well, a *sonde* (= sounding) is lowered into the well. The sonde is a capsule containing sophisticated instruments which record data from known depths as it is slowly pulled up from the bottom of the hole.

- The higher the porosity the higher the possible oil and natural gas content of a reservoir rock.
- Fluids can be identified – oil, gas or brine.
- Gamma ray spectroscopy counts the gamma rays emitted due to natural radioactive decay.
- Potential source rocks such as black oil shales and mudstones give a high gamma ray count whereas sandstones and limestones produce a low count.
- Resistivity is a measure of electrical resistance. Water, a conductor, gives a low resistance, as does coal. Covalent hydrocarbons show a very high resistance.
- Density is measured by emitting gamma rays directed at the rock and recording the reflection. Coal gives a low density.

Stratigraphic correlation

Marker bands are distinctive beds, for example with very high radioactivity or unusual lithology.

Microfossils such as pollen grains or foraminifera are usually less than 1 mm in size and so have a much higher chance of being found in a core sample. They are used in the same way as other zone fossils to correlate strata in different boreholes and as palaeoenvironmental indicators to identify possible source rocks.

Recovery of oil and natural gas

Primary recovery

Once the production well is established, it is quickly capped off to prevent *blowouts* and oil spills.

Initially the oil gushes to the surface under natural pressure due to:

- Gases in the oil coming out of solution
- Expansion of gas above the oil
- Hydrostatic pressure of the water in the pore space beneath the oil.

After the natural pressure is released, oil is pumped to the surface using submersible pumps or beam pumps called 'nodding donkeys'. Typically only 20 to 30% of the oil in the reservoir rock can be recovered by **primary recovery**.

Porosity and permeability of the reservoir rock are key factors, determined by the grain size, rounding and sorting, by the amount of matrix or cement and by structures such as joints. Oil is thick and viscous with a high surface tension and tends to stick to the grains in the reservoir rock. At depth, the rocks are warmer so the oil has a lower viscosity, but the rocks are likely to be more compacted, so have less pore space.

Secondary recovery

- *Secondary recovery* aims to extract more oil from reservoir rocks.
- *Water flood drive*: water is injected beneath the oil to maintain the pressure.
- *Gas cap drive*: natural gas, carbon dioxide or nitrogen gas is injected into the reservoir rock above the oil to maintain the pressure. Some gases also dissolve in the oil, lowering its viscosity.
- *Steam injection* increases the temperature and lowers the viscosity of the oil.
- Detergents and other chemicals reduce the surface tension of the oil to loosen it from the grains.
- Bacteria digest and break down large hydrocarbon molecules to decrease their viscosity.

Even after secondary recovery, 20–30% of the oil remains unrecoverable. If technological innovations provide a way of extracting this oil, reserves will increase.

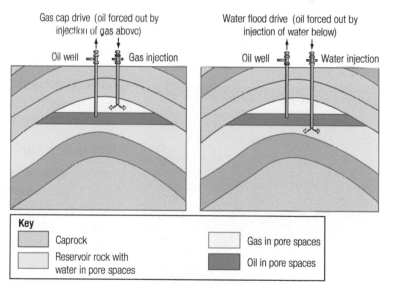

▲ **Figure 3** Secondary recovery techniques for oil

QUESTIONS

1 Describe the primary and secondary recovery methods used to increase the percentage of oil recovery.

2 Explain why 20–30% of the oil remains unrecoverable after secondary recovery.

3 Describe how geophysical exploration techniques can be used to find oil and natural gas.

4 Explain the advantages and disadvantages of exploration drilling compared to a geophysical survey.

Traditionally, oil and gas wells are vertically drilled. Advanced drilling techniques can extract more oil from one well head, reducing costs and the environmental impact.

- **Multiple-well** or **multilateral drilling**: directional drilling allows dozens of wells to branch out from one main well to tap several sources over a wide area and at different depths.
- **Extended reach drilling** allows sources to be tapped more than 10 km from the drilling rig and from reservoirs where a vertical well cannot be drilled, such as under towns or environmentally sensitive areas.
- **Horizontal drilling**, similar to extended reach drilling, starts with a vertical well that is made to turn horizontally within the reservoir rock to access more oil. This method can produce up to 20 times more oil and gas than a vertical well.
- **Complex path drilling** follows many twists and turns to try to access many sources from one drilling site.

△ **Figure 1** Burning oil shale

How long will crude oil last?

Reserves are defined as the amount of a resource that can be extracted at a profit using existing technology. This means that reserves can change, either up because of new discoveries, new technology to exploit existing reservoirs and a rise in oil prices, or down because the reservoir is exhausted, calculations of reserves are incorrect and if smaller oilfields become uneconomical.

Crude oil and natural gas are *non-renewable* resources. We consume four barrels of oil for every new barrel discovered. Estimates suggest global oil production will peak before 2030. Forty billion barrels of oil and gas have been produced by the UK, up to 2015, and estimates suggest that there are 20 billion barrels remaining. In 2020, UK production could still meet 40% of our demand.

Unconventional petroleum

Global petroleum reserves partly depend on whether we allow the exploitation of unconventional sources such as oil shales, tar sands and coal-bed methane. Historically, these sources have been unpopular due to higher production costs and greater environmental impacts.

Oil shale

This is a general term for any fine grained sedimentary rock containing significant amounts of *kerogen* that did not undergo enough maturation to produce petroleum. Oil shale can be processed by steam injection underground to produce oil that can be pumped out. Alternatively, it can be mined and the kerogen converted into synthetic oil by chemical processing, or it can be burned as a low-grade fuel. In the 1850s, oil was extracted from oil shales in the Midland Valley of Scotland and used in street lamps in Edinburgh. The Kimmeridge Clay is the richest oil shale in the British Isles.

Athabasca tar sands of northern Alberta, Canada

These cover an area of 141 000 km^2 of sparsely populated boreal forest and muskeg bogs. Reserves are estimated at 165 billion barrels and in 2015 annual production was 840 million barrels. Environmentalists point to negative impacts already being seen on the Athabasca River and the increase in carbon dioxide and other greenhouse gas emissions.

Orinoco Oil Belt of Venezuela

This contains 90% of known global reserves of 'extra heavy' crude oil, which has a lower viscosity than tar sands. It may be mixed with water or other types of crude oil to make various fuels.

Frozen gas hydrates

These are present in ocean floor sediments and permafrost so could also be a future energy resource. They produce large amounts of methane gas when melted.

Hydraulic fracturing

The first attempts, in the 1860s, to increase the flow of oil used dynamite to fracture the rocks at the bottom of the well. Later, very high pressure water was used to increase the flow from conventional oil wells. The term hydraulic fracturing, or fracking for short, is now commonly applied to the extraction of gas from impermeable rocks such as shale and coal which cannot be extracted using a conventional vertical borehole. Fracking of shales began in the 1960s with more powerful techniques introduced in the 1970s. Horizontal drilling developed during the 1980s, appropriate for accessing horizontal strata. In impermeable beds, the borehole has to go to the gas rather than the gas coming to the borehole. Several horizontal bores can radiate from the one vertical well.

Hydraulic fracturing allows the exploitation of huge reserves of shale (natural) gas, especially in the USA and Canada, but in Europe it is not clear how much it will contribute to our energy needs. In Britain, the Carboniferous Bowland Shales across Northern England are estimated to hold about 1400 trillion cubic feet but the amount which can actually be extracted depends on economic, social as well as geological factors. Basin analysis suggests that the Jurassic shales in southern England may not be mature enough to contain significant amounts of gas. The Carboniferous rocks of the Midland Valley of Scotland have potential but the extractable amount will be affected by old coal workings and faults.

The only way to test the viability of these reserves is – as with conventional methods – to drill into and to fracture the shales.

Extracting the gas

A vertical borehole is drilled down into the shale. The bore is lined for strength and to prevent leakage of gas. The bore is now manoeuvred to take a horizontal path through the shale.

The shale is fractured to release the gas using very high pressure water which is forced into the shale through perforations in the well casing. The water pressure must be higher than the pressure that the rocks are under due to overburden, perhaps up to 100 megapascals, using water at a rate of 250 litres per second. The force of the water opens up bedding planes and joints in the shale. The deeper the shale, the greater the volume of fracturing fluids that is needed. To keep the fracture open, sand (or grains of aluminium oxide or ceramic) is injected with the water.

Added to the water is a mixture of up to twelve different chemicals which help to reduce water turbulence and friction with the pipe, to prevent pipe corrosion, to increase the viscosity of the water to keep the sand in suspension and to eliminate bacteria. Each operation uses approximately 180 000 litres of chemical solutions. A typical mixture is 90% water, 9.5% sand and 0.5% chemicals.

When the water pressure subsides, gas flows under hydrostatic pressure into the borehole and up to the surface where it is collected at the well head.

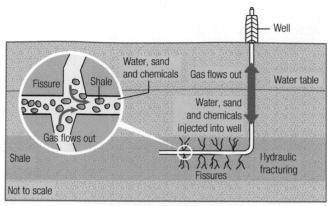

▲ **Figure 2** Hydraulic fracturing for shale gas

CASE STUDY: WYTCH FARM – AN ENVIRONMENTAL SUCCESS STORY

Wytch Farm is Europe's largest onshore oilfield, comprising three oilfields that underlie Poole Harbour in Dorset. It was discovered in 1973 and has estimated reserves of 480 million barrels of oil making it one of the top 10 UK oilfields. The oilfield is run by Perenco with permission to operate until 2037. It has 103 wells and in 2015 production was 20 000 barrels per day. Extended reach wells extend 10.7 km horizontally from the wellhead.

Oil is extracted from a faulted block with two main reservoir rocks; Jurassic Bridport Sandstone and Triassic Sherwood Sandstone. A source rock is Lower Jurassic Blue Lias clay.

The oilfield is located in an Area of Outstanding Natural Beauty. A strict environmental protection policy is in force:

- Careful consideration to the siting of operations, including on Furzey Island in Poole Harbour.
- Landscaping and screening operations with trees.
- Regularly monitoring air and groundwater quality.
- Ecological surveys to assess any environmental impacts.
- Fracking operations are not permitted.

Most people visiting Poole Harbour have no idea that a large-scale commercial oil exploitation operation is going on right under their feet! The area is a haven for wildlife including rare species such as sand lizards, smooth snakes, warblers and red squirrels.

QUESTIONS

1. Explain why locating a trap does not guarantee the discovery of economic quantities of oil and natural gas.

2. Describe the disadvantages of exploiting oil from oil shales. How can environmental concerns about fracking be ameliorated?

3. Research hydraulic fracturing in the UK on the British Geological Survey website.

4. Calculate how long the Athabasca tar sands should last using 2015 figures.

MATHS BOX: CALCULATING RESERVES

The reserves of an oilfield are estimated to be 730 million barrels. If the rate of oil production is 200 000 barrels per day, calculate how many years the reserves will last.

$730\,000\,000 \div (365 \times 200\,000) =$
$730\,000\,000 \div 73\,000\,000 = 10$ years

A *geohazard* is a geological condition that is dangerous or potentially dangerous to the environment and the people who live within it. Geohazards are natural or artificial, long term or short term, large scale or small scale, frequent or infrequent. Understanding geological processes helps to predict geohazards and to make suggestions about mitigating their effects.

Earthquakes

What happens at the focus

Earthquakes can be dangerous because of the physical effects produced by ground acceleration. The relative movement of the bedrock on either side of a fault applies stress to the rock in the fault zone which undergoes strain. This process transfers energy and increases the elastic strain energy stored in the rock. When the fault suddenly ruptures, elastic strain energy is reduced and the energy is released as movement on either side of the fault, as heat and as seismic waves. The energy transferred by the P, S and L waves is around 1% of the work done by the earthquake.

The greater the energy, the greater the amplitude of the earthquake waves.

Attenuation

When a pebble is tossed into a pond, the ripples spread out in circles. As the ripples increase in diameter, they spread out along a greater length. The **amplitude** of the ripples gets smaller. This is called geometric dispersion.

The amplitude also dissipates by **attenuation**. As the wave propagates through the rock, some elastic energy is transferred, near grain boundaries, by friction and is transferred to thermal energy. Scattering of the waves by the grains accounts for some dissipation. The wave eventually disappears altogether.

High frequency waves are more strongly attenuated because there are more oscillations per second for energy to be transferred. The earthquake becomes dominated by low frequencies. Attenuation is less for surface waves than body waves. Surface waves always have higher amplitudes than body waves, especially at long distances.

Competent rock such as granite or limestone allows the vibrations to pass through easily and attenuation is negligible. Weaker rocks such as clay or poorly cemented sandstone absorb some energy. Unconsolidated sands and gravels vibrate and move easily and attenuation may be ten times greater than in competent rocks.

Attenuation is an important factor in compiling the hazard maps used for seismic risk analysis. The further away the earthquake, the less damage is likely to be caused.

Earthquake magnitude

The amplitude of the wave and therefore the amplitude shown on the seismogram depends on energy. Seismologists assess the amount of energy (E) released by the earthquake by measuring the maximum amplitude recorded on the seismogram. This is converted to a magnitude (M) scale to indicate a size for the earthquake.

Energy is measured in joules. A joule is the energy transferred when a force of one newton acts through a distance of one metre. For earthquakes, this is a very, very small unit. A small earthquake may transfer $1\ 000\ 000\ 000 = 10^9$ joules. On a logarithmic scale, this value for E is shown as 9, a much easier number to deal with. It is now just a number on a scale, with no units.

Early magnitude scales were empirical, based on taking very many readings from very many earthquakes. A general relationship was established and used to define the term magnitude.

$M \propto \frac{2}{3}\log E$ which may be written as $\log E \propto 1.5\,M$

The relationship between the magnitude of an earthquake and the energy released is

$$M = \frac{2}{3}\log E - Q$$

Q is a constant which takes local variations into account, such as the type of seismogram used. In the OCR exam specification, Q is taken to be 6.0, giving

$$Mw = \frac{2}{3}\log E - 6.0.$$

There is no theoretical upper limit to the amount of energy which an earthquake can release, and therefore no upper limit for the scale. However, the maximum recorded value so far is 9.5. A negative value on a log scale indicates that the figure is less than one. So a magnitude of –1 represents a numerical value of 0.1.

The amount of energy released by an earthquake shows its potential for damage to the built environment.

Magnitude scales

Different magnitude scales depend on the model of seismogram, the distance from the epicentre and the depth of focus. In 1935, Charles Richter devised his local magnitude scale (M_L) which applied to one type of seismograph measuring earthquakes in California at a distance of 600 km. In this case, the constant $Q = 2.9$. Modern digital recording seismographs are calibrated with each other and the magnitude is calculated by computer. Different scales do not always give the same answer, but are reasonably consistent with Richter's original M_L.

In terms of energy released: $M = \frac{2}{3}\log E - 2.9$

Using the example of $E = 10^9$ joules: $M = \frac{2}{3} \times 9.0 - 2.9 = 6 - 2.9 = 3.1$

For an earthquake where $E = 2 \times 10^{15}$ joules: $M = \frac{2}{3} \times 15.3 - 2.9 = 10.2 - 2.9 = 7.3$

Fault geometry and seismic moment (Mo)

Magnitude scales are based on observation rather than maths. They do not measure a particular property of the earthquake. In 1979, a new scale, *moment magnitude* (Mw), was introduced which gives the most reliable magnitude for very large earthquakes because it is derived from factors which can be measured – the size of fault rupture and the displacement – as well as the energy released.

We can think of the slip along the fault in terms of rotation, about a pivot on the fault plane, of rock moving in opposite directions, like bicycle handlebars. This represents a lever which is measured, in maths and physics, by **moments.** The sum of these rotational effects is the *seismic moment* (Mo).

The orientation of the fault and the direction of movement (the *fault geometry*) are determined from the waveforms recorded at different seismographs. Waves set off by a point source, such as a nuclear bomb test, radiate out equally in all directions. If the fault displaces rock towards the north, the wave pattern received in the north is slightly different from the pattern received in the south, and very different from the pattern which spreads to the east and west.

Seismic Moment (Mo) is computed from the formula $Mo = \mu Ad$

    μ – shear strength of the rock

    A – area of the fault

    d – average displacement

Moment magnitude (Mw) is defined in terms of the seismic moment:

    $Mw = \frac{2}{3}\log(Mo) - 10.7.$

And in terms of energy released:

    $Mw = \frac{2}{3}\log E - 6.0$

It is not surprising that the Richter scale is still used by the media and the general public, even though not by seismologists!

CASE STUDY: MODELLING EARTHQUAKES

You can model an earthquake by hitting a solid object with a hammer and feeling the vibration. You will get very different results when hitting concrete, sand, a metal plate or a wood block. Try putting model houses on your different materials and observe the effect.

CASE STUDY: THE WORLD'S STRONGEST EARTHQUAKES

Location	Year	Magnitude
Chile	1960	9.5
Alaska	1964	9.2
Northern Sumatra	2004	9.1
Sendai, Japan	2011	9.0
Kamchatka	1952	9.0
Concepcion, Chile	2010	8.8
Coast of Ecuador	1906	8.8
Rat Islands, Alaska	1965	8.7
Northern Sumatra	2005	8.6
Assam – Tibet	1950	8.6
Northern Sumatra	2012	8.6
Andreanof Islands, Alaska	1957	8.6
Southern Sumatra	2007	8.5
Banda Sea, Indonesia	1938	8.5
Kamchatka	1923	8.5

QUESTIONS

1 How much more energy is released by a magnitude 8 than a magnitude 3 earthquake?

2 On a map, plot the position of the world's strongest earthquakes. Is there a pattern?

3 Research the magnitude of recent earthquakes.

CASE STUDY: 1985 MEXICO

The epicentre of the earthquake, off the coast of Mexico, was several hundred kilometres from Mexico City where the damage was greatest. Unconsolidated silt and clay of a drained lake bed amplified the ground movement to five times greater than elsewhere. The high water table encouraged **liquefaction**. The earthquake wave frequency matched the natural **resonant frequency** of 20-storey buildings so these were damaged more than taller or shorter structures. Some 20-storey buildings were destroyed, but the buildings of different heights on either side remained unscathed. Four hundred buildings collapsed and thousands were damaged.

CASE STUDY: 1906 SAN FRANCISCO

Every major pipeline breakage in the 1906 San Francisco earthquake occurred in areas of ground failure, severely hampering attempts to fight the fires that ignited during the earthquake. Ground failure movement of just two metres indirectly led to the destruction of 85% of the city.

CASE STUDY: 1989 LOMA PRIETA, CALIFORNIA

The 1989 Loma Prieta Earthquake lasted 15 seconds and measured 6.9 on the moment magnitude scale. The earthquake killed 63 people in central California and injured 3756. The epicentre was in the Santa Cruz Mountains. It caused severe damage throughout the San Francisco Bay Area, 95 km from the epicentre. Major property damage occurred due to liquefaction of ground fill on which properties had been built. The worst affected area was the collapsed double-decker freeway in Oakland.

As an earthquake propagates away from the focus, any buildings in its path are likely to be affected. This effect depends partly on the nature of the wave and partly on the nature of the soil, sub-soil and bedrock. Damage is usually caused by vibrations of the rock as its particles oscillate in response to the waveform.

Earthquake waves

P, S and L waves oscillate in a horizontal motion. For R waves, the oscillation is vertical. P and S waves have a higher frequency and shorter wavelength. L waves are lower frequency, longer wavelength, larger amplitude, travel greater distances and cause most of the shaking and damage. Buildings are built to withstand vertical forces from gravity, so R waves are less damaging.

L waves impose extreme horizontal forces on buildings due to the lateral acceleration, which we call G-forces (especially in theme park rides). A magnitude 6.7 earthquake produces 1 G, setting up large stresses on the building.

If the upward acceleration of the ground is greater than gravity ($9.81\,\text{m s}^{-2}$), loose objects are thrown upwards and great damage is caused.

Subsurface rocks

Competent rocks such as granite withstand the effects of earthquakes because the properties which make the rock competent are those which help the propagation of seismic waves; rigidity, incompressibility and shear strength. Little energy is transferred to the granite so it does not deform and buildings are not damaged. Weaker rocks such as clay or poorly cemented sandstone absorb energy, causing deformation and damage to buildings because the ground shake is amplified.

Effects on the built environment

The built environment

Movement of the ground separates one part of the structure from another.
- Bricks and stonework separate along the mortar, causing walls to collapse.
- Floors separate from supporting walls and 'pancake' on top of each other.
- Sectional bridges separate from supporting piers.
- Sections of utility pipes separate from each other and leak.

On steep slopes made unstable by high rainfall, the vibration may trigger landslides and mudflows, partly assisted by liquefaction. These cause loss of life by burial, but also hamper rescue attempts.

The main movement along a fault releases most of the energy, but subsequent movements, minutes, hours or even days afterwards, cause aftershocks. Structures weakened by the main shock may collapse during this period, causing further damage and added difficulties for rescue workers.

Earthquake intensity – the Mercalli scale

In the days before sophisticated instruments, earthquakes were classified by the noticeable effects of surface vibrations in the area, based on an earthquake's local acceleration and duration.

These effects depend on the:
- Strength of the earthquake at its focus
- Distance from the focus
- Nature of the sub-surface rocks and sediments
- Standard and design of buildings.

At the end of the nineteenth century, a scale of **intensity** was devised by Guiseppi Mercalli. Mercalli's ten-point scale was itself modified into the twelve-point scale used today. It is an arbitrary scale, meaning that it depends on the opinions of the observer and has no mathematical base. Some witnesses of the earthquake might exaggerate just how bad things were and you may not find witnesses who agree on what happened.

	Intensity	Modified Mercalli intensity scale: summary descriptions	Approximate Magnitude
I	Instrumental	Not normally felt. Birds and animals uneasy	1.0–2.0
II	Feeble	Felt only by a few persons at rest	2.0–3.0
III	Slight	Vibrations like large lorry passing. Felt by people at rest	3.0–4.0
IV	Moderate	Felt indoors by many. Cars rock	4.0
V	Rather strong	Sleepers wakened. Some windows broken	4.0–5.0
VI	Strong	Small bells ring. Trees sway. Loose objects fall	5.0–6.0
VII	Very strong	Difficult to stand up. People run outdoors. Walls crack	6.0
VIII	Destructive	Partial collapse of ordinary buildings. Chimneys fall	6.0–7.0
IX	Ruinous	Ground cracks, buildings shift off foundations. Pipes break	7.0
X	Disastrous	Landslides, buildings damaged. Many buildings destroyed	7.0–8.0
XI	Very disastrous	Few buildings stand, bridges destroyed, landslides	8.0
XII	Catastrophic	Damage total. Waves formed on ground surface	> 8.0

The greatest energy is at the focus and greatest effect is at the epicentre, directly above the focus. The epicentre can be located by plotting the intensities observed at many different localities on a map. Isoseismal lines, which join areas of equal intensity, are constructed. Intensity is greatest in the centre of the isoseismal line pattern, the epicentre.

Magnitude remains unchanged with distance from the earthquake focus. Intensity, however, decreases with distance.

Liquefaction

Seismic vibrations in bedrock pass into overlying superficial deposits such as sand or silt. If the deposit is saturated due to a high water table, the water separates from the solid particles and rises to the surface. If the water cannot drain away, the pore pressure between the grains increases and if this pressure is able to support the overlying deposit, the layer of sand temporarily behaves as a viscous liquid. Houses built on alluvial deposits suddenly find themselves standing on water and severe damage is caused. Liquefaction helps salad cream come out of the bottle when you shake it vigorously.

Effect of liquefaction on the built environment

Liquefaction is destructive if accompanied by ground displacement or ground failure.

Flow failures displace large masses of soil for many metres, even kilometres on a long slope.

Lateral spreads displace large blocks of soil down gentle slopes due to liquefaction of a layer underneath. The blocks break into fissures and scarps. The spreads disrupt foundations, cut pipelines and buckle bridges. In the 1964 Alaska earthquake, 200 bridges were damaged by spreading floodplain deposits.

Ground oscillation happens when liquefaction at depth causes overlying soil to separate, taking the form of ground waves, accompanied by the opening and closing of fissures and fracture of pavements and pipelines.

Loss of bearing strength allows buildings to settle and tip. During the 1964 Niigata (Japan) earthquake, several four-storey buildings tipped by as much as 60°. Buried tanks may rise buoyantly through the liquefied soil.

KEY DEFINITIONS

Liquefaction describes saturated or partially saturated unconsolidated material losing strength and rigidity in response to an applied stress, usually an earthquake.

Resonant frequency the frequency at which the amplitude of the oscillation is greatest.

Intensity a measure of the surface shaking and damage caused by an earthquake. The Mercalli scale measures the intensity of an earthquake. Intensity is based on an earthquake's local accelerations and how long these persist.

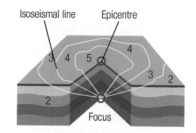

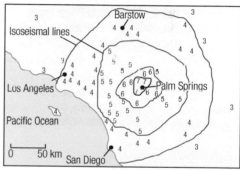

▲ **Figure 1** Isoseismal lines around the epicentre

QUESTIONS

1 Within the same area there may be cemented sandstones and unconsolidated sands. Will the intensity of an earthquake be the same or different on the two rock types? Explain why.

2 What are the differences between earthquake intensity and magnitude?

CASE STUDY: TRANSAMERICA PYRAMID, SAN FRANCISCO

The 260 m high Transamerica Pyramid has a system of cross-bracing, used above the first floor, which supports both vertical and horizontal loading and resists torsional forces. Interior frames reach up to the 45th floor. During the magnitude 7.1 Loma Prieta event in 1989, the top storey of the pyramid swayed more than 30 cm from side to side, but there was no damage.

◗ **Figure 1**
Transamerica Pyramid, San Francisco

Living with earthquakes

It is often said that 'Earthquakes don't kill people, buildings do'. Fatalities and injuries are caused by falling buildings and glass, fires and later disease often caused by lack of clean drinking water.

It is thought possible to build an earthquake-proof building but the cost would be prohibitive. Instead, the aim is to build earthquake-resistant buildings which, although severely damaged, do not collapse but allow people to escape. The ultimate design might be for a building which deforms during an earthquake but then returns to its original shape afterwards.

Planning

A risk assessment will forecast the likely number and magnitude of future earthquakes. Local authorities may ban buildings from the fault line itself, from areas of alluvial deposits which may suffer liquefaction, and from areas liable to landslides.

Building design

The aim is to counteract the effects of both the horizontal and vertical stresses which act on a building.

- Choose the height of the building so that its natural **frequency** does not match the expected local frequency of seismic waves.
- Avoid irregular or asymmetrical designs because they are more susceptible to *torsion*, or twisting. In symmetrical buildings, stresses are distributed equally throughout the structure.
- Avoid ornamentation, such as cornices or fascia stones, which can be dislodged.
- The building should be able to dissipate some energy by undergoing plastic deformation. Wooden structures are flexible and accept a certain amount of strain. The steel rods in steel-reinforced concrete make it more ductile. Pre-formed beams, angles and plates made of structural steel are even more ductile, allowing the buildings to bend without breaking. Brick and concrete buildings absorb very little energy which instead goes into fracturing the structure.

Foundations may be reinforced by pumping liquid cement into a network of micropiles drilled into the ground.

Ground or base isolation systems

The aim is to separate the vibrating ground from the building. When an earthquake strikes, the inertial mass of the building keeps it stationary whilst the ground vibrates beneath, or allows the building to move without damage. The building's horizontal acceleration is reduced and there is much less deformation and damage.

Various types of 'isolators' are used; lead-rubber bearings show the principle. A solid lead core makes the bearing strong in the vertical direction and it absorbs energy. The wrapping of alternating layers of rubber and steel make the bearing flexible in the horizontal direction. Other devices include rollers, rubber pads, springs or sliders coated with non-stick Teflon™.

Older buildings sometimes have this system retro-fitted for future protection. After the 1989 Loma Prieta earthquake damaged 100 000 buildings, San Francisco City Hall was retro-fitted with more than 500 base isolators which will safely allow the building to sway horizontally by up to 60 cm.

⬢ **Figure 2** Mechanisms to prevent bridge collapse

Resisting shear forces

The framework of the building is strengthened by diagonal cables or rigid girders. Large open spaces such as atria are avoided, and floors are fixed to the walls to add rigidity and prevent 'pancaking' (where floors collapse on top of each other rather like a pile of pancakes). Shear walls without any openings, extending the full height of the building, are installed, for example, around lift shafts or stairwells, to add to the rigidity.

Shear walls limit the flexibility of the building's design, so moment-resisting frames may be used instead. Columns and beams are allowed to bend, but the joints or connectors between them are rigid so that the whole frame moves in response to a lateral force.

Absorbing sway

Tall buildings are designed to sway, absorbing the energy through flexible supports and materials made, for example, of rubber. Hydraulic systems, sometimes computer-controlled, dampen the movement, like shock absorbers in a car. Flexible connections between different parts of the building help to counter movements. *Active mass damping* uses a heavy mass mounted at the top of a building and connected to dampers. When the building begins to oscillate, the mass moves in the opposite direction, reducing the amplitude of mechanical vibrations.

The *controlled rocking frame system* consists of steel frames, steel cables and shock diffusers. During an earthquake, steel frames rock up and down and the energy gets directed downward to the diffusers. The steel teeth in the diffusers grind together and may even fail, but the frame itself remains intact. Once the shaking has stopped, the steel cables in the frame pull the building back into an upright position. Any damaged diffusers are easily replaced.

Services

Broken gas mains and power lines can cause fires, and fractured water mains prevent the fires from being put out. These problems can be avoided by using flexible piping to allow for movement.

Natural frequency

Seismic waves make the ground vibrate at a particular frequency which makes buildings vibrate at the same frequency.

All structures such as buildings and bridges vibrate at their own natural frequency. A ruler clamped to the desk will vibrate, when twanged, at a particular frequency depending on its length, thickness and material.

The natural frequency of a building is given by the formula:

$$f = \frac{1}{2\pi}\sqrt{\frac{k}{m}}$$

f = natural frequency (measured in Hertz)
k = the stiffness of the building, which depends on its construction
m = the mass of the building, which depends partly on its height

The greater the stiffness, then the higher the frequency; the greater the mass, the lower the frequency. Taller buildings are usually more flexible, so they have lower natural frequencies compared to shorter buildings.

Buildings sway because their foundations move sideways but the top storeys lag behind. If the natural frequency of the building matches the frequency of the earthquake's vibration, **resonance** is set up which amplifies the oscillation, increasing the amount of movement and damage.

▲ **Figure 3** Cross-bracing, Japan

CASE STUDY: TAIPEI 101, TAIWAN

The 508 m high Taipei 101 tower has a 660 tonne active mass damper fixed between the 88th and 92nd floors by eight steel cables each connected to a damper. If the building sways, the dampers counteract the motion, reducing vibrations that could cause stress on the structure.

◀ **Figure 4** Building 101, Taipei

QUESTIONS

1 Research other building designs which help to reduce damage during earthquakes.

2 Look at photographs of damaged buildings and try to work out why they collapsed.

Earthquake zones run through some of the most populated areas of the world. Over 500 000 people have died in the last decade due to earthquakes, mostly in the developing world. In densely populated areas, earthquakes can kill a large number of people within a few minutes. Such large losses justify the effort of identifying potential hazards and carrying out seismic risk assessments. Earthquakes are, of course, **geological hazards**.

Seismic risk

Seismic risk refers to the possibility of death, injury or damage that may occur within a certain period of time. It is assessed using seismological and historical records and geological observations. Risk depends not only on the hazard but also on a numerical model chosen to describe the occurrence of earthquakes. High seismic hazard does not necessarily mean high seismic risk, and vice versa. An earthquake-resistant building in a seismically active area may be at lower risk of collapse than a brick-built building in a seismically quieter area. Seismic modelling can give general probabilities for casualties or economic damage. The actual risk for individual buildings varies considerably making seismic risk estimation an expensive business.

Each organisation has its own priorities for allocating funds for mitigating risks:

- Local government – protecting public buildings and emergency planning.
- Business – protecting buildings to avoid loss of equipment and production.
- National government – maintaining water, electricity and communications.

Seismic risk can be reduced by:

- Enforcing up-to-date building codes
- Improving emergency response and basic infrastructure
- Preparing for emergencies which will arise after an earthquake.

Seismic risk assessment

It is the job of the geologist to carry out seismic risk assessments and to draw up hazard maps. These are passed on to other experts in building, engineering and logistics to compile building codes and to prepare emergency response plans, which may apply to other hazards, too, such as floods. Qualified experts in disaster planning have a role in communicating these plans through public education programmes, for example using the Internet.

Notification of immediate risk may be passed to the public by the emergency services via alarm sirens, loudspeakers, radio, television and the social media.

Information from historical records is added to a map, showing areas with many earthquakes and those with a few, what depth they originated and where the maximum intensity has occurred. This shows what happened in the past but it does not tell us what will happen for an earthquake which is not the same as the previous ones. Adding one degree of intensity on the Mercalli scale to all the values on the map provides a safety margin.

Other maps show the geological features, including faults, rock types, soil type and thickness, groundwater conditions, steep slopes and filled areas. Areas where earthquakes have occurred in the past can be matched with other areas with similar geological settings which have not yet experienced earthquakes but which might in the future.

KEY DEFINITIONS

Geological hazard a geological condition that is dangerous or potentially dangerous to the environment and the people who live within it.

Seismic risk the possibility of suffering harm or social and environmental loss because of a seismic event that may occur in a specified period of time.

Probability a measure of the likelihood that an event will occur. Probability is quantified as a number between 0 (impossibility) and 1 (certainty).

Return period the average length of time for an earthquake of a given magnitude to occur again or to be exceeded. (Sometimes called the recurrence period.)

KEY FACT

Exceedance is the probability that an earthquake will generate a level of ground motion that exceeds a specified reference level during a given exposure time.

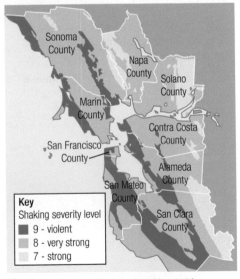

Figure 1 Hazard map showing shaking severity, California

Figure 2 Flow diagram for risk analysis

HOW SCIENCE WORKS: GEOGRAPHIC INFORMATION SYSTEM

Geographical information systems (GIS) are sophisticated computerised maps which store, edit, analyse, share and display any information you want to include. As well as the usual topographical information, a seismic analyst would add faults, rock and soil types, landslips, filled areas, hospitals, gas pipes, power lines – anything to do with earthquakes.

The map can be divided into layers to show particular features. It can show relationships, for example, between sites where injuries might be expected and the location of hospitals. Maps can be produced showing, for example, areas liable to liquefaction or routes which may be followed by tsunamis.

GIS provide a powerful tool for hazard planning, for example by identifying the best tsunami evacuation routes and where to locate earthquake shelters.

What's the risk?

How much money and effort should go into mitigating seismic hazards? In Greece, for example, ten people a year are killed by earthquakes but over 2000 on the roads. Where is money better spent?

When you cross a road, do you consider the risk? The authorities do it on your behalf, in stages, ranging from doing nothing (which may incur extra medical costs after an accident) through pedestrian crossings to railings to prevent you crossing the road at all. Does the initial outlay save money in the long run and make life better for the community?

The number of accidents on a stretch of road will give the **probability** of subsequent accidents. Values for probability range from 0 (impossibility) to 1 (certainty). Of course, along with the probability (P) that an event will happen, there is the complementary probability (Q) that it won't happen. In statistical, terms $P + Q = 1$.

Return period

The **return period** is the average length of time for an earthquake of a given magnitude or more to happen again. It is a statistical measurement based on historical data giving an estimate of the likelihood of an earthquake event happening. It assumes that the probability of the event does not vary over time and is independent of past events. It is used for risk analysis when designing structures to withstand an earthquake of a chosen magnitude with a certain return period.

For example, an earthquake with a 50-year return period for magnitude 6 or more means that such an earthquake should occur at that location on average only once every 50 years. The fact is that an earthquake has occurred once in this period of 50 years – the rest is statistics.

The formula for the return period is $T = \dfrac{n + 1}{m}$

n – number of years on record;

m – number of recorded earthquakes

The return period (T) is the inverse of the probability (P) that an earthquake of a given magnitude or more will happen again, given as $T = 1/P$

For example, an annual probability (P) of 0.1 (10%) implies a return period (T) of 10 years. This is called a 10-year earthquake. Conversely, if an earthquake occurs once in 10 years, the probability of it occurring in a one-year period is $1/10 = 0.1 = 10\%$.

A 50-year earthquake has a $1/50 = 0.02 = 2\%$ chance of occurring in a one-year period. It has the same chance of 2% of occurring the next year without waiting the full 50 years – but it probably won't happen. It may not happen at all in 50 years – but it probably will.

In the USA, the most commonly used probability is a 10% chance of an earthquake happening within a period of 50 years. This equates to a 0.2% probability of an earthquake happening in one year, and a return period of 475 years. Buildings should last at least 50 years and so are designed for a '475-year earthquake'. Important buildings are built for a 2500-year earthquake, with a 2% chance of an earthquake in 50 years.

MATHS BOX: RETURN PERIOD AND PROBABILITY

If there is a probability (P) that an earthquake will happen, there is a complementary probability (Q) that it won't happen. Statistically $P + Q = 1$. Re-arranging this: $P = 1 - Q$ and $Q = 1 - P$

For a return period of T years, the probability that the earthquake won't occur is: $Q = (1 - P)^T$

For example: for a 100-year earthquake $T = 100$ and $P = 1/T = 0.01$.

$$Q = (1 - 0.01)^{100} = 0.366$$

So there is a 37% chance that the 100-year earthquake will not occur during a 100-year period.

Conversely, the probability (P_T) that it will happen during the T-year return period is given by

$$P_T = 1 - Q = 1 - (1 - P)^T$$

For the 100-year earthquake: $P = 1 - (1 - P)^T = 1 - (1 - 0.01)^{100} = 0.634$

There is a 63% probability that a 100-year earthquake will happen during a period of 100 years, and a 37% chance that it won't.

For example: for a 10% probability (P_T) over a 50-year time period (T).

$$P_T = 0.1 = 1 - (1 - P)^{50}$$

This calculates as $P = 0.0021$ and $T = 1/P = 475$ years

The annual probability (P) $= 0.0021 = 0.2\%$, with a return period (T) of 475 years. Some calculations are based on a 30-year period, being the average length of a house mortgage.

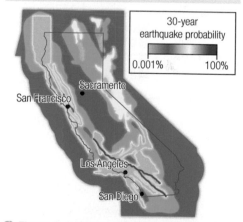

🔺 **Figure 3** Uniform California Earthquake Rupture Forecast (2015). Hazard map of M > 6.7 for 30 years

QUESTIONS

1 Research the earthquake drills carried out in Japanese schools.

2 Show that a 2500-year earthquake has a 2% chance of an earthquake in 50 years.

3 Research the resilience programme of the San Francisco 'Association of Bay Area Governments' on http://resilience.abag.ca.gov/earthquakes/.

Figure 1 This earthquake shelter in Japan is an open space away from tall buildings

PGA (g)

0.005–0.02	0.1–0.12
0.02–0.04	0.12–0.14
0.04–0.06	0.14–0.16
0.06–0.08	0.16–0.18
0.08–0.1	

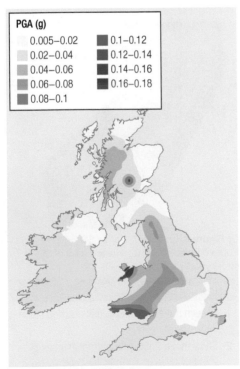

Figure 2 Seismic Hazard Map for the UK building regulations, showing levels of peak ground acceleration (PGA)

We cannot control earthquakes. The best we can do is to mitigate the effects by:

- Forecasting the areas where earthquakes may happen
- Predicting the possibility of a seismic event
- Looking for precursor signs leading up to an event.

Forecasting and predicting

The terms **forecast** and **prediction** are different, although they are sometimes used interchangeably. We can forecast with a high degree of certainty that you will take an exam in Geology within the next two years. It is more difficult to predict what grades you will get! We can forecast with a high degree of certainty that earthquakes will happen along plate boundaries. It is more difficult to predict where, when and how strong the earthquakes will be.

- Forecasting refers to the probability of an earthquake event occurring.
- Predicting tries to determine when and where the event will take place.

Whilst predicting the exact time and location of an earthquake is not possible, seismologists are increasingly able to forecast the likelihood that an earthquake might occur in a certain area within a certain time.

Whether you want a forecast or a prediction (or both) depends on your point of view:

- Home-owner
- Local authority emergency planner
- Skyscraper architect
- Nuclear power plant company
- Insurance company.

Forecasting using data

Earthquake forecasting is the probabilistic assessment of an earthquake hazard, with reference to the frequency and magnitude of damaging earthquakes in a given area over a period of years.

Consider a source area where the probability of a magnitude 5 (M5) earthquake occurring within 50 years can be computed from global and local seismic monitoring, historical records and geological field work. The area affected by such an earthquake is represented by a circle. A town is concerned about the effects of a Mercalli intensity 7 ground movement caused by the M5 earthquake. The area liable to be affected is represented by another circle.

Do the two circles overlap?

- If they do not, the probability of an intensity 7 ground movement affecting the town due to an M5 earthquake is zero – at least in theory.
- If they overlap by 20%, the probability of the town being affected increases by 20%.
- If they overlap completely, an intensity 7 event certainly will happen within 50 years.

If the exercise is repeated for all possible magnitudes, and then for different intensities and time periods, a complete seismic hazard map for the town is built up, contoured according to the value of the probability.

Probabilistic forecasting takes into account the full range of possible earthquakes, their location, frequency, magnitude, and the propagation of waves. Uncertainty in each of these elements is taken into account. This method is widely used, but it has been criticised for lacking a physical and mathematical basis. Its use could lead to either unsafe or overly conservative engineering design. It involves a degree of subjectivity and judgement when constructing the initial model. Civil engineers base their designs on good science. If there is no physical basis for a hazard forecast, then there are problems of understanding and communication between civil engineers and seismologists.

Predicting using science

It is not currently possible to make deterministic predictions of when and where an earthquake will happen and how strong it will be. We cannot yet identify the seismic, biological, physical or chemical changes which would almost certainly indicate that an earthquake will happen.

A useful question to ask is 'what is the worst-case scenario?' To do this, we:

- Find the nearest active fault
- Calculate the largest earthquake that could happen on this fault
- Estimate the return period
- Assume the epicentre is closest to the site
- Calculate what the ground motion will be.

This method looks at real faults, is easy to do and it gives a value that maximises safety. On the other hand, the answer may be too safe and mitigating measures may be taken which are unnecessary. Deterministic analyses do not consider the probability of the earthquake hazard, just what will happen if it does strike.

Instead of modelling for the 'maximum' earthquake, a theoretical 'expected' earthquake can be used, or one which is agreed to be 'sufficient' for design purposes, depending on whether it is for a house or for a nuclear power station. However, the maximum earthquake could still occur, even though the chances are small. On the other hand, if a very large earthquake is so likely, you may as well just model it straight away. It's a question of judgement. We don't know what the maximum earthquake on a fault can be, until it happens. And we don't know how the ground will actually respond to an event, even if we know the magnitude and the distance from the epicentre.

Who needs to know?

The 'maximum considered earthquake' (MCE) for an area is a 2500-year earthquake, with a 2% probability of being exceeded in 50 years. In general building codes, non-essential buildings are designed for 'collapse prevention' where the building remains standing, even if damaged, to allow for safe evacuation of people.

Another version of MCE stands for 'maximum credible earthquake', which is used in designing skyscrapers and larger civil infrastructure where failure could lead to catastrophic consequences.

Social consequences of earthquake prediction

It is one thing to predict an earthquake, but quite another to do something about it. One problem is the uncertainty of predicting the exact date, place and magnitude. Another is human nature. Consider your reaction to the order that you (and a million others) should evacuate your homes because there may be an earthquake next week. Consider the panic if the warning was just one hour. If the earthquake failed to strike and you lost a day's earnings, what then? If an earthquake struck without warning, who would you blame? And who would you expect to come to your assistance?

Seismologists provide hazard maps and risk assessments for local authorities so as to minimise risk. They have to decide if a public warning is appropriate. If the likelihood is very low, a warning may do more harm than good. This is vital work which seismologists will only undertake if their role is clearly defined and if they feel confident that they can offer advice without fear of retribution.

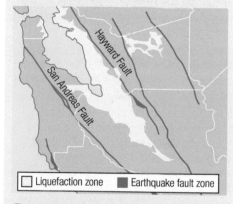

Japan's Earthquake Early Warning system uses a network of 1000 seismometers. The instrument nearest the focus immediately detects a disturbance and automatically sends warnings, which are flashed on television, showing the area most at risk. The seismic waves took 100 seconds to reach Tokyo and the warning was given 60 seconds before the earthquake struck, giving people long enough to take action. The arrival of the faster P waves also gives invaluable seconds to evacuate a building before the more damaging S waves strike.

These earthquakes were the third (M9.2) and fourth (M9.0) largest earthquakes since 1900 when accurate records began. Both were due to mega-thrusting along a subduction zone, the Indian Plate under the Burma Plate, and the Pacific Plate under the Eurasian Plate, causing vertical displacements of up to 15m of the sea floor along several hundred kilometre stretches of the plate boundaries.

These conditions were ideal for the formation of the highly destructive tsunamis which devastated coastal areas, leaving more than 250 000 people dead or missing in 15 countries.

It is difficult to plan for events which are far stronger than anything known previously. The Fukushima Daiichi Nuclear Power Plant was designed to withstand an M7.5 earthquake and was protected by a 6-m high sea wall. It was not expected that the entire coastline would subside by almost a metre, lowering the seawall and allowing the tsunami to flood the power stations, putting the generators out of action, leading to the explosion and the leak of radioactive material.

▲ **Figure 1** Epicentre for the 2004 earthquake

No method for predicting earthquakes, which is based on physical principles, has yet proven successful. All the suggested effects can occur without being followed by an earthquake.

Most methods of predicting earthquakes (and volcanic eruptions) look for change, but first you must identify the norm. This involves expensive long-term monitoring, which is easier for some countries than others.

Physical properties

Large earthquakes are sometimes preceded by a series of smaller ones, increasing in frequency before the main event. If these smaller earthquakes disperse the strain energy, there will be no major earthquake. In central Italy, only 10% of M>3 earthquakes are followed by a larger event within 30km and 48 hours.

P wave velocities may decrease, then increase again before the earthquake. Laboratory experiments suggest that velocities change when rock expands slightly close to its fracture point.

Coloured lights in the sky immediately before an earthquake may be caused by changes in the electrical properties of quartz and other minerals under stress.

Stress

Many minute cracks develop within a rock under stress, increasing its volume and allowing the inward percolation of water and gases. The water in the micro-cracks increases the electrical conductivity of the ground, lowering its resistivity. P wave velocities decrease then increase again.

The area around the earthquake focus may tilt due to deformation, and swell slightly due to the micro-cracks. Tiltmeters, using laser technology and GPS, measure the slope of the ground very accurately. Strain gauges in boreholes measure deformation and therefore any increase in stress.

Groundwater percolates into the micro-cracks, lowering the level of water in wells. The levels return to normal as the water is replenished, before the earthquake occurs.

Animal behaviour

Research in China confirms that animals show disturbed behaviour just before an earthquake. Pigs squeal and dogs bark (as they do for other reasons too), ground-living birds perch in trees and animals such as snakes leave their 'burrows'. Animals may feel the P waves which arrive well before the higher amplitude S waves. Alternatively, the behaviour may be in response to low magnitude foreshocks or to slight changes in the Earth's magnetic field.

Radon emissions

Radon, a radioactive decay product of uranium in granite, is easily detected and its short half-life (3.8 days) makes radon levels sensitive to short-term fluctuations. As a heavy gas, radon accumulates in water wells. An increase in the amount of radon suggests that the gas is percolating through new micro-cracks and an earthquake is imminent, for example before the 1995 Kobe (Japan) earthquake. Overall, there is little correlation between radon levels and earthquakes.

Seismic gap model

Where two plates slip past each other, every section of the fault slips, but not all at the same time. Different sections will be at different stages in the cycle of the build-up and release of strain.

In the seismic gap model, the 'next big earthquake' is expected in the gap between the recently slipped sections, where there is more stored strain energy. If the pattern is regular, the next earthquake may occur in the quiet sections where the fault is locked and the stress is steadily building up. Those sections of the fault which move frequently dissipate their elastic energy in many small, less destructive earthquakes.

This model can be used in long-term forecasting but statistical analysis shows that it does not lead to good predictions. Patterns are not regular enough within the human time frame. Movement, frictional forces and other properties are not constant. However, if the seismic gap theory suggests that an earthquake is due, then it is time to monitor the area more closely.

Seismic warnings

A warning alerts people to an event which has already started.

Tsunami early warning

A tsunami (= harbour wave) is the movement of a body of water caused by the displacement of a large volume of water by a landslide, volcanic eruption, meteorite impact, and most commonly, by uplift of the sea floor. A sudden, vertical movement of the ocean floor displaces a column of sea water upwards. As the surge of water settles back, it collapses in on itself, sending a series of wave forms radiating outwards in all directions.

The movement of the crust that causes the tsunami also sets off an earthquake which provides an early warning system. The global network of seismographs quickly identifies the focus and strength of the earthquake and the type of crustal movement which caused it. Tsunamis are associated with high magnitude earthquakes, M>6.5, usually M>8, but not all high magnitude earthquakes are associated with tsunamis. The displacement of the sea floor must be vertical and large.

P waves travel at about $8\,\mathrm{km\,s^{-1}}$ and tsunamis at up to $0.2\,\mathrm{km\,s^{-1}}$. Except for the communities closest to the epicentre, this gives minutes or hours for action to be taken between the arrival of the P waves and the tsunami.

Tsunami hazard management

Saving lives:
- An efficient early warning system by siren, radio and loudspeakers
- Signs showing evacuation routes to high land
- Education that a retreating sea indicates an approaching tsunami and that there may be a series of waves over several hours due to the long wavelength of the wave train.

Saving property relies on reducing the energy of the wave by:
- Maintaining coral reefs in good condition
- Maintaining coastal trees and vegetation
- Prohibiting building along the coast, although unpopular with hotels and fishermen
- Designing buildings with no permanent accommodation on the ground floor.

▲ **Figure 3** Tsunami siren, Hawaii

CASE STUDY: 1985–1993: PARKFIELD, U.S.

Earthquakes of about M>6 regularly occur on the Parkfield segment of the San Andreas Fault: 1857, 1881, 1901, 1922, 1934 and 1966. Excluding 1934, they occur every 22 years, ± 4.3 years. Seismologists predicted a 95% chance that the next earthquake would hit in 1988, and by 1993 at the latest. One of the 'most sophisticated and densest nets of monitoring instruments in the world' was established in part to identify any precursors. Plans were made for alerting emergency authorities if necessary.

1993 came and went. An M 6.0 earthquake hit in September 2004, without obvious precursors. While the experiment in catching an earthquake is considered to have been successful, the prediction was unsuccessful. Later research showed several M 4.0 earthquakes had reduced the stresses, delaying the predicted M 6.0 earthquake.

CASE STUDY: TOHOKU 2011

Shortly after the Tohoku earthquake, the Pacific Tsunami Warning Centre issued warnings for the entire Pacific Ocean. Countries were warned to expect several waves over a twelve-hour period. They evacuated people, cancelled trains, closed schools and sewage works and patrolled beaches. Ships put to sea where the amplitude of the tsunami is low.

▲ **Figure 2** Setting for the Tohoku 2011 earthquake

HOW SCIENCE WORKS: DETECTING A TSUNAMI

In open water, tsunamis travel as low amplitude (1 m), very long wavelength (10 to 500 km), very fast (up to 700 km per hour) waveforms.
- Tide gauges, with filters to negate the effects of wind-waves, measure sea-level changes within 1 mm accuracy.
- Satellite altimeters accurately measure the height of the ocean surface directly by the use of electro-magnetic pulses.
- Sea-bed pressure recorders detect the passage of a tsunami by the change in height of the water column. Data are transmitted to a surface buoy via sonar. The surface buoy radios the information to the Pacific Tsunami Warning Centre on Hawaii, via satellite.

QUESTIONS

1   Using a map of the Indian Ocean, research and draw circles showing the arrival times, at hourly intervals, of the 2004 Indian Ocean tsunami.

2   Explain why Hawaii is often significantly affected by tsunamis.

Due in part to our recent glacial history, a large proportion of the British Isles is covered in clays. Approximately 25% of the land area has a moderate or significant risk of damage by swelling and shrinking of clays. It is on these glacial deposits that we erect dwellings and other buildings, construct transport links and rely upon for agriculture.

Classification of clay minerals

Layer type	Group	Examples of clay minerals in group	Structure and properties	Industrial use
1:1	Kaolin-Serpentine	**Kaolinite** Serpentine	Non-expanding Low shrink-swell capacity	Ceramics and porcelain. Filler for paint, rubber and plastics. Paper industry to produce glossy finishes.
2:1	**Smectite**	**Montmorillonite** Saponite	Expanding structure High shrink-swell capacity	Drilling mud for oil industry. Protective liners. Catalytic processes. Facial powder. Fillers. Fulling. Cat litter.
	Vermiculite	**Vermiculite**	Limited expansion Medium shrink-swell capacity	[When heated it expands to be a low-density medium] Refractory processes/fireproofing. Insulation (when added to cement). Packing material with high absorbency. Gardening use. Cat litter.
	Mica	**Illite** Muscovite Biotite	Non-expanding Low shrink-swell capacity	Ceramics and stoneware. Fillers.

The table shows a simplified classification of common clay minerals. Those shown in bold are the most common industrial clay minerals.

Sedimentary clays are mostly mixtures of illite and kaolinite, with some montmorillonite. Shales are dominated by the clay mineral illite, and this is the most widely distributed clay mineral in marine argillaceous sediments.

Clay crystal structures

It is the arrangement of layers and the method of bonding between the layers that gives the variety of clay minerals and their peculiar and useful properties.

All clays consist of mineral sheets packaged into layers, and most can be classified as either 1:1 or 2:1. These ratios refer to the proportion of tetrahedral sheets to octahedral sheets.

▲ **Figure 1** The hazard potential due to swelling and shrinking clays (BGS)

Key
- Low to nil
- Moderate
- Significant

KEY FACTS: DIFFERENT CLAY MINERALS

Clay minerals are hydrous aluminium phyllosilicates. They form flat hexagonal sheets, with a small particle size of less than two micrometres.

Smectites are 'swelling clays' with an expandable crystal structure for water to be contained. Montmorillonite is a smectite clay mineral, although there are others.

Kaolin is known as China Clay, and is largely made up of kaolinite, along with a mixture of other clay minerals. This is formed by the breakdown of feldspar by the action of water and carbon dioxide; the weathering of granite.

Another useful clay, bentonite, is composed mainly of montmorillonite, along with some other clay minerals. Bentonite is formed by the weathering of ash deposits from volcanoes. Bentonite is not a mineral in its own right but is recognised as a useful clay deposit, with a common name of 'Fuller's earth'.

CASE STUDY: KAOLINITE

A common clay mineral is kaolinite ($Al_2Si_2O_5(OH)_4$). Kaolinite is known as a 1:1 type clay mineral, which means it comprises one tetrahedral sheet linked to one octahedral sheet. Tetrahedral sheets consist of $Si_2O_3(OH)_2$ and the octahedral sheets are $Al_2(OH)_6$. All tetrahedral sheets have the apices of their tetrahedra facing in the same direction. The aluminium is present in the octahedral sheets, which has bonds known as hydroxyl groups, holding the layers together. The relatively strong bonding means that the lattice is fixed and no other cations or water can get between the sheets. The result is low ion adsorption, low **plasticity** and low swelling and shrinking. Kaolinite does not expand when it comes in contact with water and is the preferred type of clay for the ceramic industry.

CASE STUDY: SMECTITE

These clays are more complex, forming a 2:1 type structure and the most common example is Montmorillonite (Ca or Na).

The unit has an octahedral aluminium layer sandwiched between two tetrahedral silica layers. In this case the units are only weakly bonded by those H^+ ions and attract cations. Water molecules can enter the clay by moving into the spaces between the units. These clays have high plasticity, high cohesion and high swelling and shrinkage.

Montmorillonite is quoted to have many different formulae, given from different sources. The one we use here is the general formula (either Na or Ca montmorillonite):

$(Na, Ca)_{0.3} (Al, Mg)_2 Si_4 O_{10} (OH)_2 \cdot nH_2O$

The nH_2O in this formula denotes the expandable nature of the clay as n can be variable.

Smectite, when fully expanded, can increase in volume by up to 1500%. This is the preferred clay mineral for drilling muds, creating a protective clay liner for hazardous waste landfills and is a great absorbent.

For completeness, you should be aware that there are some 2:1 clays that do not expand and many mixed clay deposits.

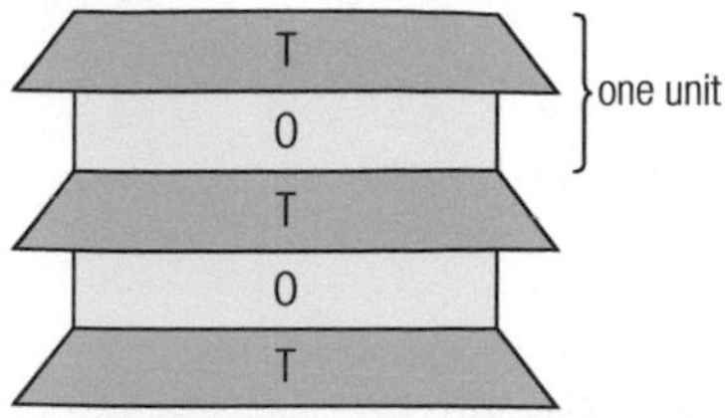

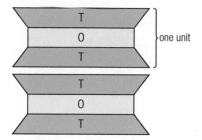

▲ **Figure 2** Generalised kaolinite structure. The sheets of silica tetrahedral (T) are interleaved with the aluminium octahedral sheets (O). Side view.

▲ **Figure 3** The three-layer unit of the Montmorillonite structure. Side view.

Technical and engineering problems

Expansion of lattice

Expansion of swelling clays produces unwanted forces. This is enough to crack and lift footings, foundations and floor slabs.

Shrinkage of lattice

The shrinkage of the heavy clays in the east and south-east of England caused subsidence that cost £250 million during the 1975–76 drought. It does not have to be a substantial outcrop to cause problems; if a soil contains at least 5% of clay minerals by mass, it could have the ability to shrink and swell. Of course, this is dependent upon the clay minerals present. Shallow pipes buried in the zone of seasonal water table fluctuation, are exposed to enormous stresses by shrinking soils. If water or sewage pipes break, the resultant leaks can cause swelling damage to nearby structures. Hence the mineralogy of soils is of importance to engineers.

Engineering geology mitigating shrink-swell hazards

It is essential to have a very detailed geotechnical investigation of the proposed site for any major structure. At this point it should be known if swelling clays are going to be a problem. The structure should have foundations deep enough to be below the zone of any fluctuations in the water table. Clays have a reasonable compressive strength but poor shear strength which affects the way piles are supported by friction. A continuous, reinforced raft foundation will help prevent damage to smaller structures. It is a good idea to avoid the contrast between the dry clays under the centre of foundations and the watered clays at the edges due to rain or irrigation. An effective drainage system can be a lot cheaper than underpinning a house. The water table can be artificially lowered by pumping or kept high by irrigation. Trees can be removed but as the ground re-saturates it may 'heave' and cause damage anyway.

Changing the properties of clays

There are soil chemical treatments available designed to alter the clay mineralogy and reduce the expansion potential. Treatment with 'lime' or calcium oxide is the most traditional method. The expansion of an Na-smectite is 1500%, whilst that of a Ca-smectite is only 100%.

Given the importance of clay minerals in soils (where they hold organic matter and nutrient elements) and their possible role in removing harmful cations from the environment, much research has been carried out on the effects of oxidation and reduction on the properties of clays. The exciting bit is that this can be done in situ, e.g. by using bacteria to reduce the Fe^{3+} to Fe^{2+}. In smectites, this decreases the tendency to swell and increases the negative surface charge. It is thought that this also increases interlayer cation fixation so the clay particles will hang onto K, most undesirable for nutrient supply to plants. The manipulation of the iron oxidation state causes large changes in chemical and physical behaviour and because the changes can be made in the field, there is great potential to benefit humanity.

QUESTION

1   Give two economic uses for each type of clay mineral. Ensure that you link the properties of the clay to the uses you describe.

The shrinkage and swelling of clays is one geological reason why buildings and other structures can become unstable. Other geological reasons such as subsidence may be much more dramatic and dangerous than clays shrinking. A detailed understanding of the local geology in these cases can help avoid chaos!

Subsidence due to mining

Any part of the country underlain by coal measures could be affected by subsidence resulting from longwall mining – now increasingly rare, as deep mining in the UK has been phased out (2015). Shallow coal mining in old bell pits and pillar and stall workings have left unstable voids underground. At the time of working, no records were kept and so collapse was often a surprise. Later, deep mines led to more predictable subsidence and, eventually, systems of compensation for those affected were put into place.

In longwall mining the excavation of the seam is followed by the collapse of the roof as the process moves forward. The rock above the seam will be heavily fractured (to about 2.5 times the seam thickness) and the resulting subsidence spreads out to affect an area as much as 1.4 times the height from the workings to the surface. The angle at which the subsidence spreads out towards the surface depends upon the strength of the strata and the depth of cover to the coal seam. Subsidence at the surface will be less than the thickness of coal removed as the fallen rock underground will still have lots of voids following collapse.

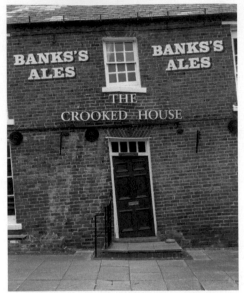

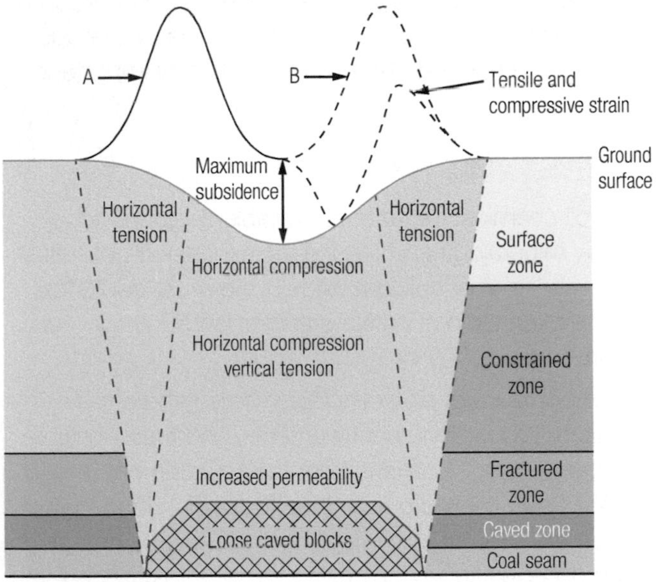

Figure 1 Subsidence in the UK is a real problem especially in areas where there has been underground coal mining.

Subsidence is measured in mm, cm or even m. If the subsidence is vertical, then there is little cause for concern; there is not much to complain about if your whole house is 100 mm closer to sea level. The problem is that the subsidence does not occur all at once, but as a wave with some parts subsiding before others. It is not very good if *part* of your house is 100 mm lower! This results in horizontal extensions and then compressions, the horizontal displacements can be greater than the vertical subsidence. **Tilt** is calculated as the change in elevation between two points divided by the distance between those points.

Faults can extend the duration of the hazard. Subsidence following longwall extraction is usually completed within weeks and is generally complete within 4 months. However, the reactivation of faults can extend the subsidence hazard for years after mining has ceased.

Salt extraction by drilling and pumping down hot water leaves large caverns under parts of Cheshire held up by pressurised brine. There are some instances of the brine escaping, followed by subsidence. There is even a Cheshire Brine Subsidence Compensation Board.

A = Tilt
B = Horizontal movement

Figure 2 Stresses and strains due to longwall mining. Vertical scale exaggerated. Note that the tilt, movement and strain curves are only shown for half the zone of subsidence; they repeat across the centre. Also note that tensile strain changes to compressive strain as it passes the **inflection point** on the ground profile. This is also where the maximum tilt and horizontal movement occur.

Crown holes and sinkholes

The more dramatic of the subsidence hazards; suddenly a deep hole opens up in a road or under houses with catastrophic consequences. Although limited in the area affected they are becoming more common as climate change produces more prolonged and intense periods of rainfall. Unusually wet weather in 2014 resulted in a dozen or so sinkholes in urban areas. Most affected were the south-east, the Isle of Wight and parts of Yorkshire.

Crown holes (denes, meaning a 'hole' or 'valley') are the result of human activity such as the gypsum mining in Yorkshire. Heavy rain finds its way into these voids and begins to enlarge them. The cavity formed holds more water, causes more solution and increases in size, until it moves to the surface as its roof collapses.

In limestone country, the continual collapse of cavities leads to a 'pock-marked' landscape of depressions known as 'Karst topography'. **Sinkholes** result from the natural dissolution of bed rock. If you live above salt or gypsum you are at risk, but carbonates are the chief culprits, as even the usual mild acidity of rainwater (pH ~5.7) is sufficient to attack chalk and limestone. Joints and bedding planes are enlarged, they allow in more water and more dissolution takes place leading to the formation of cave systems. As the rock dissolves, the soil is carried down leaving a funnel-shaped depression on the ground above. This depression collects more rainwater and the process continues. If there is a lot of clay in the superficial layers, the sinkholes may end up holding water providing an unusual circular pond or lake; a weird sight in limestone country.

Sometimes these caves collapse suddenly, and the cavity breaks the surface. What triggers their collapse? A sudden increase in surface water adds to the load on the roof of the cavity or, during droughts as water tables fall, the previous support of the water is lost.

The sinkholes that make the news are usually formed in built up areas and are commonly caused by leaking sewers or water pipes. In August 2015, a 12 m crater opened up on the Mancunian Way, Manchester, which took a very long time to repair. This sinkhole is tiny compared with the 662 m deep Xiaozhai tiankeng ("heavenly pit") in Chongqing, China. This limestone sinkhole is a famous site for the sport of base jumping.

Engineering geology to the rescue?

There are few preventative measures to consider in relation to sinkholes. Avoiding Karst terrain for major structures would help. Attention to drainage around the area will be essential. Techniques for protecting structures from mining subsidence include using reinforced raft foundations or isolating the foundations from the movement of the surface with PVC or sand layers. Traditional ground survey techniques are used for monitoring subsidence impacts. Satellite radar imaging can provide the level of accuracy required for analysis, but it is costly.

HOW SCIENCE WORKS: ENGINEERING GEOLOGY – COST-BENEFIT ANALYSIS

A huge amount of engineering geology is carried out for mining operations. Imagine trying to test and then apply the strengths of the different rocks above the coal seam to work out their capacity to bridge the voids resulting from mining! This complex process is routinely carried out to predict where damage to buildings, other structures or water resources might occur.

The simplest solutions are to limit the width of the extracted panels, leaving more coal to support the rock above or to reduce the worked seam thickness. This has to be balanced against the profitability of the mine. This compromise requires intensive geological investigations, rock mechanical testing and numerical computer modelling of alternative mine layouts and dimensions. The aim is to minimise tensile strain and tilt or where protection of surface waters and aquifers is the primary concern, to limit vertical crack propagation. A typical 2.4 m thick seam would require at least 120 m of overburden cover to protect overlying water bodies and would dictate a maximum panel width of 75 m. Some mines backfill with waste to reduce subsidence or to grout between separating beds in the roof.

KEY DEFINITIONS

Tilt calculated as the change in elevation between two points divided by the distance between those points (the first derivative of the subsidence profile). The maximum tilt occurs at the point of inflection in the subsidence trough. Tilt is usually expressed in millimetres per metre.

Inflection point a point on a curve at which the curve changes from being concave to convex, or vice versa.

Sinkholes hollows or holes in the ground caused by collapse of a surface layer, usually by dissolution of limestone.

CASE STUDY: SUBSIDENCE DUE TO GROUNDWATER EXTRACTION

Central London, a comparatively expensive area, sank almost 300 mm due to over-extraction of groundwater from the confined chalk aquifer. That artesian water used to rise to within 3 m of the surface in 1820 but by 1936 had fallen to 90 m below, as the water was extracted faster than it was recharged, and the pressure dropped. The mass of the overlying rocks and the city were partly supported by that water pressure and lowering it caused the aquifer to lose porosity and subsidence occurred. Levels continued to fall to up to 1965 when the abstraction of water from the aquifer was restricted and recovery began. This caused alarm in the 1980s as the rising water table (3 metres per year in the centre of the basin) threatened the tunnels and massive structures put in place when the levels were low! A policy of abstraction to keep the aquifers at a stable level has been successful.

HOW SCIENCE WORKS: SINKHOLES OR CROWN HOLES?

The media use the term sinkhole to describe any hole which suddenly opens up in the ground.

QUESTIONS

1 Describe why subsidence occurs following underground mining.

2 Research several mining accidents that occurred in the United Kingdom during the 20th century. Try to ascertain the reasons for these accidents.

3 Research the geology of where you live, to determine the risk of subsidence. Is there a history of sinkholes or mining? Are there any other geological risks?

Figure 1 The Greek village of Ropoto after a catastrophic slope failure

KEY DEFINITIONS

Talus the debris accumulating at the foot of a slope due to erosion of the rock face above.

Translational slide when material begins to move as an entity, e.g. several beds sliding down slope. It usually breaks up as it gathers speed and the energy increases. Contrast the rotational slide that marks failure of incompetent, clay-rich rocks.

Isotropic clay having equal properties in all directions. This is not the same as being homogeneous, which conveys being made of the same material throughout. A slate can be homogeneous in being made of metamorphosed mudstone but is not isotropic in that it has a strong cleavage.

Tsunamite a tsunami deposit which is a sedimentary unit deposited onshore during the inundation phase. The offshore deposits as the water retreats are different.

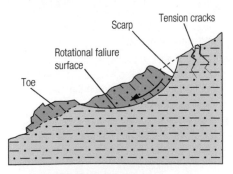

Figure 3 Rotational slide or slump in isotropic clay-rich rocks. The over steepened scarp may cause another slump to occur.

Mass movements are slope processes, ranging from very slow soil creep, to virtually instantaneous rock falls. They include landslips, slides, slumps, flows and falls. They are usually much less spectacular than volcanic eruptions, earthquakes, fires and floods, but annually cause far more damage.

In 1970, a debris avalanche, buried the town of Yungay, Peru, killing 20 000 people. More than 50 million m³ of debris slid 15 km downhill at an angle of about 14° at speeds up to 272 m s⁻¹.

Ropoto was a large village with around 300 families in residence, but a landslide in 2012 meant that the village had to be abandoned. Ropoto is still sinking, and the half standing structures are still moving.

Causes and effects

The mass of rock on a slope is held in place by frictional forces. Increase the mass, increase the slope or decrease the friction and movement will be the result, down the slope.

It is easy to increase the mass: just add water or a building. Decreasing the friction is also a regular occurrence: just add water to pump up the hydrostatic pressure on that incline, or to reduce the shear-strength of clays. Changing the slope can be the result of erosion by rivers or waves; or the result of excavations to build roads, tunnels or buildings on slopes.

Types of mass movement

Mass movements are often classified by the *velocity* of the material on the move.

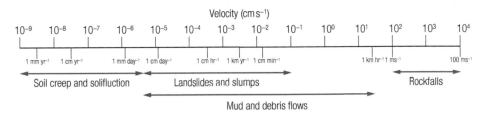

Figure 2 Land instability based upon the rate of movement

They are also classified by the type of material that moves. If water is a major component we have a flow; if clays and water are involved it is a mud flow or mudslide.

Competent rocks form steeper slopes but have discontinuities (bedding planes, joint sets, foliation). If the planes of weakness are horizontal or dipping into the face, it may be possible to sustain vertical faces. When these fail, the result is a rock fall which will create a **talus** slope at the base of the cliff. Competent rocks failing along planes of weakness may result in **translational slides,** in which the rock moves as an entity down the slope. As it gains speed it will break up and the rocks will separate, bouncing off each other, to form a more fluid-like landslide.

Incompetent rocks may be more **isotropic** and form less steep slopes but contain a high proportion of clay minerals, which have a low shear strength, especially when wet. When these materials fail, the resulting slump (a rotational slide) occurs along a spoon-shaped surface as the material moves down and out to create a toe of debris.

Landslides also occur underwater. The 'Storegga Slides' are considered to be amongst the largest known landslides. They occurred at the edge of Norway's continental shelf causing a very large tsunami in the North Atlantic Ocean. This collapse involved an estimated 290 km length of coastal shelf, with a total volume of 3500 km³ of debris.

What triggers mass movements?

It's important to recognise the triggers, as many people live at the base of slopes. In mountainous terrain, valley floors offer the only flat ground for buildings, roads, railways and agriculture. It is vital to avoid actions which increase the load by adding water or building structures up the slopes, yet this often occurs. Clear-felling slopes drastically increases the available water in the sub-surface and reduces the stabilisation provided by root systems.

The triggers are not always the direct result of human actions and most slope failures are the result of heavy rainfall. Rainfall raises the pore water pressure in the sub-surface. Extreme weather, due to climate change, is not improving the situation. Less obviously, sudden falls in the water table can trigger failures, e.g. sudden drops in river or reservoir level cause shear forces.

Earthquakes are notorious for starting mass movements. The seismic waves pump up the pore water pressure and reduce that crucial friction, as in liquefaction. The complex accelerations of the various waves can be enough to reduce the load forces acting on the surfaces and there is an interaction with the topography that seems to focus these effects at ridges, giving maximum slope failure.

A likely triggering mechanism for the Storegga Slides is thought to be an earthquake causing a calastrophic expansion of methane hydrates. A cubic metre of solid hydrate expands to 164 cubic metres of methane.

▲ **Figure 4** Rock fall warning sign. Road cuttings over-steepen slopes and increase the danger that ice or water will trigger rock fall.

HOW SCIENCE WORKS: SLOPE STABILISATION

- Slope modification. The slope is reduced to a lower angle to increase its stability.
- Retaining wall. Usually constructed of concrete and used to support the sides of road cuttings.
- Gabions. Wire mesh boxes filled with rocks and placed as lateral toe-support at the bottom of slopes to prevent failure by slumping.
- Rock bolts. Steel rods several metres long are drilled and cemented into rock faces then tensioned. They pin loose blocks of rock to the sound rock behind and prevent rock falls. Rock bolts can only be used in competent rocks.
- Rock drains. Addition of water is a common cause of slope failure, so pipes or drains of broken rock can be constructed, especially in clay, to remove water and reduce the pore fluid pressure.
- Wire netting.
- Shotcrete. Concrete is sprayed at high pressure onto rock surfaces. Shotcrete increases strength, reduces permeability, and protects the surface from weathering.
- Vegetation. Plants 'fix' soil in place and reduce infiltration of water. It is one of the few effective stabilisation strategies for incompetent rocks such as clay.

CASE STUDY: GEOLOGICAL EVIDENCE FOR TSUNAMIS

The Shetland Islands have evidence of three North Sea tsunamis between 8000 and 1500 years ago. The oldest is the well-known Storegga Slide off the Norwegian continental slope, around 8100 years ago. Sea level was around 10 to 15m lower at the time, and deposits from this tsunami are over 9 metres above current sea level, meaning run-up exceeded 20m. Deposits contained rip-up clasts, sand layers, redeposited sediments and diatoms.

There was a younger tsunami dated at 5500 years, with a run-up of 10m.

A further tsunami was dated at 1500 years, with sediments from this event found at two sites, 40 km apart.

These deposits are called **tsunamite** or *tempestite* (storm deposit).

Tsunamis

Tsunamis are normally triggered by shallow-focus earthquakes, above magnitude 6.5; displacement of the sea floor causes a series of high velocity, low amplitude waves in the entire depth of the water.

As these waves approach land and slow down, so the amplitude builds. This is especially serious if the waves are entering a restriction on the coast line. The sea withdraws then waves with amplitudes of up to 30m race onshore engulfing everything before them.

Coastal dwellers in the UK do not live in fear of these horrific events, which are considered a hazard for those living closer to destructive plate margins but they are not always triggered by earthquakes and they have affected these islands in the past.

TSUNAMI HAZARD ZONE

In case of alert go to high ground Or assembly area

▲ **Figure 5** Warning sign from Taupo Bay, New Zealand

QUESTIONS

1. If the slope angle increases from 30 to 50 degrees, what is the % increase in the force acting to cause a slope failure?
2. Why is it better if a road cut is made at right angles to the strike of bedded rock?

10.10 Rock strength

In engineering geology, it is essential to know that the foundation rock used will not fail once loaded. Ask anyone living in a valley below a large dam whether it matters. Large loads can be exerted on samples of rock in the laboratory to test their strength. The strength of a rock can be defined as its ability to resist **stress** without large-scale failure.

The mechanical properties of a rock depend on its mineral composition, the arrangement of the mineral grains/crystals (**fabric**), and any cracks that may have been introduced along the way. This means that there will be variation, not just between different rock types, but also between different samples of the same rock. The large number of variables that affect rock strength does not stop us from trying to gain some idea in order to safely build tunnels and structures. It is better to do the tests and then allow a very safe margin.

Some necessary basics

Common stress units for working with rocks are MegaPascals (MPa = 10^6 Pa). The Pascal (1 Nm^{-2}) is very small; atmospheric pressure is 101325 Pa at sea level.

When a stress is applied, it causes the rock to deform. The amount it deforms is called the **strain** and is defined as, for example in a tensile stress, the change in length divided by the original length so that it is dimensionless. The definitions will be slightly different for the application of shear stress which changes the shape of the rock or the application of **confining pressure** which results in a change in volume.

The way in which rocks respond to stress is also important. You will know from structural geology that the response depends not just on the size of the stress and the rock type but also on temperature, confining pressure and time. Rocks at depth undergoing high temperatures and pressures are more likely to deform in a **ductile** manner to give folds, especially if the strain happens over a very long time. If the opposite conditions apply and the cold, near-surface, low pressure rock is deformed rapidly, it will fail in a **brittle** fashion, and faults and joints result.

Measuring rock strength

The availability of core samples recovered by diamond drilling has encouraged the measurement of intact, homogeneous, cylindrical rock samples. In the most commonly applied test, Uniaxial Compressive Strength, these cylinders are placed between the jaws of a hydraulic press. The load is very slowly (e.g. 1 µm per second) increased and recorded along with the distance moved until the sample fails at peak strength. It is called uniaxial because the **principal stress** is down the axis of the cylinder. The other stresses are negligible.

In graphs of the results (Figure 1), it is usual to show the stress applied (force divided by cross-sectional area of the cylinder) and the strain that results (divide by the original length of the cylinder). This is an unconfined test, there is no confining pressure. Tests which can measure strength *and* apply confining pressure are called triaxial.

There are hydraulic rigs for applying all the types of stress to rocks (and soils). Shear strength involves applying force in opposing senses and is more often used for soils or to measure the strength across discontinuities.

KEY DEFINITIONS

Stress defined as the force per unit area acting on the rock. It can be *compressive stress* that tends to compress or shorten the material acting normal to the stressed area or *shear stress* that acts in a plane parallel to the stressed area or *tensile stress* that tends to stretch or lengthen the material; acting in the opposite sense to compressive stress.

Fabric describes the spatial and geometric configuration of all the 'components' of a rock: the mineral crystals and their sizes, shapes and relative orientations or the mineral grains, the voids between and their relative sizes and dispositions. Close to the term 'texture' for sedimentary, igneous and metamorphic rocks.

Strain the response of a system to an applied stress. When a material is loaded with a force, it produces a stress, which then causes a material to deform. Engineering strain is defined as the amount of deformation in the direction of the applied force divided by the initial length, volume or shape of the material (it has been normalised).

Confining pressure the combined lithostatic and hydrostatic pressure. At depth all principal stresses are equal ($\sigma 1 = \sigma 2 = \sigma 3$).

Ductile deformation occurs when a rock suffers large strains without large-scale fracturing. It bends and flows.

Brittle deformation causes the rock to fracture (possible after some elastic deformation has occurred).

Principal stresses a useful way to describe what is going on. $\sigma 1$ is the maximum, and $\sigma 3$ being the minimum and $\sigma 2$ is the intermediate principal stress direction. All three principal stresses are perpendicular to each other.

Lithostatic pressure the vertical pressure due to the mass of the rock only. It is also referred to as the overburden pressure.

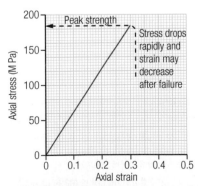

Figure 1 Graph of a uniaxial compression test on a granite sample.

Interlocking crystals and cementation

The common rock-forming minerals are tough. The rocks made from them are often not as strong. Look at the table of differences in the strengths of a metaquartzite with intergrown quartz crystals and an orthoquartzite sandstone. Both metaquartzite and orthoquartzite are composed of the same minerals but the grains of the orthoquartzite are more easily separated. The strengths of a rock will depend on its fabric and so it is only possible to give broad ranges of values for rock types.

Rock type	Uniaxial compressive strength (UCS) (MPa)	Uniaxial tensile strength (MPa)	Shear strength (MPa)	Classification based on lower limit of UCS
Granite	100–250	7–25	14–50	Very high strength
Dolerite	100–350	15–35	25–60	Very high strength
Basalt	100–300	10–30	20–60	Very high strength
Quartzite	150–300	10–30	20–60	Very high strength
Sandstone	20–170	4–25	8–40	High strength
Shale	5–100	2–10	3–30	Low strength
Limestone	30–250	5–25	10–50	High strength

The obvious trends are that compressive strength is much greater than shear which is greater than tensile strength. This reflects the relative difficulty of crushing crystals together compared to shearing off parts of them. The easiest action is to separate the crystals or grains by pulling them apart. This has implications for shear strength across discontinuities with rough surfaces.

Rock density

In order to calculate the **lithostatic pressure** we will need to know rock density. Measured in SI units of kilograms per cubic metre this large amount of rock is troublesome for experimental purposes so you will often find traditional measurements quoted in grams per cm^3 (this gives the same value as tonnes per cubic metre).

Rock density does depend on the constituent minerals but that does not result in a great range as the common rock-forming minerals have relatively similar densities.

The quoted ranges are for the compositional differences, e.g. in the plagioclase solid solution series. There are some higher densities quoted for olivine but these may be representative of high-pressure forms.

The way in which these mineral components are arranged is also very important. Any voids, even if water filled, result in a large reduction in density. Any voids will be lost as depth increases, due to increasing confining pressures, but in engineering geology we are mostly dealing with near-surface materials.

The pressure due to the rocks above will be significant in mining and tunnelling. Lithostatic pressure is the vertical pressure due to the weight of the rock only. At depths greater than a kilometre, the stresses act just like hydrostatic pressure. Lithostatic pressure then has all the principal stresses equal ($\sigma1 = \sigma2 = \sigma3$) and is equal to $\rho \times g \times h$ where ρ is the density, g is the acceleration due to gravity and h is the depth of rock involved.

Weird and dangerous things happen under these conditions; it is possible for the floor of the tunnel to explode inwards in a rock burst!

There are other pressures at work such as the pore pressure; that of the fluids within any voids. The hydrostatic pressure gradient is as shown in Figure 2, but pore pressure may be higher, particularly if high temperatures are involved.

MATHS BOX: ROCK DENSITY

A quartz sandstone has a porosity of 15% (and it is highly permeable). What is the rock density of the dry rock and what is it when it is saturated?

Simple if we assume that the density of air is negligible in this scenario. The porous rock will be 85% of quartz ($2650\,kg\,m^{-3}$) i.e. $2252\,kg\,m^{-3}$.

Make it easy on yourself – just work out the mass of one cubic metre.

15% of 1000 litres ($1\,m^3$) is water, so that is $150\,kg$.

85% of that cubic metre is quartz at $2650\,kg\,m^{-3}$ which is $2252\,kg$.

The total for the saturated rock is $2402\,kg\,m^{-3}$.

Mineral	Density (ρ) $kg\,m^{-3}$
Quartz	2650
Plagioclase	2620–2730
Muscovite mica	2820
Olivine	3220–4390
Pyroxenes	3260–3310
Amphiboles	2850–3440
Biotite mica	3200
Calcite	2710

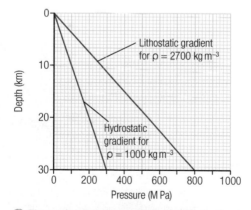

▲ **Figure 2** Increases in pressure with depth

QUESTIONS

1 Describe the differences between the following pairs of terms:
   - Stress and strain
   - Lithostatic pressure and confining pressure
   - Plastic deformation and ductile deformation.

2 A granite is composed of 30% Quartz ($2650\,kg\,m^{-3}$), 45% Orthoclase ($2560\,kg\,m^{-3}$), 15% Plagioclase ($2620\,kg\,m^{-3}$) and 10% Biotite ($3200\,kg\,m^{-3}$). What is the rock density?

Samples from cores when tested in laboratory conditions yield consistent results. The trouble is that rock is usually not that homogeneous and the samples in the laboratory may not be representative. There may be foliation or lamination, and at an outcrop scale, the rock will contain discontinuities, which are places where the properties suddenly change. These could be actual fractures such as joints, faults, or bedding planes where the rock type changes. These represent potential weaknesses in the rock and there is a high probability that the rock will fail at these weak points.

In engineering geology, it is essential to analyse the discontinuities in terms of their severity, frequency and orientation before assessing whether the site is suitable. The testing of unconfined samples usually provides the lower limit of the rock strength, as rock at depth has confining pressure.

Strength across discontinuities – the basics

Failures at discontinuities are usually reactions to shear stresses; those massive blocks you hoped wouldn't move down an inclined surface. Shear strength measurements give different results from compression in that, as the stress increases, there will be some elastic behaviour (O–A–B) until the point where the material holding the discontinuity together fails (C). This could be due to the roughness of the surface and the **asperities** (uneven surfaces) being forced over each other or crystals fracturing or some combination of the two. With further displacement the stress falls (C–D) until it reaches a **residual strength** which remains no matter how large the movement.

KEY DEFINITIONS

Asperity the term mostly used to describe the roughness of the surface of a discontinuity.

Residual strength the remaining resistance to movement after the rock has failed and been displaced.

Joint sets are fractures across which there is little displacement. They are mostly to dissipate the residual stresses left after folding. As they are the result of regional stresses they tend to form sub-parallel sets.

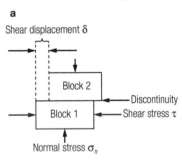
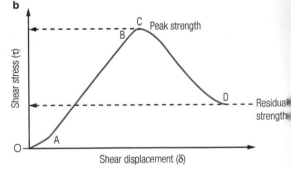

Figure 1 **a** Displacement due to shear stress. **b** Stress-displacement curve for rock with a discontinuity under shear stress.

Joints

The strength of a jointed rock mass depends on the strength of the rock and on the type, orientation and spacing of the **joint sets**.

Joint types are:
- Tensional, producing an angular discontinuity which may resist shear stresses
- Shear, which are smoother and less likely to resist stress.

A jointed rock mass is stronger if:
- The joints are infilled with secondary minerals, forming veins
- The joints are narrow or tight rather than wide and open
- There are fewer joints per unit length.

Figure 2 Sea cliffs at Portovenere, Italy, with failing faults and joints. The cliffs in the foreground are more unstable as bedding dips towards the face, in the distance the opposite is true.

Fractures allow water to enter the system. The resulting hydrostatic pressure acts against the area of the joint surface no matter how close the joint, decreasing frictional forces.

Some joints arise due to unloading. The internal stresses due to burial are released when the rock is excavated or exposed by erosion and uplift. This produces dangerous sets of joints parallel to the exposed surface. These may need grouting to improve their stability.

Faults

Faults are discontinuities which pose a range of stability hazards. They may contain gouge, which is likely to have a clay content, and lose strength when saturated. They may allow passage of water, leading to dissolution and removal of fine material. These problems are magnified for dams and reservoirs where the hydrostatic pressure can be high enough to cause failure.

Faults are likely to have different rock types on either side. These will have different properties and therefore react to loads unequally, e.g. settling and compression.

Bedding planes and foliation

Bedding planes mark a time when deposition temporarily ceased. During that hiatus, fine material could settle out and that is often clay rich. This forms a weak layer especially when wet.

Alternatively, the plane marks a change in rock type and therefore properties which could lead to a concentration of stresses at that boundary. If the plane marks a sudden decrease in permeability it may encourage water to permeate downwards to accumulate at that level.

Foliation appears to be less of a problem, as metamorphic rocks are recrystallised *but* a strong anisotropy can lead to unexpected failure. The Malpassat thin-arch dam near Freyjus in France failed in 1959. The increased hydrostatic pressure on the gneiss it was built into lifted a huge block of rock; all the impounded water was released in a single wave killing 400 people.

Other geological structures

Severe folding can be a nightmare for a geotechnical investigation, as there are rapid changes in orientation and the jointing. The Tebay stretch of the M6 motorway was an unexpected nuisance and required a massive investment in rock-bolting to hold the sides of the contorted cutting together. The Tebay road cutting is a classic piece of geology which is studied by many undergraduate geologists on a parallel road away from the motorway (A685).

Weathering of rocks and sediment

Weathering of sediment will change its properties and may adversely affect its suitability for engineering purposes. Occasionally, it can be useful as the finer crystalline igneous rocks have improved bonding with bitumen if the surfaces are weathered.

Weathering alters rock strength and if not taken into account could cause catastrophic failure. Often the weathered rock is removed at great expense until sound and strong rock can be found to serve as a foundation. It only needs one susceptible mineral out of all those that make up the rock to destroy the competence.

As an example; in a set of sandstone samples the 'partially weathered' unconfined compressive strength (UCS) values averaged 9.78 MPa but in 'distinctly weathered' samples the average was only 3.72 MPa. Even tunnels may be affected as weathering can be very deep if it is following major joints sets or faults, so proper sampling of the site is essential. Geophysics can provide useful coverage as seismic velocities are much lower in weathered rock.

MATHS BOX: FORCES ON A SLOPE

The force due to the mass of the block (shown as **mg**) can be resolved into the forces acting to move the block down slope (**mg sinθ**) and the force acting normal to the surface which helps it stay in place (**mg cosθ**).

The graphical method of resolution is shown by the dashed rectangle. The small arrows acting normally to the joint surface are representing hydrostatic pressure. Notice that it acts to reduce the normal frictional force and to increase the downslope forces when the dip of the joints is orientated in this way.

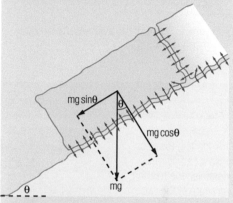

🔺 **Figure 3** Forces acting on a block on a slope

MATHS BOX: STRESS AND SHEAR STRENGTH

The higher the normal stresses (those acting at right angles to push the surfaces together), the higher the stress needed to get the movement going. If you plot up the results of peak shear stress versus normal stress there is a simple linear graph and its intercept 'c' is known as the **cohesion**.

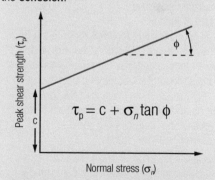

🔺 **Figure 4** Normal stress vs peak shear strength, produces a straight-line graph. The Mohr-Coulomb equation (shown) describes the relationship and shows that shear strength is made up of two components: the cohesive (c) and the frictional (σ_n tanφ).

QUESTION

1 The Malpassat dam was 60m high. What was the hydrostatic pressure in the rocks at its base?

Figure 1 Some of the symbols used by the BGS in drift mapping

- Peat
- Alluvium
- River terrace deposits
- Alluvial fan

Before the building of any major structure, there has to be a survey by engineering geologists. Without a careful site investigation, the results can be embarrassing, expensive or worse, cause loss of life. If a new building sinks it is a problem; if a structure fails, then it is lethal, as when more than 2000 lives were lost in the Vajont Dam disaster.

There are usually five stages to the investigation:

1 A preliminary research into published information.
2 A detailed geological survey.
3 Geophysical surveys.
4 Boreholes and excavations at critical points.
5 Testing of soils and rocks to assess their mechanical properties.

Desk study

The BGS mapping of the UK shows **drift** as well as bedrock and structure. We are very fortunate in that the 1:63 360/1:50 000 maps of England and Wales and of Scotland are available to consult and that for professional use, up-to-date digital data can be purchased. The BGS also hold data on the results of previous boreholes which can provide information on how the geology changes with depth. They may also help with a range of geohazard mapping such as ground stability and radon potential. Not all countries have such easy access to data.

Studying aerial and satellite photographs can be useful to analyse terrain.

Any data on the site will need research; for example, there may be previous workings known beneath the site.

The desk study will include some records of how the water table has fluctuated in the past. Once again the BGS has records of groundwater levels in many areas.

Site surface mapping

The map will be at large scale and of a cleared site, as when covered in soil and vegetation the information is limited. All changes in earth materials are mapped. This is not just about rock types and weathering but often includes subsoils, which may contain clay and are left in place.

All structures are mapped to a detail required, which is dependent on the scale of the project and size of the proposed 'load'. They include faults and their displacement, joint sets and their frequency and openness. The type and orientation of joint movement could be vital. It would be helpful to know which layers of the strata are affected by faulting so that the most recent activation is known. The dip of beds and other discontinuities is an obvious structural measurement as it affects stability and fold axes are often sites of increased jointing.

Geophysical surveys

For small sites this could include ground-penetrating radar to detect near-surface anomalies and changes. Larger areas could use resistivity surveys (both electrical drilling and constant spaced electrode mapping). Resistivity surveys are good at finding voids, they have a very high resistivity! Reflection seismic surveys may be needed where deeper changes in rock types are mapped, for example in tunnelling.

Simple data, such as the depth of the weathered layer, can be cheaply assessed using a **seismic refraction survey**, as the weathered rock will have a significantly reduced velocity. Such surveys also provide the seismic velocity of the upper layers.

CASE STUDY: INSTABILITY OF BUILDINGS

It is easy to quote the leaning tower of Pisa as a failure to establish the bearing strength of the two different clays it was built across. That structure was started in 1173 and the masons were aware of ground stability problems in the area. Estimated at 14 500 metric tonnes the tower stands 56.7 m on the high side with the centre displaced sideways 3.9 m at the top.

Figure 2 Il Torre Pendente, Pisa (leaning tower)

Built on inadequate foundations it started to tilt by the time the second stage was built. Work stopped and over time the compressed clays strengthened. Some of the tilt was compensated in the building of the tower but the dangerous 5.5° lean had to be corrected by extracting material from under the high side between 1990 and 2001.

It is less easy to excuse San Francisco's $350 million Millennium Tower completed in 2009. An ambitious structure of 58 storeys (197m) of glass and concrete, its foundations are supported by concrete piles up to 27m deep. Unfortunately they are supported by their friction with the surrounding material, in this case muds and sand. These are not ideal and by 2016 the building had sunk over 40cm causing (so far) a 15cm tilt at the top of the tower. If piles had been driven down to bedrock (end-bearing piles) they would have needed to be 60m deep and would have prevented the subsidence.

Site subsurface mapping

Test pits in softer earth materials can be dug to sample all critical areas likely to be loaded and to provide background information for the site as a whole. It is relatively cheap to excavate a shear-sided pit large enough for someone to climb into and record the profile and extract samples. Some soil mechanical properties can be tested in situ. However, there are health and safety factors to consider in this practice especially in pits deeper than a metre, as collapse is a possibility.

Cores should be drilled at intervals likely to represent the area and especially under loaded foundations. Cored material will provide samples for laboratory analysis and breaks in core recovery will show up weak layers. Drilling will also reveal hydrogeological information. The water table level should also be known across the site.

Rock and soil property measurements

Samples of rocks and soils are analysed for strength and composition. For soils, simple penetration tests give an idea of the strength, whereas for rocks it could be necessary to carry out compressive and shear strength testing to establish how they will react to loading. Testing is very important on weathered rock, as strength will vary drastically according to conditions such as water penetration into the rock.

Permeability and porosity should be assessed as the passage of water through the subsurface can change the rock and soil strength.

Geohazard mapping

In steeper terrain, gradients are routinely mapped to look at the stability of slopes, the effects of the work planned in the area and possible work needed to stabilise slopes. This used to be accomplished using contour maps, analysis using stereo pairs of aerial photographs and ground surveys. Now, in many areas, satellite altimetry is available such that it is only a matter of setting the required parameters and letting some heavy-duty program do the calculations. This is not to say that experienced ground survey will not be necessary, if it can be afforded. Satellite images and air photos often show tell-tale signs of previous landslips and slumps which would alert the geologist to look for the potential hazards around the site.

Integration of data

Modern computing power makes analysis of all this disparate data much easier. GIS enables the geologist to look at all the factors needed to plan the work in map form with data represented as layers.

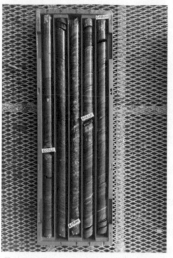
▲ **Figure 3** Diamond drilled core of dipping strata boxed in order with markers showing depth. In geotechnical investigations, the breaks in core recovery are important as they show weaknesses and levels of potential failure.

▲ **Figure 4** Tracked drilling rig for near surface sampling

QUESTIONS

1 (a) In the cross-section, there are proposed drill core points marked. After drilling at **J**, it is decided to next drill at **H**, explain the decision.

(b) How many boreholes are required to fix the position of the fault?

(c) Due to expense it is decided to drill only four boreholes. Which positions would give most information?

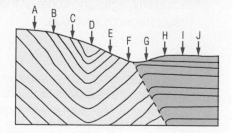

KEY DEFINITIONS

Drift the superficial deposits such as glacial and fluvioglacial material. The 'solid' maps show an overlay of drift. Alluvium refers to the recent deposits by rivers.

Seismic refraction surveys can be made using a simple dropped mass or large hammer as the energy source. Waves are refracted by changes in velocity such as at the base of the weathered layer.

Earth, the early years

Well-preserved ancient rocks are needed for an analysis of plate tectonics in the early part of Earth's history. Isotope studies (hafnium) of Greenland rocks have shown a marked change 3.2 billion years ago. Before that time the mantle was comparatively hot and processes were dominated by mantle plumes. The world was almost entirely ocean with a few small continents with exotic names such as Vaalbara (now part of the South African **craton**). After 3.2 billion years, the mantle had cooled enough to allow stable convection cells to operate, and the tectonic processes we are familiar with today took over.

One of the earliest supercontinents, Kenorland, formed around 2.7 billion years ago, and had a major influence on climate. It provided shallow seas in which photosynthetic bacteria could gain dominance over anaerobic deep-water bacteria. Even though the solar energy was much weaker (<85% of present levels) this marks the start of the **Great Oxygenation Event**. When Kenorland began to break up, around 2.6 billion years ago, it caused major change in the climate. Increased rainfall caused increased weathering which removed CO_2, and the resulting temperature drop led to the first **Snowball Earth**.

Columbia formed between 2000 and 1800 Ma, breaking up some 500 million years later. The next to be formed was Rodinia around 1000 million years ago, which contained most of Earth's continents. The break-up of Rodinia caused the second Snowball Earth to be triggered, by several mechanisms:

- There was increased rainfall which caused extensive weathering using up lots of CO_2.
- Atmospheric oxygen rose from 0.1 to 1% during this time.
- The **albedo** of continents is higher than oceans so the region cools.

A combination of these factors started the formation of the ice cap and set up the feedback of: more ice equals more reflectivity and more cooling. The resulting glaciation lasted 60 million years! Rodinia was separated by rifting which was caused by a mantle super-plume and the continent broke into eight micro-continents around 600 million years ago. Those fragments may have formed an intermediate, short-lived supercontinent **Pannotia**. The continents eventually collided to form the familiar supercontinent Pangaea; a single land mass surrounded by the ocean Panthalassa.

How can we know any of this? The geology of the continents, even when eroded, reveals a patchwork of old cratons with younger rocks, known as greenstone belts, joining them together. The rocks can be analysed to get radiometric ages and some of them yield up their past global positions using palaeomagnetic methods.

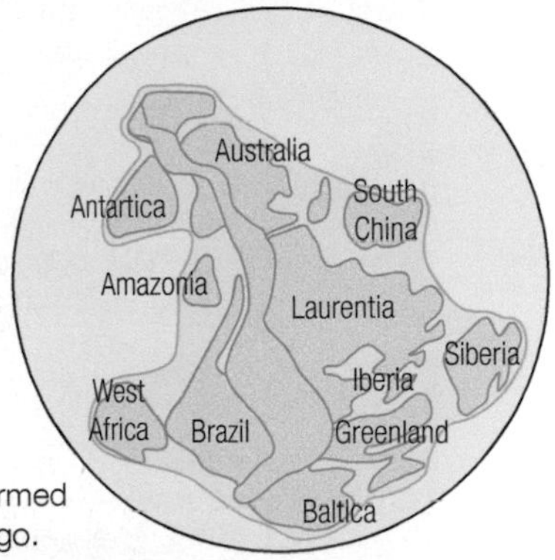

KEY DEFINITIONS

Cratons stable portions of the continental crust no longer tectonically active, often ancient and described as continental shields. Mostly crystalline basement rocks, they are thicker parts of the crust.

Great Oxygenation Event (GOE) was the biologically induced appearance of oxygen in the Earth's atmosphere around 2.3 billion years ago.

Snowball Earth describes a condition in which the planet is either entirely frozen or is frozen to very low latitudes. The hypothesis has been resisted by some, on the grounds that there is no clear mechanism to reverse the condition.

Albedo the fraction of solar energy (shortwave radiation) reflected from the Earth back into space. It is a measure of the reflectivity of the Earth's surface. Ice, especially with snow on top of it, has a high albedo: most sunlight hitting the surface bounces back towards space.

▶ **Figure 1**
A representation of Kenorland, a Precambrian supercontinent formed 2.7 billion years ago.

▲ **Figure 2** View centred on the south pole of the supercontinent Pannotia may have existed during the Ediacaran period (about 600 million years ago) and towards the end of the Precambrian (about 541 million years ago). After that, it split into smaller continents. The small ones were Laurentia, Siberia and Baltica, with the main land mass Gondwana to the south of them.

HOW SCIENCE WORKS: THE WILSON CYCLE

It is now known that plate tectonics is cyclical. Continents rifted apart to form new oceans which were balanced by the destruction of other oceans by subduction. The closure of the oceans ends with continental collision. J. Tuzo Wilson was one of the first to recognise that this cycle must have been occurring throughout Earth history. Palaeomagnetic data now confirms that the assembly of successive supercontinents takes place approximately every 500 Ma.

One model attributes the breakup of supercontinents to the action of mantle superplumes; the consequent volcanic emissions and the wild swings in climate have caused or helped cause mass extinctions such as the Permo-Triassic.

Pangaea

Pangaea, the whole Earth, was a concept named by Wegener in 1920. The southern half of this supercontinent was **Gondwanaland,** formed by collisions in the Late Precambrian. It incorporated South America, Africa, Arabia, Madagascar, India, Australia, and Antarctica. Gondwanaland existed before Pangaea formed and remained intact after Pangaea broke up. In the Carboniferous, Gondwanaland collided with **Laurasia** (North America, Greenland, Europe, and Siberia) resulting in the **Variscan orogeny** and the supercontinent of Pangaea.

The detail of the formation of Pangaea is complex, and new evidence and ideas are still emerging. The land mass that was to become Laurasia was initially drifting away from Gondwanaland creating the proto-Tethys ocean. Laurasia itself rifted creating new oceans including the well-known Lower Paleozoic ocean, Iapetus.

In the Early Ordovician, the microcontinent of Avalonia, a land mass that would become the north-eastern United States, Nova Scotia, and England, broke free from Gondwanaland and began its journey to join Laurasia.

A number of collisions of the northern continents occurred and by the end of the Ordovician, Avalonia was joined to a minor northern supercontinent as the Iapetus closed. Meanwhile Gondwanaland was approaching the South Pole.

By Permo-Carboniferous times (around 299 Ma) the three main blocks (Laurentia, Baltica and Gondwana) were assembled and then various late arrivals were joined on at the edges. Although Pangaea was essentially complete by the early Permian, Gondwanaland moved 3500 km westward, relative to the northern land masses. This brought Africa next to North America by the Late Triassic, producing the classic Pangaea configuration.

The breakup of Pangaea

Pangaea began to break up about 200 million years ago during the early part of the Jurassic.

First Gondwanaland split from Laurasia. Then, about 150 million years ago, Gondwanaland itself broke up. In the Cretaceous, India moved away from Antarctica, heading north at relatively high speed. Then in the Middle Cretaceous, the South Atlantic Ocean developed, separating Africa and South America. The third major phase of the break-up of Pangaea occurred in the early Cenozoic when North America/Greenland (Laurentia) broke free from Eurasia at about 60–55 Ma. The Atlantic and Indian Oceans continued to expand, closing the Tethys Ocean.

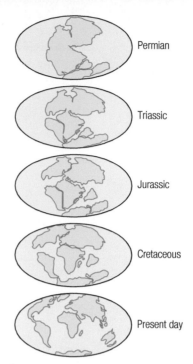

Permian

Triassic

Jurassic

Cretaceous

Present day

🔺 **Figure 3** Evolution of continents from the Permian supercontinent of Pangaea, to the present day

KEY DEFINITIONS

Pannotia a short-lived supercontinent that formed at the end of the Precambrian.

Gondwanaland also called Gondwana; formed between 570 and 510 Ma ago. It included most of the southern land masses, including Antarctica, South America, Africa, Madagascar, and the Australian continent, as well as the Arabian Peninsula and India.

Laurasia combines the names of Laurentia, the name given to the North American craton, and Eurasia. It contains most of the present northern hemisphere continents.

Variscan orogeny (or **Hercynian**) is a mountain-building event caused by late Palaeozoic continental collision between Euramerica and Gondwana to form the supercontinent of Pangaea.

HOW SCIENCE WORKS: CLIMATE

The position of the continents has major effects on global climate. The present distribution of continents encourages the ocean and atmosphere currents to redistribute heat energy around the globe. If the continents were over the poles, it would encourage accumulations of snow and ice, which reflect radiation, so reducing the temperature (albedo effect). There would be a greater contrast in temperature between the poles and the Equator.

CASE STUDY: FORMATION OF CONTINENTS

The main Archaean, Proterozoic and Phanerozoic supercontinents and their ages are shown in the table.

Approximate age (Ga)	Name of continents	Description of continent
3.5	Vaalbara	Single small continent
3.0	Ur	Half size of Australia
2.7–2.6	Kenorland	Supercontinent
1.8	Columbia	Third of present land area
1.0–0.6	Rodinia	Supercontinent
0.65–0.56	Pannotia	Short-lived supercontinent
0.30–0.17	Pangaea	Appropriately 'all the Earth'

QUESTIONS

1 What is the difference between Laurasia and Laurentia?

2 List the geological evidence that suggests that the Gondwanaland supercontinent existed in the past.

3 Which mass extinction is associated with the breakup of Kenorland?

We live in turbulent times, there are concerns that we have destabilised the climate and poisoned the oceans, by altering atmospheric composition. In September 2016, carbon dioxide levels reached 400 parts per million (ppm) for the first time in many millions of years.

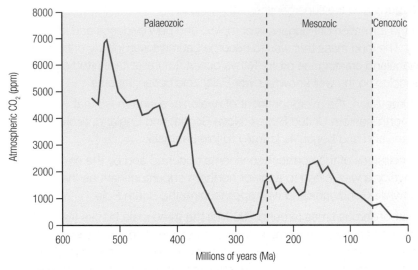

Figure 1 Calculated atmospheric CO_2 throughout geological time

Evolution of the very early atmosphere

As the solar system formed, the atmosphere was probably H and He, but soon those light gases escaped. The primitive atmosphere was composed of gases released from volcanoes and as a result it was dominated by carbon dioxide, similar to the atmosphere of Venus today. There would have been some water vapour, nitrogen, ammonia and methane but only traces of oxygen. Hawaiian volcanoes today emit gases that are thought to be similar to those emitted in the geological past; mostly composed of water vapour and carbon dioxide. The water vapour condensed to help form the oceans about 4 billion years ago. This would leave carbon dioxide as the dominant gas in the atmosphere. This stage lasted about a billion years, but it was early life in those oceans that produced radical changes in atmospheric chemistry.

Evolution of oxygen in the early atmosphere

Around 3.8 billion years ago, photosynthetic cyanobacteria began to produce oxygen. At first this was used up by oxidising Fe^{2+}, which is water-soluble, to Fe^{3+} which precipitated in the oceans. When all the iron was used up, the oxygen began to leak into the atmosphere and from about 2.3 billion years ago, atmospheric oxygen began to build up. Two billion years of photosynthesis had also worked to reduce atmospheric CO_2. The bacteria transformed the atmosphere paving the way for the Cambrian explosion of aerobic life.

Changes in climate – icehouse and greenhouse

On a geological timescale there are periods where the planet is hotter and colder than average. We refer to the warm periods as Greenhouse Earth and the cold as Icehouse Earth.

We currently live in an icehouse where large continental ice sheets exist at both poles. This began in Antarctica 34 million years ago and in the Arctic about 2 million years ago. At least three times during Earth's history, the planet has been in a true icehouse state, when ice sheets extended from the poles to the tropics.

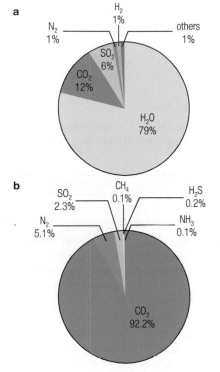

Figure 2 The gas composition of **a** emissions from Hawaiian volcanoes and **b** the primitive atmosphere, an estimate due to volcanism

Figure 3 Modern stromatolites in Western Australia. Blue-green algae trap fine sediment in shallow waters and build up to form these finely layered mounds. Similar organisms were responsible for oxygenating the primitive atmosphere.

Icehouse events are characterised by lower temperatures, ice caps and glaciers. The ice sheets from the last period of glaciation, called the 'Ice Age', are still present in Antarctica and Greenland. The huge increase in ice coverage then increases the drop in global temperatures by reflecting more of the Sun's radiation back into space (albedo effect).

Greenhouse events are characterised by a lack of ice coverage and an overall increase in global temperatures. They can be caused by an increase in the amount of solar radiation reaching the Earth or a change in the concentration of gases in the atmosphere.

Volcanic events contribute to both factors; the particulates and sulfate aerosols reflect incoming solar radiation but in the longer term the greenhouse gases emitted may raise temperatures. The eruption of Pinatubo in 1991 caused a surface cooling in the Northern Hemisphere of up to 0.6°C. Major temperature cycles are linked to plate tectonics and the Wilson cycle. The aggregation of continents leads to icehouse events; the breakup is associated with greenhouse conditions.

Changes in sea level in the Phanerozoic

Effects of climate change on sea levels

At the time of maximum glaciation (about 12 000 years ago), people were living on the floor of what is now the English Channel and the North Sea. Rising sea level gradually separated Britain from the rest of Europe and submerged forests are seen around the coast of England and Wales.

Today, if all of the Greenland ice cap melted, it would create a sea-level rise of 7 m, while the melting of the ice of Antarctica could create a 75 m rise.

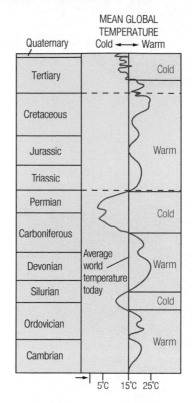

▲ **Figure 4** Mean global temperatures throughout geological time; the centre line marks the mean surface temperature today.

CASE STUDY: VAIL SEA-LEVEL CURVES

The seismic interpretation of sea-level curves was pioneered by a team at Exxon, the oil company, led by Peter R Vail, and later revised by Haq. Oil companies need to understand sea-level changes for successful oil exploration. Graphs produced using this technique are known as the Vail sea-level curves.

From the Triassic to the present, these sea-level curves identified more than 100 global sea-level changes. When looked at over millions of years the combination of all the data on sea-level changes shows patterns at many levels. If the graph is smoothed a little, it shows a pattern over tens of millions of years of frequent transgressive and regressive events, known as second-order cycles. If the graph is smoothed further, it shows a broad cycle lasting hundreds of millions of years as first-order cycles, which coincide with times of major continental plate break-up.

The high sea level that existed in the Cretaceous occurred at the same time as the opening of the Atlantic Ocean when the rate of seafloor spreading was high. Sea-level oscillations have continued with a rough periodicity of 100 000 years linked to astronomical cycles.

CASE STUDY: ISOSTASY AND EUSTASY

These are two mechanisms for changes in sea level:

Isostatic sea-level changes are due to uplift or subsidence of the continental crust. Crust often sinks when loaded with ice or sediment, rising again when such loads are removed. These are only seen in the affected region. Isostatic rebound is evident in Scotland today where raised beaches can be seen.

Eustatic sea-level changes are due to changes in the volume of the ocean basins or volume of water in them. These changes are seen worldwide.

QUESTIONS

1 Compare the primitive composition of the atmosphere with that of the present day. Explain the major differences.

2 If sea levels rose by 110m during the Jurassic period what is the overall rate of sea level rise in mm per year? How does this compare to the 2007 eustatic rise of 3 mm per year? Suggest an explanation for the difference.

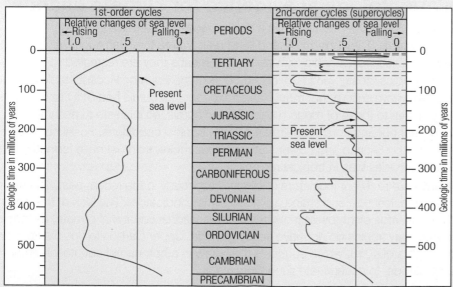

▲ **Figure 5** Changes in sea level through the Phanerozoic

Lithological evidence for climate change

The UK has Carboniferous coal deposits and Permo-Triassic desert sandstones and *must* have experienced northerly drift since they were deposited. This is the essence of *uniformitarianism*.

There are a number of rock types associated with different climates and the latitudes in which those climates dominate today.

Coal

In order to produce peat in sufficient thicknesses to be compressed into economic coal seams, there has to be a highly productive ecosystem. Rapid plant growth requires high rainfall and high temperatures such as in the tropical rainforests found in equatorial regions.

Desert sandstone

Sands exposed to the air develop a red colour as the iron hydroxides oxidise to hematite. Desert sands are fine-grained, very well sorted, with well-rounded grains consisting mostly of quartz. They are transported by wind to produce spectacular dunes. Dune-forming desert latitudes are generally around 20–30° south or north of the Equator.

Evaporites

These deposits form where rainfall is low and evaporation is high, for example, in hot deserts. Evaporites today are found at latitudes of about 20–30° south and north, so the same is likely for ancient deposits.

Tillites

These are ancient boulder clay deposits so are glacial deposits, which are most commonly formed at high latitudes of more than 60° north or south of the Equator. However, there are glacial deposits in low latitudes as small areas of high-altitude glaciers.

Reef limestone

Today, reef limestone is built mainly of colonial corals which are restricted to latitudes of less than 30° north or south of the Equator, so reef structures most likely formed at these latitudes.

🔺 **Figure 1** **a** Coal showing remains of plant structure.
b Sands exposed to air for about 5000 years forming dunes, the sandstones of the future.
c Fossils of colonial corals in reef limestone.

Palaeontological evidence for climate change

- **Corals** Modern reef corals live in a narrow range of water depths and temperatures. They only thrive and build reefs where they can successfully exploit their symbiotic relationship with algae. This means they are restricted to the warm tropics mostly within 30° N and 30° S of the Equator. Recent stable isotope analyses have shown that most Palaeozoic tabulate corals had $\delta^{18}O$ to $\delta^{13}C$ ratios very similar to those of modern scleractinian corals, so it is safe to assume that fossil reef building corals required the same tropical conditions. Fossil coral reefs are found in Silurian and Lower Carboniferous rocks, so must have formed when the British Isles was in tropical regions.

- **Plants** are excellent indicators of climate, especially in the recent past. The use of tree rings, leaf size and shape are good climate indicators. The lack of tree rings in Carboniferous plant fossils indicates a non-seasonal equatorial climate, as those in current day tropics do not have seasonal rings. Upper Carboniferous plants grew to heights of 40 m, which suggests they grew in a hot, humid equatorial climate and latitude, to accumulate such biomass.

Isotopic evidence for climate change

There are changes in Rare Earth Element (REE) ratios and more obvious correlations with salinity and pH, as temperatures change and ice forms, but the common markers for temperature are provided by stable isotope ratios.

Oxygen isotopes and temperature change

Water may not appear to change in any obvious way, but the atoms of oxygen within it can vary. Oxygen has three stable isotopes, ^{16}O, ^{17}O and ^{18}O. The majority (99.76%) of the world's oxygen is ^{16}O, but a noticeable percentage (0.20%) is ^{18}O.

The lighter ^{16}O isotope is removed more easily by evaporation, so tends to leave more ^{18}O behind in the ocean, altering the ratio of the two isotopes.

Under normal circumstances the ^{16}O in the evaporated water is returned to the ocean after falling as rain or snow and draining or melting through the river system. In periods of glaciation, the light oxygen is trapped in the ice caps and glaciers. As temperatures drop globally, the ratio of ^{18}O to ^{16}O changes. More ^{18}O is present in the oceans in colder temperatures than in warmer temperatures. Once the ice caps have melted and the ^{18}O to ^{16}O ratio has returned to normal, the only evidence of the change remains in the oxygen of the calcium carbonate shells of bivalves, belemnites and some microfossils such as foraminifera.

Carbon isotopes and temperature change

Isotopes of carbon have been used in a similar fashion. The common ^{12}C accounts for the majority (98.9%) of the carbon in the global cycle, but there is a small amount of the heavy isotope ^{13}C (1.1%).

The temperature-dependent process using carbon isotope ratios is similar to that of oxygen, so that delta ^{13}C values from pelagic and benthic fossils can be correlated with climate change. Plants preferentially take up ^{12}C. Thus, during glacial periods, when the terrestrial biomass was greatly reduced, the ocean appeared relatively depleted in ^{13}C.

Carbon isotopes are more useful than oxygen isotopes for stratigraphic purposes in the remote past up to 10 million years ago because they are more resistant to diagenetic change.

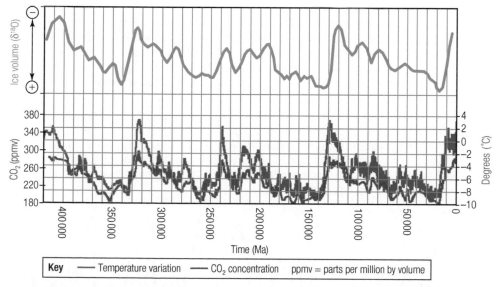

Figure 3 Oxygen isotope (the values are expressed as delta ^{18}O, which is the difference in the ratio compared to the world standard) compared to CO_2 and temperature measured in ice cores based on data from the Vostok station in Antarctica

QUESTION

1 Describe and explain how oxygen and carbon dioxide isotopes can give information about climate.

CASE STUDY: THE NORTHWARD DRIFT OF THE BRITISH ISLES

The evidence in the British Isles includes:

- Tillites are found in Precambrian rocks in the west of Scotland, suggesting a glaciation.
- The colonial corals in the reefs of the Silurian Wenlock Limestone suggest shallow tropical seas.
- Reef limestones of the Lower Carboniferous in the Pennines again suggest tropical seas.
- The Coal Measures of the Carboniferous formed in equatorial rainforests growing on a huge delta, extending from Scotland to Kent.
- The Permian and Triassic rocks include red desert sandstones and evaporites of the Cheshire Basin, which formed within 30°N of the Equator.
- Some colonial corals in the Jurassic mean tropical seas north of the Equator.
- The Chalk Seas of the Cretaceous represent temperate conditions as the British Isles move northwards into cooler climates.

Period	Diagnostic rock types	Approximate palaeolatitude
Quaternary	Glacial deposits	55°N
Tertiary	Palms and tropical plants	40°N
Cretaceous	Chalk	35°N
Jurassic	Rare colonial corals	30°N
Triassic	Desert sandstones	31°N
Permian	Desert sandstones, evaporites	12°N
Carboniferous	Reef limestones, coals	0
Devonian	Desert sandstones	20°S
Silurian	Reef limestones	30°S

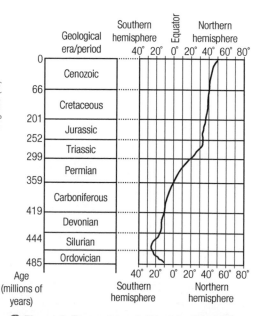

Figure 2 The northward drift of the British Isles

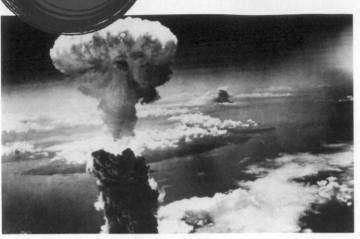

Figure 1 The start of the Anthropocene?

If alien geologists landed on a future Earth dominated by arthropods and started to work out the pattern of events in the past, they would doubtless come to the same conclusions about where to divide up the timescale (they *are* geologists). They would use major events such as mass extinctions or rapid climate changes. Some scientists think that they would use the present as the beginning of a new epoch, the **Anthropocene**. In August 2016, it was recommended to the International Geological Congress that 1950 should mark the end of the **Holocene** and the start of the Anthropocene. The working group say that although it is a very short period in geological terms, the changes involved are irreversible.

Environmental changes

Are the environmental changes of such a scale as to stand out as a marker, given the climate changes that have previously occurred? If the present variations are part of a repetitive cycle, then we are not at the start of something new and it would provide ammunition for those that deny anthropogenic climate change.

2016 marks the first time that atmospheric CO_2 has stayed above 400 ppm in the last 15 million years. Rising from 280 ppm before the industrial revolution, it is currently increasing at the fastest rate since the end of the Cretaceous. Increases in greenhouse gases can be matched with increases in temperature as we move back through the geological record. Scientific evidence suggests that the current rise in mean surface temperature is linked to the anthropogenic release of CO_2 and that rising sea levels will follow as land-ice melts. The human population has also doubled the amount of methane in the atmosphere while driving up carbon dioxide concentrations due to our activities.

HOW SCIENCE WORKS: TEMPERATURE CHANGE

Geologists often take a long-term view; this can be a very valuable standpoint alongside industries and modern-day companies who benefit from short-term gains and have vested interests. On climate change they could point out that we are currently in the middle of a glacial period, the third major cold cycle in the Phanerozoic. Average temperatures have been falling for the past 80 million years. In comparing present-day changes to the past it should be noted that there have been some sharp thermal spikes such as the Paleocene-Eocene Thermal Maximum (PETM).

Geologists could point out, on a shorter time-scale, that ice-core data shows a series of cycles which have been correlated with Earth's orbital variations. However, this does not reduce the seriousness of the change occurring now: In the past century alone, the temperature has risen 0.7°C, roughly ten times faster than the average rate of the post ice-age warming in the last million years. Our civilisation has been based on the relatively stable climate of the last 11 700 years, the Holocene, and it seems that we have disrupted that equilibrium.

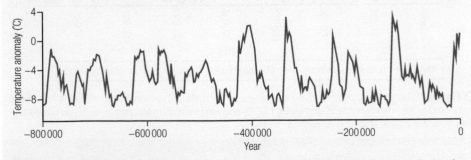

Figure 2 Temperature anomalies for the past 800 000 years (data from the European Project for Ice Coring in Antarctica Dome C (NOAA))

Possible biological changes

We, as humans, are most likely causing the sixth major mass extinction in the Phanerozoic. This puts humanity on a par with a 10 km diameter asteroid strike or similar to a major volcanic eruption resulting in basalts covering 2 million km^2.

Species have always become extinct: changing environments or competition used to result in the loss of one to five species a year. Biologists estimate the extinction rate is presently 1000 to 10 000 times that rate. Amphibians, with us since the late Devonian in some form, are disappearing at 25 000 to 45 000 times the background rate. Globally, it is estimated that, excluding bacteria, there are 8.7 million species living on Earth, so at a rate of 50 000 a year we could have a planet with only bacteria in 174 years. More reasoned estimates are that we are on course to lose 75% of species in the next few centuries unless we reverse current trends. This is on a par with some of the mass extinctions in geological history.

From a palaeontological point of view, the domestic chicken is a serious contender to be a fossil that defines the Anthropocene for future geologists. It has become the world's most common bird, fossilised in thousands of landfill sites. The Anthropocene could also be marked by strata that contain the remains of cities. Rising sea levels will inundate a good proportion of our coastal cities, preserving the lower parts of buildings in marine muds.

Which marker horizon is suitable?

To define a new epoch we need a marker that can be recognised in the stratigraphic record. These are 'global standard stratotype-section and points' (GSSPs); geologists call these 'golden spikes'. These are locations where the strata show clear evidence on a worldwide scale, such as a particular zone fossil or global event. Once ratified, the golden spikes are set as the boundary of a time period even if the absolute age has to be changed. For example, the Cretaceous-Tertiary has the 'spike' of higher iridium levels marking the impact of the asteroid.

Similar geochemical markers could be the unique combination of radionuclides distributed around the world by nuclear tests of the mid-twentieth century. One famous set of nuclear tests was at Bikini Atoll where 23 nuclear devices were detonated by the USA between 1946 and 1958. Many of the dangerous radionuclides such as Caesium-137, Strontium-90 and Iodine-131 have relatively short half-lives but the Plutonium-239 has a half-life of 24 400 years and would make a long-lasting marker as it was first made in 1941 (altogether, about 3 tonnes was distributed into the atmosphere).

Some regard the effects of agriculture as a better world-wide marker: we have added nitrogen and phosphorus as fertiliser, ploughing to create unique soil profiles. This would require setting the start of the Anthropocene back a thousand years or so. The plastics in our oceans and rivers break down into microscopic granules, and will form a recognisable layer for thousands of years, out of the reach of UV light. We recognise past burning events by the soot layers and our industrial efforts have created one such layer visible in sediments and glacial ice.

Anthropocene – an acceptable term?

The term was proposed partly to draw attention to the permanent damage being done to the environment and to life on Earth. In this it has achieved its purpose. The 35 scientists on the working group voted to designate the Anthropocene and will decide both which markers are the strongest and on a location which contains the recognised boundary. It remains to be seen whether the International Commission on Stratigraphy will adopt the term.

STUDY TIP

Weather and climate are different. The day-to-day changes in temperature, precipitation or sunshine are familiarly understood as weather. *Climate* is a longer-term synthesis of weather for a particular place. Some use a 30-year average of weather variables to define climate. Climate changes can be seen by looking at weather records for the past few hundred years and then pushed back hundreds of thousands of years by analysing ice cores. For longer-term data we have to look at the geological record.

QUESTIONS

1 Research the Milankovitch astronomical cycles which can influence (force) the changes in climate. These are eccentricity, obliquity and precession. Which of the cycles fits the long-term period found above?

2 If Strontium-90 with a half-life of 28.8 years was released by testing in 1945, what proportion would be around in 2029?

Evolution

Evolution is the process by which the inherited characteristics of a population are passed on from parent to offspring. The English naturalist, Charles Darwin, is the best-known contributor to several theories of how this has resulted in the species we know today. His work was mostly based on living species and could be summarised as a gradual change resulting from natural selection. Alfred Wallace should also be famous, as he jointly proposed the theory with Darwin.

They both reasoned that some of the random variations resulting from cross-breeding and mutations are advantageous. The individuals who were more able to adapt would be more likely to pass these characteristics to their offspring and so the changes would spread in the population. The phrase 'survival of the fittest' described the mechanism well. Eventually there would be so many small changes that a new species would have to be acknowledged. A species is an organism that can interbreed to produce fertile offspring. The technical description for this view of evolution is **phyletic gradualism** which theorises that speciation is slow, uniform and gradual.

Evidence from the fossil record has been sought to prove this theory, and is not as easy as it first appears. Preservation as a fossil is a rare circumstance, and the geological record itself is full of gaps. Also, the fossils that are found may not show important changes, such as in the soft tissues, as these are only rarely preserved.

Many biological ideas of speciation are built upon knowledge of soft tissues and genetic coding, which is absent from the fossil record. Most soft tissues are not preserved due to decay and DNA is only stable for a short time, quickly becoming fragmented in the fossil record, over thousands of years, rather than millions.

KEY DEFINITIONS

Phyletic gradualism model of evolution which states that most speciation is slow, uniform and gradual.

Genes sections of DNA that code for a protein. A gene is made up of two alleles, one from each parent, if the organism reproduces sexually.

Adaptive radiation a process when organisms diversify rapidly into many different forms, usually as a response to a change in the environment or to exploit a new ecological niche.

HOW SCIENCE WORKS: THE OLDEST HUMAN GENOME

Scientists have sequenced the oldest human DNA ever, extracted from 430 000-year-old samples of fossilised tooth and a thigh bone, found in Spain's Sima de los Huesos, which translates to 'pit of bones'.

In doing so, the team from Germany has found evidence that the ancient ancestors of modern humans must have split from the ancestors of Neanderthals hundreds of thousands of years earlier than we thought, which means it might be time for us to redraw the human family tree.

[Excerpt from www.sciencealert.com 15th March 2016]

CASE STUDY: EXTINCTION AND RECOVERY

The Cretaceous–Tertiary boundary is well known for the extinction of the dinosaurs. This mass extinction, described elsewhere in this book, had a massive effect on evolution of life on Earth. The extinction of so many dominant groups allowed others to take their place, and there was a series of adaptive radiations early in the Tertiary. Animals such as insects underwent diversification and became more dominant, including butterflies and ants. The dinosaurs were replaced by mammals, which evolved quickly to fill the niches that were vacant. Mammals did not appear on Earth at this boundary, as their earliest ancestors evolved far earlier than this, originally from a group of reptiles called the *therapsids* in the Permian.

CASE STUDY: THEORIES OF EVOLUTION

There are several *theories* of evolution, summarised here:

Phyletic Gradualism

This is the process of gradual evolutionary change over time, as understood by the Darwin–Wallace theory. This involves 'creation' of new characteristics or **genes** due to cross-breeding and mutations. Evolution is seen as smooth and continuous.

Punctuated equilibrium

This reinterprets fossil evidence. It suggests that for long periods of time, very little change occurs in the population as it expands, known as 'stasis'. At the limits of its environment, the population is forced to adapt and a new species, which is more successful than its parent stock, takes over in a short time. There is no need to hypothesise that the intermediate stages in gradual evolution have been lost, the 'missing links' were never there. One species splits into two distinct species, rather than one species gradually transforming into another.

Genetic drift

This is especially apparent when there are few selective pressures on the population; the random sampling of parent genetic material in the offspring can eventually lead to the elimination of some characteristics or genes. The genes passed to the next generation are derived from 'lucky' individuals, not necessarily healthier or better.

Gene flow

This involves the exchange of genes between populations, for example, by migration in animals or pollination in plants. If gene flow carries genes that did not already exist within that population, then this will most likely result in genetic variation.

Fossil ranges

Geological time has been marked by periods of extinction, followed by **adaptive radiation**. This can be seen by the changing numbers of groups of different fossils throughout geological time, shown in the charts.

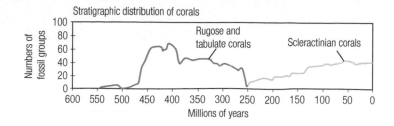

Stratigraphic distribution of corals

Rugose and tabulate corals

Scleractinian corals

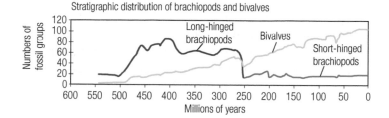

Stratigraphic distribution of brachiopods and bivalves

Long-hinged brachiopods

Bivalves

Short-hinged brachiopods

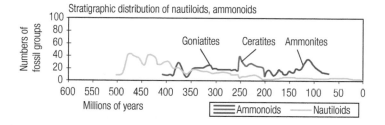

Stratigraphic distribution of nautiloids, ammonoids

Goniatites   Ceratites   Ammonites

Ammonoids ▭▭▭   Nautiloids ▭▭▭

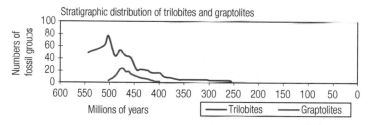

Stratigraphic distribution of trilobites and graptolites

Trilobites ▭▭▭   Graptolites ▭▭▭

🔺 **Figure 1** Stratigraphic ranges and relative abundance of fossil groups

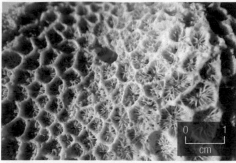

a Coral

b Brachiopod

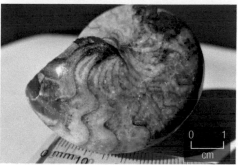

c Goniatite

d Trilobite

🔺 **Figure 3** Photographs of four major fossil types.

CASE STUDY: THE DISCOVERY OF DNA AND EVOLUTIONARY BIOLOGY

James Watson and Francis Crick first produced a model of DNA, a painstaking job of analysing X-ray data produced by another researcher, Rosalind Franklin. Watson and Crick realised that the DNA molecule was arranged like a twisted ladder, a double helix.

Evolutionary biology was revolutionised by the discovery of DNA, when it was realised that minor changes in a set of genes, often due to a mutation, are duplicated in the offspring. These mutations conveyed some kind of selection advantage to an individual, allowing them to live and reproduce more successfully. Successive generations may replace earlier versions in the gene pool. Genetic coding has allowed scientists to establish the interrelationships between living organisms by looking at individual bases.

Of course, when we look in the fossil record we have no idea of the genetic interrelationships between extinct organisms. Instead we rely on morphological similarities and differences to establish interrelationships.

🔺 **Figure 2** DNA double helix. DNA is 2 nanometres wide with 10 base pairs per turn.

QUESTIONS

1 What is the evidence for gradualism and punctuated equilibrium. What evidence does the fossil record give us for each of these theories?

2 Study the fossil distribution charts. Can you notice any patterns by comparing these charts? Can you account for any similarities or differences?

Classification

Trilobites belong to the Phylum Arthropoda, which includes familiar creatures such as lobsters, crabs, insects and spiders. It also includes some extinct forms such as eurypterids (sea scorpions) and, of course, the Class Trilobita.

Trilobites lived in the oceans of the Cambrian and Ordovician periods, beginning around 540 million years ago, becoming extinct at the Permian mass extinction. The morphological diversity actually peaked in the Ordovician. To have survived for nearly 300 million years is a testimony to the successful design and adaptability of trilobites. They were probably the first organisms on Earth to have good vision with an advanced design of eyes.

KEY FACTS: THE EXOSKELETON

- Comprises a cephalon, thorax and pygidium.
- Made up of three layers. A waxy outer layer and two further layers of chitin and protein. The middle layer may be impregnated with calcium carbonate for increased strength.
- Provides protection and support and attachment areas for muscles.
- May have had sensory hairs emerging from the exoskeleton to detect chemicals or changes in the environment.

KEY FACTS: FEATURES OF TRILOBITES

- Had complex compound eyes (if not blind).
- Each pleuron had a pair of jointed appendages (legs) for movement and gills for gaseous exchange.
- Trilobites underwent ecdysis (or moulting) to grow, as do modern day arthropods such as crabs and spiders, splitting along lines of weakness called sutures.
- Some could enroll into a ball to protect their softer tissues, legs and gills from predators.

KEY DEFINITIONS

Morphology describes the shape of the organism.

Benthonic means living on the bottom of the sea or a river, maybe in the sediment or on top of it.

Epifaunal means living on top of the sediment or substrate.

Substrate the sediment or rock at the bottom of the sea.

Pelagic means floating in the water column, usually the open ocean.

Nektonic means actively swimming in the water column.

Infaunal means living in the sediment, maybe a burrower.

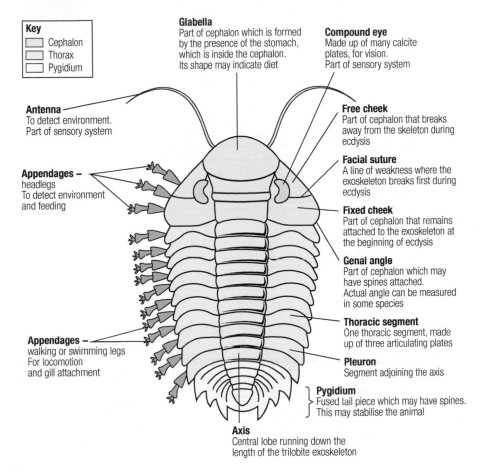

Key
- Cephalon
- Thorax
- Pygidium

Glabella
Part of cephalon which is formed by the presence of the stomach, which is inside the cephalon. Its shape may indicate diet

Compound eye
Made up of many calcite plates, for vision. Part of sensory system

Antenna
To detect environment. Part of sensory system

Free cheek
Part of cephalon that breaks away from the skeleton during ecdysis

Facial suture
A line of weakness where the exoskeleton breaks first during ecdysis

Appendages –
headlegs
To detect environment and feeding

Fixed cheek
Part of cephalon that remains attached to the exoskeleton at the beginning of ecdysis

Genal angle
Part of cephalon which may have spines attached. Actual angle can be measured in some species

Thoracic segment
One thoracic segment, made up of three articulating plates

Appendages –
walking or swimming legs
For locomotion and gill attachment

Pleuron
Segment adjoining the axis

Pygidium
Fused tail piece which may have spines. This may stabilise the animal

Axis
Central lobe running down the length of the trilobite exoskeleton

▲ **Figure 1** Trilobite morphology

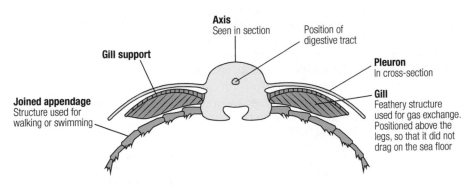

Axis
Seen in section

Position of digestive tract

Gill support

Pleuron
In cross-section

Joined appendage
Structure used for walking or swimming

Gill
Feathery structure used for gas exchange. Positioned above the legs, so that it did not drag on the sea floor

▲ **Figure 2** Cross-section of one thoracic segment showing position of legs and gills. The legs and gills are not usually preserved.

▲ **Figure 3** Trilobite showing eight thoracic segments and long genal spines

Adaptive radiation

Trilobites evolved to be a variety of shapes and sizes (different **morphologies**), probably as a response to the selection pressures in the environments where they lived.

BENTHONIC (E.G. *CALYMENE*)

These trilobites have little streamlining and possess complex compound eyes (360-degree vision). They had many pleura, which meant they could enroll for protection. Each pleuron had a set of gills for respiration and so the trilobite probably had an active metabolism. The mode of life can be inferred as an active hunter, **benthonic**, **epifaunal,** living on the **substrate**. As bottom dwellers, they crawled on the sea floor looking for food and preying on smaller invertebrates or scavenging. As well as giving them a good depth and field of vision, the eyes would have been sensitive to movement. This would have helped them detect food and predators, an advantage over other bottom dwellers in the same environment.

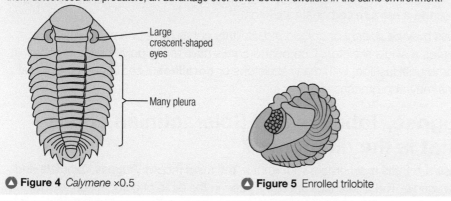

Large crescent-shaped eyes

Many pleura

▲ **Figure 4** *Calymene* ×0.5   ▲ **Figure 5** Enrolled trilobite

BURROWING (E.G. *TRINUCLEUS*)

These trilobites lack eyes and have a modified shovel-shaped cephalon with spines, probably to increase the surface area of the animal to avoid sinking into the sediment. The cephalon had pits running along its margins which have been interpreted as a sensory system to detect the environment (possibly chemicals, movement, currents or temperature) and the pits may have housed sensory hairs for this purpose. The inferred mode of life is benthonic, living on the substrate, as epifaunal or **infaunal** organisms. They may have dug shallow burrows for protection or to feed on the organic-rich sediment. Eyes would have been useless as these trilobites lived at depth, where light does not penetrate.

Large cephalic shield

Sensory pits

Genal spine

▲ **Figure 6** *Trinuclus* ×1

CASE STUDY: TRILOBITES ACROSS OCEANS

During the Ordovician, trilobites had evolved to exploit different environments, adapted to both different water depths and temperatures. At the time, Laurentia and Eastern Avalonia were separated by the Iapetus Ocean. Laurentia lay at or near the Equator, whilst Eastern Avalonia lay to the North (boreal). The deep-water trilobites showed similarities at species level, the medium depth water dwellers showed similarities at generic level. Those that occupied shallow waters were very different.

CASE STUDY: PLANKTONIC TRILOBITES

There are probably no examples of any truly planktonic trilobites. However, the small growth stages of trilobites when juvenile are thought by experts to have had a planktonic mode of life. By the time they became adults they were truly benthonic or nekto-benthonic.

Agnostus has been interpreted as having a planktonic lifestyle but is now believed to be nekto-benthonic. In fact, most experts now believe that it was not a trilobite at all, and may have been a stem crustacean.

CASE STUDY: PELAGIC (E.G. *OPIPEUTER*)

This genus probably lived high up in the water column. It had large complex compound eyes that could see backwards, forwards, up and down. It was small with a streamlined body. It had separated pleura and an inflated glabella, which may have helped it stay buoyant in the water column. It probably had strong muscles and was a very good swimmer.

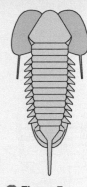

▲ **Figure 7** *Opipeuter* ×3

QUESTIONS

1 List the morphological differences between a benthonic and nektonic trilobite.

2 Explain why some trilobites have wide cephalic fringes and are blind.

3 Describe the likely mode of life for the trilobite shown.

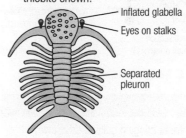

Inflated glabella

Eyes on stalks

Separated pleuron

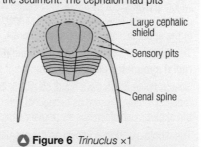

EQUATORIAL   Iapetus Ocean   Sea level   BOREAL

illaenid-cheirurid community   Pelagic community, e.g. *Opipeuter*   Pelagic community, e.g. *Cyclopygae*   neseuretus community

nileid community

olenid community   olenid community   raphiophorid community

LAURENTIA SPITSBERGEN/N.AMERICA   GONDWANALAND E. AVALONIA

Similarities at species level

Similarities at generic level

▲ **Figure 8** Lower Ordovician trilobite communities. This shows different trilobites which lived in the same water depths, but in different latitudes.

Corals belong to the phylum *Cnidaria*, which also includes sea anemones and jellyfish. Corals belong to the Class *Anthozoa*. Corals have a calcium carbonate skeleton, which is of immense importance in the geological record.

You should remind yourself of coral morphology, found earlier in this book.

Modern coral morphology

The polyps consist of two layers, an outer ectoderm and an inner endoderm, with stinging cells called **nematoblasts**. They have a larval stage where they are free swimming which attach themselves to a hard substrate and the soft **polyp** begins to secrete a calcium carbonate skeleton.

Corals may be *solitary* or *compound*. Solitary corals have one polyp which secretes a single skeleton. Compound corals have many polyps living together, in a communal fashion, with many skeletons or **corallites** fused together. Colonial forms may also branch.

Rugose, Tabulate and Scleractinian corals – what is the difference?

These are three main orders important in the fossil record: *Rugosa*, *Tabulata* and *Scleractinia*. The differences are highlighted in the table of comparison, below.

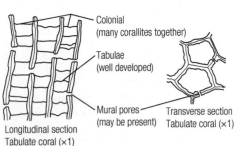

Figure 1 Septal growth in rugose and scleractinian corals

Scleractinian coral   Rugose coral

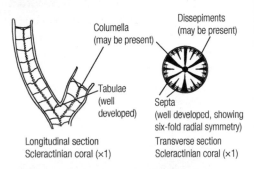

Colonial (many corallites together)

Tabulae (well developed)

Mural pores (may be present)

Longitudinal section Tabulate coral (×1)

Transverse section Tabulate coral (×1)

Columella (may be present)

Dissepiments (may be present)

Tabulae (well developed)

Septa (well developed, showing six-fold radial symmetry)

Longitudinal section Scleractinian coral (×1)

Transverse section Scleractinian coral (×1)

Figure 2 Tabulate and scleractinian corals

HOW SCIENCE WORKS: MAIN WAYS TO DISTINGUISH CORALS

Rugose corals are distinguished by having one plane of symmetry, known as **bilateral symmetry** which is picked out by the distinct septa. When the corals started to grow, there were initially six primary septa. New septa were only added in four of the resulting spaces. This complex arrangement of the septa is diagnostic of rugose corals.

Tabulate corals have small corallites and **mural pores** may be present in some species. They show **radial symmetry**, but the development of septa is either poor or absent.

Scleractinian corals have small corallites, with no mural pores between the colonial forms. They show radial symmetry, known as hexagonal symmetry. This symmetry is picked out by the distinct septa in the skeleton which shows six primary septa and evenly developed secondary septa between them.

The whole skeleton is known as the **corallum** in solitary corals.

KEY DEFINITIONS

Nematoblasts stinging cells to drive away enemies or catch prey.

Polyp the soft-bodied part of the coral that sits inside the calice.

Corallite a cup-like calcareous skeleton of a single coral polyp.

Bilateral symmetry means one plane of symmetry, with two identical halves.

Mural pores connections between adjacent corallites, perhaps for communication.

Radial symmetry where many planes of symmetry can be seen.

Corallum the whole skeleton of a solitary or colonial coral.

Extant describes a species that still exists today.

	Type of coral		
	Rugose	**Tabulate**	**Scleractinian**
Geological range	Ordovician to Permian (extinct)	Cambrian to Permian (extinct)	Triassic to Recent (**extant**)
Tabulae	Always present	Present and well developed	Always present
Corallites	Large	Many small	Many small
Dissepiments	Sometimes present	Absent or sometimes reduced	Always present
Symmetry	Bilateral	Radial	Radial
Columella	Always present	Absent	May be present
Septa	Major septa at six points, with four sets of minor septa	Sometimes present but reduced or poor	Major septa at six points radially
Colonial/solitary	Colonial or solitary	Always colonial	Colonial or solitary
Mural pores	None	Mural pores may be present	None
Example	*Lithostrotian* sp.	*Halysites* sp.	*Thecosmilia* sp.

The symbiotic relationship

Modern species have a type of algae that lives inside their tissues, in a **symbiotic relationship**. These algae, a type of dinoflagellate, are called **zooxanthellae**. The zooxanthellae can photosynthesise, as they have many chloroplasts inside of their cells. They take the carbon dioxide, phosphates and nitrates produced as waste from the coral and use them to produce oxygen, water, amino acids and sugars. The coral uses the waste from the zooxanthellae and they, in turn, use the coral waste. These organisms cannot survive without each other and the success of coral reefs is dependent upon this relationship.

Soft tissues

The soft tissues are not preserved in fossils, but we assume that fossil corals had similar soft tissue to modern corals. Soft-bodied polyps sit on top of the hard-secreted skeleton, or calice. The polyp itself has tentacles which it extends for feeding, usually at night. Food particles in the water or zooplankton are extracted. Sometimes they paralyse prey with the aid of the nematoblasts, or prey may become trapped in the mucus which is secreted from the polyp. Food is passed to the mouth and into a primitive gut for digestion. Undigested material and waste are removed through the mouth, as they do not have an anus. This sort of feeding is in addition to the nutrients supplied by the zooxanthellae.

Conditions needed for good coral growth

Some solitary corals can be found in all seas and oceans, regardless of temperature and to some depth. However, reef building corals require more specific conditions, which are considered tropical, to grow and survive. They need:

- To be at or just below sea level. Water depth is important as light is filtered out at depth. Few reef building corals are found below 30 m.
- Clear waters, as they need sunlight for the symbiotic algae to survive.
- The water should be free from particles of mud or sediment that may clog the polyps, tending to be offshore and far from rivers or other sources of sediment.
- High energy levels or wave action, to incorporate more oxygen into the water and circulates the correct level of nutrients by upwelling.
- Fully marine environment, with a salinity of 30 to 40 parts per thousand.
- A temperature between 23 °C and 29 °C.

HOW SCIENCE WORKS: MODERN-DAY CORAL PROBLEMS

It has been found that modern-day corals are very sensitive to environmental change. Very small changes in temperature can cause **coral bleaching** to occur. There are of course many modern-day threats to the coral reef, such as global changes (El Nino and La Nina), over-fishing, dredging and pollution.

HOW SCIENCE WORKS: GLOBAL DISTRIBUTION (MODERN-DAY REEFS)

Most coral reefs are generally found close the Equator, between 30° north and 30° south of the Equator, where conditions are good for coral growth. Good examples of modern coral reefs are: the Great Barrier Reef, Australia; many of the islands in the Caribbean, such as the Bahamas and the Maldives in the Indian Ocean. We assume that the distribution of fossil reefs was controlled by the same factors as modern reefs.

HOW SCIENCE WORKS: MODERN CORAL GROWTH

The coral growth needs to be fast enough to keep pace with subsidence. 'Massive' modern corals are the slowest growing species, increasing between 5 and 25 mm per year. Branching and staghorn corals can grow much faster, increasing in size by as much as 20 cm per year.

KEY DEFINITIONS

Symbiotic relationship where two organisms live together for mutual benefit, neither of which can successfully live without the other.

Zooxanthellae a type of dinoflagellate (algae) that lives symbiotically inside modern coral polyps.

Coral bleaching occurs where a small increase in temperature (or pollution) causes the polyps to expel the algae within them and they die. This kills the reef and the dead corals look white.

KEY FACTS

Types of modern coral reefs:
- Fringing reefs meet the land and some parts may be above sea level at low tide.
- Barrier reefs are further out to sea, with a lagoon separating the land from the reef.
- Atolls or coral islands are ring-shaped reefs found far offshore and are associated with volcanic hot spot activity.

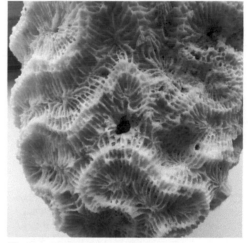

▲ **Figure 3** Modern-day coral ×1

QUESTIONS

1 Give one characteristic of rugose, tabulate and scleractinian corals that may be used to distinguish between each order.

2 Research and find the global positions of fossil reef systems (Carboniferous, Devonian and Silurian). Compare this distribution with the distribution of modern coral reefs today.

Rhynchonellid (×1)

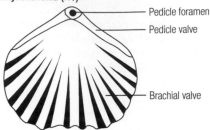

Pedicle foramen
Pedicle valve
Brachial valve

🔺 **Figure 1** Rhynchonellid brachiopod showing strong ribbing and a large pedicle opening (pedically attached). It was adapted to live in a high-energy environment.

Spiriferid (×0.5)

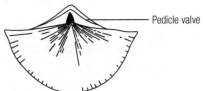

Pedicle valve

🔺 **Figure 2** Spiriferid brachiopod, showing a long straight (**strophic**) hinge line, was not attached but free lying on the sea floor. It was adapted to live in a low-energy environment.

Productid (×0.5)

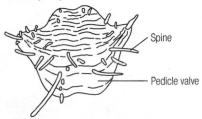

Spine
Pedicle valve

🔺 **Figure 3** Productid brachiopod bearing many slender spines for anchorage in the sediment and a straight hinge line. Not attached to the sea floor, it was adapted to live in a low-energy environment.

Terebratulid (×1)

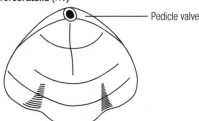

Pedicle valve

🔺 **Figure 4** Terebratulid brachiopod showing a curved (**astrophic**) hinge line and smooth valves. This had a large pedicle opening (pedically attached). It was adapted to live in a high-energy environment.

Adaptive radiation

Brachiopods evolved to be a variety of different shapes and sizes, probably as a response to the environment they lived in and the selection pressures in that environment. Specific adaptations with their possible explanations are shown below.

Adaptations linked to turbulent water

Adaptation	Possible reason for adaptation
A large pedicle opening (foramen)	To support a large pedicle for secure attachment to the substrate.
Strongly ribbed valves	To strengthen shell against wave action.
A folded or zigzagged **commisure**	To reduce the amount and size of sediment moving into the shells when the valves are open.
A thick and heavy shell	To provide extra stability on the substrate and prevent rolling in the current.

Adaptations linked to quiet water and/or soft substrate

Adaptation	Possible reason for adaptation
May have a median **fold** and **sulcus**	To separate currents of water entering and leaving the animal. Prevents mixing of fresh water and waste.
Valves flat with a large resting area	To provide a large surface area to prevent sinking into the sediment.
May have extension of the valves to form wings	To provide a large surface area to prevent sinking into the sediment (quiet waters are often muddy environments).
Smooth or weakly ribbed valves	No need to be robust in quiet conditions.
No pedicle opening	Pedicle not needed for attachment.
One margin of the shell may be turned upwards, away from the sediment	To ensure that some part of the shell remains out of the sediment for feeding.

Adaptations linked to a hard substrate

Adaptation	Possible reason for adaptation
Elongate ventral valve, cemented to the substrate at the base	For attachment in a high-energy environment.
May have spines from the ventral valve	To stabilise, by extending these to the substrate; prevents overturning by a current.
Small brachial valve	Acts as a lid to open for feeding and respiration.

Brachiopods as reef builders

Fossil reef systems were formed by many organisms including: algae, stromatolites, sponges, corals, bryozoans and more rarely, brachiopods. The dominant *reef building* organisms within reef systems vary throughout the Phanerozoic. For example, there are algal reefs in the Carboniferous, coral reefs in the Silurian and brachiopod reefs in the Permian. These Permian reefs are common in Texas, Southeast Asia and Sicily and brachiopods are considered the main reef builders. Brachiopods are preserved in life position within many reef systems of the lower Palaeozoic.

Similarities between brachiopods and bivalves

Brachiopods and bivalves are similar as they both have two calcareous valves, umbos, growth lines, ribs and they filter feed.

Differences between brachiopods and bivalves

Morphological features	Brachiopods	Bivalves
Shell shape	Two different size valves (larger pedicle valve and smaller brachial valve).	Two identical size valves (left and right).
Shell symmetry	Bilaterally symmetrical about a plane from the umbo to the anterior margin. Each valve is symmetrical about a line from the umbo to the anterior margin.	Bilaterally symmetrical about a plane passing between the two valves. Each valve is asymmetrical about a line from the umbo to the ventral margin.
Opening and closing mechanisms	Adductor muscles contract to close the valves. They leave a pair of muscle scars in the pedicle valve and two pairs in the brachial valve. Diductor muscles contract to open the valves. Two pairs run from the floor of the pedicle valve to the cardinal process in the brachial valve.	A pair of adductor muscles contract to keep the valves closed. When muscles are relaxed, the external ligament pulls the valves open. They leave a pair of muscle scars at the anterior and posterior.
Foot	None.	Found at the posterior end used for movement and digging.
Pedicle	Pedicle for attachment to rocks.	None.
Pallial line and sinus	None.	Clearly seen around the margin of the shell.
Feeding	Lophophore to gather food particles and pass these to the mouth.	Siphons to take in water and food particles.
Oxygen for respiration	Mantle and lophophore.	Gills.
Teeth and socket	Two teeth within the hinge apparatus of the pedicle valve. Sockets on brachial valve (this is where the teeth fit).	Dentition (collective term for teeth and sockets) all along the hinge plate. Cardinal teeth under umbo and lateral teeth beyond umbo or many similar teeth and sockets all along the hinge plate.

HOW SCIENCE WORKS: ATTACHMENT OF BRACHIOPODS

Valves are directly cemented onto a hard substrate or attached with a pedicle during **spatfall**. Some species of brachiopod may be cemented or attached by a pedicle when young, but detach when the brachiopod increases in size.

Some pedicle foramen are very small and only support juveniles, becoming closed when adult.

QUESTIONS

1 Describe the morphological features that suggest a brachiopod lived in a turbulent sea.

2 Research the types of brachiopod that are alive today. What is their global distribution?

3 Describe and draw a diagram to show how brachiopods were attached to the sea floor.

KEY DEFINITIONS

Strophic means a straight hinge line. These are sometimes extended as 'wings'.

Astrophic means the hinge line is not straight, may show a shallow or highly curved pattern. Sometimes known as non-strophic.

Commisure the margin between the valves at the posterior and may be curved, folded or zigzag.

Fold and sulcus a fold in the central or middle part of valves.

Spatfall the settling and attachment of juveniles to a substrate.

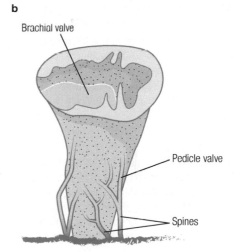

🔺 **Figure 5** Reef building richthofenid brachiopods **a** photograph showing several specimens in life position and **b** labelled diagram showing spine-like supports.

Amphibians were the first land-dwelling creatures with four legs, known as **tetrapods**. They evolved from **lobe-finned fish** in the late Devonian to early Carboniferous. They then ventured forth into **terrestrial** environments, where they would later evolve into reptiles, birds and mammals.

What are lobe-finned fish?

Lobe-finned fish are *lungfish*. These fish had the ability to breathe both in and out of water because they had evolved to have 'functional lungs', which were modified **swim bladders**. This allowed them to breathe out of water, a great advantage to start the transition to life on land. They could move from one water source to another and exploit new niches, with the absence of predators.

They had robust fleshy fins with an arrangement of central bones inside, similar to the structure of bones in a hand, which would have allowed for more mobility and support (especially on land). The only disadvantage of this design was that there was no strengthening girdle connecting the limb bones to the rest of the skeleton. This gave an area of weakness. These fins would later evolve into the limbs needed for more permanent life on land.

How do we know this about the lobe-finned fish? There are many fossil remains of these creatures found in Devonian strata, and some even earlier that show the skeletal structure. There are also live specimens that can be found on Earth today.

Another piece of evidence is that **coelacanths** were thought to be extinct. They are now known as 'living fossils'. They have morphological features that have remained virtually unchanged for 400 million years. Coelacanths are marine fish that do not have lungs, which suggests that lungfish are more closely related to the first amphibians.

Similarities between the lobe-finned fish and the early amphibians

- The four fins of lobe-finned fish and early amphibians were skeletally similar.
- Their limbs were in the same position on their bodies.
- They both lacked claws or nails.
- The skull morphology, the jaw bone and teeth of the lobe-finned fish and amphibians were very much alike.
- The amphibian's skull became increasingly slender, with the temporal and opercular bones becoming smaller in size, and the jaw bones becoming more fused together.
- The teeth of both the lobe-finned fish and amphibians were complex.
- Early amphibians still had a tail fin, suggesting they still spent a great deal of time in the water. Body shape, and presumably movement, still resembled that of the lobe-finned fish.
- Also, the early amphibians still had traces of small bony scales on the skin, a trait of fishes.

KEY DEFINITIONS

Amphibian a creature with two life styles, one in water, the other on land.

Tetrapod a creature with four limbs.

Lobe-finned fish are fish that possessed both lungs and gills and had four fleshy fins supported by bones in a similar structure to a hand, e.g. lungfish.

Terrestrial describes anything relating to land.

Swim bladder a sac-like structure, which can be filled with gas or fluid, to control buoyancy in fish.

Coelacanth a fish from an ancient lineage, closely related to lungfish and reptiles. They were thought to be extinct since the end of the Cretaceous period, but are actually extant.

CASE STUDY: THE COELACANTH

The coelacanth was rediscovered in 1938 off the Comoros Islands, living between 150 and 300 m depth. Previously, it was believed to be extinct, more than 65 million years ago.

Coelacanths were swimming in the oceans when their close relatives became the first vertebrates to venture onto land. They shared the oceans with trilobites and primitive molluscs. They were almost identical to the 'modern' coelacanth.

The limb-like ventral fin and the pectoral fins made it possibly the first fish to crawl out of the sea. That could make them the closest living relative to the ancestor of all amphibians, reptiles, birds and mammals. However, this could be the lungfish, so debate still rages. The coelacanth group reached its peak in numbers around 200 million years ago and numbers today are low.

▼ **Figure 1** Photos showing **a** lungfish and **b** coelacanth

a lungfish

b coelacanth

Adaptations to life on land

As well as having similar characteristics to those of lobe-finned fish, the newly evolved amphibians also had some other entirely unique features, which allowed them to adapt to life on land. These included:

- The development of a skeletal *girdle* connecting the limb bones to the skeleton for better movement on land.

- A more *robust* skeleton strengthening the vertebral column and rib bones, for extra support on land.

- *Eyelids* formed to help keep eyes moist, as it was no longer always submerged in water.

- The development of a *double circulatory* system with a three-chambered heart to pump mixed blood before and after it had been to the lungs. This allowed a more efficient gas exchange to take place, so it could provide its more active cells with the oxygen and to remove waste products more efficiently.

- A *tongue* formed within its mouth, which could be used to catch prey, as well as having a sensory role.

- *Ears* adapted so it could detect sound waves through the thin medium of air, allowing it to listen to its surroundings for prey or predators.

However, even with new adaptations to terrestrial life, the early amphibians still had to remain close to a water source. They still used their skin for gas exchange, so they had to keep it moist to allow the transfer of oxygen and carbon dioxide. They also had to lay their eggs in water, because without it their eggs would dry out, as they were only protected by a layer of jelly and not a shell. The young would hatch into aquatic larvae with gills, and then undergo metamorphosis to develop into a terrestrial adult, able to walk on land. It was not until the development of the amniotic egg that evolution proceeded to give rise to such creatures as the dinosaurs, birds and mammals.

General morphology

a

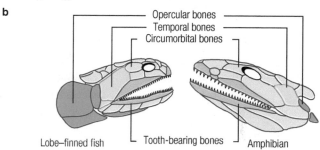

Lobe-finned fish

Early amphibian

Skull morphology

b

Opercular bones
Temporal bones
Circumorbital bones

Lobe–finned fish — Tooth-bearing bones — Amphibian

Limb morphology

c

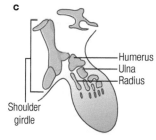

Humerus
Ulna
Radius

Shoulder girdle

Lobe–finned fish

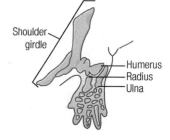

Shoulder girdle

Humerus
Radius
Ulna

Primitive amphibian

🔺 **Figure 2** Diagrams showing **a** general morphology **b** skull morphology, and **c** limb morphology of lobe-finned fish and early amphibians

---

CASE STUDY: FIRST AMPHIBIANS

At first they thought the fossil was just a fragment of a lungfish snout – interesting, but not earth shaking. In 2002, geologists went to Ellesmere Island, high in the Canadian Arctic, in search of the fish that first dragged itself out of water nearly 400 million years ago, the evolutionary forebear to all land vertebrates. That unassuming fragment of bone helped geologists of the University of Chicago to one of the most important fossil finds of the decade: several specimens of an almost perfect intermediate between fish and land vertebrates – tetrapods. All are so beautifully preserved that the researchers could see almost every detail of their skeleton. The new creature was named *Tiktaalik* – from the Inuit name for a large, shallow-water fish. Several fossil fish with amphibian-like features have been found. The fish's fins became the amphibian's rudimentary legs. The head, bound rigidly to the body by bony plates, turned into a mobile neck. The first amphibians lived at the time that the first major terrestrial ecosystems were establishing themselves. Full-grown forests with roots a metre deep were developing, where formerly just mosses and other tiny, shallow-rooted plants grew.

Precis from Issue 2568 of *New Scientist*, September 9, 2006, pp. 35–39.

HOW SCIENCE WORKS: AMPHIBIAN AND REPTILE EGGS

Amphibians lay eggs with a soft, gel-like consistency, lacking any hard shell. They have to be laid in water so they do not dry out.

Reptiles lay eggs with a tough, leathery shell. These can be laid on land as the leathery exterior prevents drying out.

QUESTIONS

1 What is a lobe-finned fish?

2 What are the similarities between lungfish (lobe-finned fish) and the early amphibians?

3 Describe and explain three adaptations of an early amphibian for life on land.

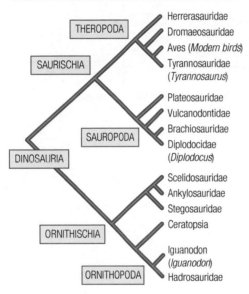

Figure 1 Evolutionary relationships of dinosaurs

The amniotic egg

Dinosaurs are believed to have laid **amniotic eggs**. This type of egg has an amnion, or a sac which is fluid filled around the embryo during development. The amniotic egg is a significant feature in reptile evolution, as it allowed for life on land without the need for a water source in which to reproduce.

The first advantage for life on land was the development of a hard outer shell, which provided protection whilst remaining porous. This allowed the diffusion of oxygen into the egg and carbon dioxide out, allowing respiration to take place. The yolk sac provided the embryo with food and the albumin supplied water and nutrients, eliminating the need for a larval stage. Instead the embryo would develop directly into a miniature version of the adult.

Evolution of dinosaurs

Dinosaurs evolved from **Archosaurs**, after the Permo-Triassic mass extinction, thrived throughout the Mesozoic and were extinct at the end of the Cretaceous (66 Ma).

The hip bones can be used to classify dinosaurs. These bones are actually three connected bones (Ilium, ischium and **pubis**), collectively called the pelvis, with a hole in the centre for the head of the femur. In **Ornithischia** the hip bone points backwards (bird hipped) and in **Saurischia** it points forwards (reptile hipped). The hip allowed an erect posture.

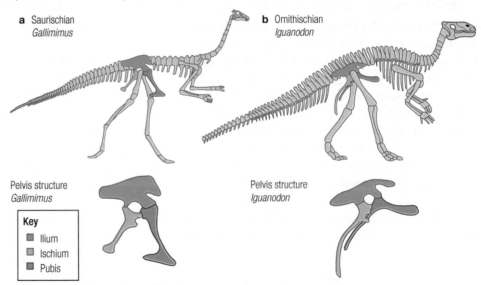

a Saurischian *Gallimimus*

b Ornithischian *Iguanodon*

Pelvis structure *Gallimimus*

Pelvis structure *Iguanodon*

Key
- Ilium
- Ischium
- Pubis

Figure 2 Arrangement of hip bones in **a** Saurischia and **b** Ornithischia

Characteristics of Saurischian dinosaurs (Theropoda and Sauropoda)

This group includes the carnivore *Velociraptor,* popularised by the film Jurassic Park and the herbivore *Diplodocus*.

Features include:

- The primitive arrangement of the hip bones is similar to reptiles, in which the pubis points forwards.
- They have long, S-shaped flexible necks, allowing rapid and precise movement.
- Hands consisted of only three digits (fingers). The digits were asymmetrical with the first digit similar to a thumb, allowing the hand to grasp and with the second digit being the longest of the three.

Characteristics of Ornithischian dinosaurs

This includes *Iguanodon* and *Stegosaurus*.

- The arrangement of the hip bones is similar to birds, in which the pubis points backwards.
- The front teeth are small or absent, replaced at the front with a horny beak which became broader as they evolved, giving them the name 'duck-billed dinosaurs'.
- Many were armoured with bony plates, such as the *Stegosaurus*. These were thought to be a defence mechanism against predators, but could have also acted as heat-exchangers. The fossilised plates have tiny grooves, which may have housed blood vessels allowing heat to be given off or absorbed.

CASE STUDY: DIPLODOCUS (SAURISCHIA SAUROPODA)

Diplodocus was an herbivore common in the Jurassic and the Cretaceous. It had a long slender skull, which was small compared to the rest of its body. It also had an extremely long neck, which it carried parallel to the ground. This allowed it to reach vegetation in forests where most other Sauropods could not venture due to their large size. Its long neck could have also helped it to search for foliage in wetlands. Peg-like teeth are only found to the front of the jaw and unlike many other herbivores' teeth, are not differentiated into grinding and tearing teeth, suggesting that it bit off vegetation and swallowed it whole. This would have made digesting plant material difficult, so it is believed that bacterial action and swallowed stones (**gastroliths**) aided the digestion of its food. Its body was long but lightly built. The spine had extra bones underneath it, these bony protrusions running both forwards and backwards (anvil-shaped), providing support and extra mobility of the neck and tail. Its tail was long and whip-like, which could have been used for defence.

CASE STUDY: TYRANNOSAURUS (SAURISCHIA THEROPODA)

The largest of all terrestrial predators, was a carnivore and lived in the late Cretaceous. Was the *Tyrannosaurus* a predator or a scavenger? There are many arguments against them being predators; for example, they had small eyes while large eyes are more suitable to spot prey. They had small arms, not suitable for grasping onto prey. The back legs on which it walked were huge, indicating slow movement, which once again is not ideal in a predator. What these large legs did prove was that it was capable of walking large distances, a possible adaptation of a scavenger. *Tyrannosaurus* also had large **olfactory lobes**, another adaptation if you are a scavenger.

However, there are arguments against scavenging. Many predators today are happy to scavenge meat if it is readily available, but most prefer it fresh. Also, many predators are successful hunters without having to use their forelimbs. Whether or not the *Tyrannosaurus* was a slow animal is difficult to determine as few traces can be identified as made by *Tyrannosaurus*. It could have been a predator, a scavenger or both. It was a meat eater as its teeth were relatively large, curved and jagged, making them ideal for tearing flesh from a carcass.

CASE STUDY: IGUANODON (ORNITHISCHIA)

Iguanodon appeared in the late Jurassic. It was a large, heavily built creature with heavy shoulders and forelimbs. Its skull was large with a horse-like snout, ending in a horny toothless beak, which was used to crop vegetation. It could then mince its food with a long series of leaf-shaped cheek teeth (resembling the teeth of living iguanas), due to a hinged upper jaw which was able to flex from side to side. This meant that the upper teeth could grind against the lower to mince vegetation. Its hands had three digits (fingers), which ended in hooves and had a large, conical thumb spike, thought to be a weapon or used to obtain food. The most outstanding feature of the *Iguanodon* was its ability to be quadrupedal, in which it used its tail as a counter balance, and bipedal to run or to rear up and swing its spike-like thumb in defence.

QUESTIONS

1. Compare the hip arrangement in both Saurischian and Ornithischian dinosaurs.
2. Research the dinosaur *Stegosaurus* to find out about its mode of life. List the evidence you discover for the mode of life.

KEY DEFINITIONS

Dinosaur a Mesozoic reptile with an upright walking position or gait.

Amniotic eggs types of eggs with shells produced by birds and reptiles.

Archosaurs a group of reptiles dominating the Triassic. These were the ancestors of dinosaurs, birds and crocodiles.

Pubis the pubic bone.

Ornithischia one of the main divisions in dinosaur classification, in which two bones in the hip point backwards. This is the same as in birds, so they are known as '*bird hipped*'.

Saurischia one of the main divisions in dinosaur classification, in which one of the bones in the hip points forward. This is known as '*reptile hipped*'.

Gastroliths stones swallowed by animals which stay in the stomach and grind the food that they eat. This is usually alongside a gizzard, which is a lining inside the stomach to protect it from damage.

Olfactory lobes part of the brain that can process smell. The larger this area, the more acute the sense of smell.

🔺 **Figure 3** *Iguanodon*

🔺 **Figure 4** A skull of *Pachycephalosaurus*, an Ornithischian dinosaur.

Evolution of birds

There are several theories to explain the evolution of birds. The most accepted of these is that birds are thought to have evolved from Theropod dinosaurs in the late Jurassic. Theropod dinosaurs, in particular the **coelurosaurs**, are thought to be the closest relatives of birds, with many skeletal similarities, for example:

- Hollow thin-walled bones, to make the bones less dense.
- S-shaped curved neck.
- Elongated arms and forelimbs, and clawed hands.
- The pubis shifted from an anterior (forward position) to a posterior (backward position).
- Large orbits (eye sockets in the skull).
- Hinged ankles (reduces the rotation of the ankle).

Archaeopteryx

Archaeopteryx is the first known 'bird-like' fossil. It is believed to be intermediate between birds and reptiles. This was a semi-arboreal animal (a creature that probably lived in trees) capable of gliding and sustaining flight, although some believe it had poor flight and simply glided. They show characteristics of dinosaurs and birds, known as **dinobirds**.

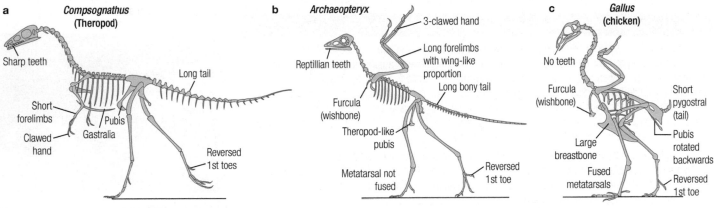

Figure 1 Comparison of skeletons
a Theropod dinosaur, **b** *Archaeopteryx* and **c** modern bird

HOW SCIENCE WORKS: DISCOVERY OF *ARCHAEOPTERYX*

The first feather from an *Archaeopteryx* was discovered in 1860 near Solnhofen, Germany. The following year (1861) the remains of the first *Archaeopteryx* skeleton were also found. A close relationship between birds and dinosaurs was first proposed soon after the discovery of this primitive bird-like specimen of *Archaeopteryx*.

Reptilian/dinosaur-like features (*Compsognathus*)

- Long, bony lizard-like tail.
- Three digits (fingers) on the wings, each digit had a claw.
- Snout with developed reptilian teeth.
- Reptilian skull and brain.
- The **sternum** was not bony or keeled.
- Gastralia ('belly ribs' that do not articulate with the vertebrae) were present.
- S-shaped curved neck.

Bird-like features (*Archaeopteryx*)

- Wings for flight.
- Feathers.
- Hollow bones.
- Legs directly underneath the body.
- A **furcula** (wishbone) was present.
- Reversed big toe.

Evolution of feathers

Reptiles are well known to have scales and it has long been thought that feathers developed as outgrowths or elongations from such scales. Scales are relatively flat and the calamus (or quill) of a feather is tubular, and so there is some uncertainty that this was indeed the way feathers originated. Instead, it is thought that feathers probably evolved separately and this is supported by the fact that the proteins (keratin) that form feathers and scales are different.

The earliest preserved feathers are found in the late Jurassic. These feathers originated as 'protofeathers', simple unbranched downy filaments, a few millimetres in length covering the body and were most likely used for insulation. This probably gave the dinosaurs a 'fluffy' look. In some dinosaurs, they grew quite long, up to 15cm, and were present on the neck, head and tail. These filament-like feathers originated nearly 100 million years before the origin of flight and are described as *plumulaceous*.

Feathers evolved to be more complex, becoming elongate, symmetrical, rather like those from modern birds and were described as *pennaceous*. Developing to a greater extent in avian-like dinosaurs.

There is some evidence of **sexual dimorphism** in plumage and colour patterns in late Jurassic and early Cretaceous dinosaurs. This tells us that plumage was probably used for display purposes and so these dinosaurs may have behaved very much like birds.

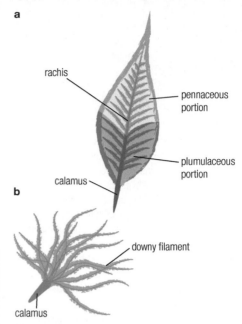

Figure 2 labels: rachis, pennaceous portion, plumulaceous portion, calamus, downy filament, calamus

⬤ **Figure 2** Early types of feather **a** pennaceous and **b** plumulaceous

CASE STUDY: EVOLUTIONARY RELATIONSHIPS OF COELUROSAURS

A recent study by Dr Stephen Brusatte and his team at the University of Edinburgh has produced a comprehensive family tree showing the interrelationships between Theropod dinosaurs (coelurosaurs) and prehistoric birds.

The team analysed the anatomy of more than 853 body features in 150 species of coelurosaurs to construct this family tree.

Troodontids (bird-like theropods) and Dromaeosaurids (raptor dinosaurs) are seen as being very closely related to *Archaeopteryx*. A clear lineage is identified between *Archaeopteryx* and modern birds, with several intermediaries.

The fact that they found features of birds in the specimens of dinosaurs, such as feathers, wings and the fork-like **'furcula'**, proves that these features first evolved in dinosaurs and their initial evolution was nothing to do with flight.

Feathers increased in frequency moving down this family tree but even the large tyrannosaurs at the top of the chart have been found to have feathers and are described as 'fluffy'.

At first the early birds in the Jurassic were indistinguishable from other dinosaurs. There was probably no moment when a dinosaur

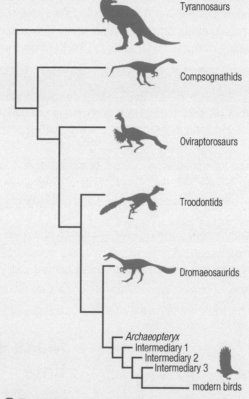

Figure 3 labels: Tyrannosaurs, Compsognathids, Oviraptorosaurs, Troodontids, Dromaeosaurids, *Archaeopteryx*, Intermediary 1, Intermediary 2, Intermediary 3, modern birds

⬤ **Figure 3** Evolutionary relationships of Coelurosaurs [Published in *Current Biology*]

became a bird, simply that they evolved gradually, with *Archaeopteryx* being an intermediary. Once the successful bird body morphology evolved, there was a rapid rate of evolution leading to the thousands of bird species today with powered flight.

CASE STUDY: COMPARATIVE EVIDENCE

The wrist became increasingly more flexible, allowing wings to fold back along the body. This was not developed in early Theropods but was well developed in later ones, like *Velociraptor*.

A reversed hallux (first toe) is considered an adaptation for roosting. Initially the hallux is small and non-opposable and doesn't even reach the ground in most dinosaurs. It was like this in early birds too, developing into the grasping structure we see today.

QUESTION

1   In what ways are birds, *Archaeopteryx* and Theropods similar? How do they differ?

CASE STUDY: PTEROSAURS

These 'winged lizards' are flying reptiles, which are often erroneously called dinosaurs. Although dinosaurs and pterosaurs had a common ancestor, their lineages split much earlier in their evolution and the two groups evolved separately. Pterosaurs lacked the characteristics of Ornithischian and Saurischian dinosaurs.

Pterosaurs are the earliest vertebrates known to have evolved flight, with membranous skin forming wings, strong muscles and an elongate fourth finger. They had hollow bones, large brains and well-developed optic lobes. It was thought that they had rather poor flight, but recent work has suggested that they were accomplished fliers, dominating the Jurassic and Cretaceous skies. They had crests on their bones to which flight muscles attached, which is consistent with powered flight (flapping). This is a case of convergent evolution.

⬤ **Figure 4** Reconstruction of **a** *Archaeopteryx* and **b** *Rhamphorhynchus*, a pterosaur

What is a mass extinction?

It is almost impossible to assess the real nature of a mass extinction, as the fossil record shows such a small and biased number of organisms preserved. How quickly a mass extinction occurs is a matter of debate. Do organisms die out gradually or catastrophically? Mass extinctions are not usually single, instantaneous events. Most seem to take place gradually, over at least 100 000 years, possibly following environmental deterioration. It is possible that events moved more quickly than this, but we have no means of measuring if they did. Unfortunately, the fossil record is not good enough to tell us the answers.

A mass **extinction** is when there is a massive decline in the number of different **species**, over a relatively short period of time, perhaps spanning several thousand or even a few million years. For any one species, extinction is catastrophic. The normal process of extinctions occurs continually, generating a regular change of all the species living on Earth, called background extinction. Sometimes, however, extinction rates rise suddenly for a relatively short time, described as a mass extinction event.

The most famous of the mass extinctions is the Cretaceous–Tertiary mass extinction (K–T) also known as the Cretaceous-Paleogene (K-Pg) extinction event. It is of particular interest as it is when the dinosaurs became extinct. This was the most recent large-scale mass extinction and has been well documented.

Mass extinction events are not rare and some environmentalists and biologists believe that we are in the middle of another major mass extinction event, fuelled by man's effect on the environment. This is due to the decline in the numbers of species of many organisms on Earth since the industrial revolution.

Extinction event	Time	Organisms most affected	% extinct of groups alive at the time
Ordovician-Silurian	443 Ma	Plankton and bottom dwellers, particularly brachiopods and trilobites. 100 families extinct, including more than half of the bryozoan and brachiopod species extinct.	27% of all families 57% of all genera
Late Devonian	372 Ma	Shallow marine ecosystems especially corals, brachiopods and trilobites.	19% of all families 50% of all genera
Permo-Triassic	252 Ma	Marine organisms, 95% of all marine species. Most brachiopods, corals, ammonoids, trilobites. Land reptiles, with 50% of all animal families and many trees die out.	57% of all families 83% of all genera
Triassic-Jurassic	201 Ma	Sea and land animals (amphibians and reptiles). Many ammonoids. 35% of all animal families die out. Most early dinosaur families became extinct.	23% of all families 48% of all genera
Cretaceous-Tertiary	66 Ma	Bivalves, belemnites, dinosaurs, ammonites, pterosaurs, plesiosaurs, mosasaurs, many families of fish, bivalves and many others.	17% of all families 50% of all genera

Mass extinctions through time

Mass extinction events have occurred throughout geological time, with remarkable frequency. There have been five major mass extinctions and many more minor ones.

Above is a table summarising the 'big five' extinction events.

Each of these extinctions affected well-established **genera**. Some losses affected one main type of habitat, such as shallow marine dwellers, whilst others were more widespread with global implications.

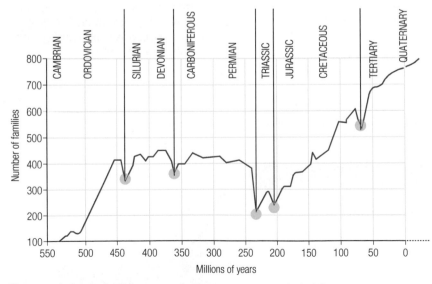

▲ **Figure 1** Graph showing mass extinction events over geological time

Ordovician-Silurian mass extinction

During the Ordovician, most life existed in the sea which was dominated by trilobites, brachiopods and graptolites. The relatively warm temperatures during the Ordovician, gave way to a cooling event at the Ordovician-Silurian boundary. An ice age has been identified as the most likely cause for this cooling, as a huge ice sheet extended over the southern hemisphere. The presence of an extensive ice sheet would have caused:

- The climate to become cooler, by the **albedo effect**.
- The sea level would fall, as water would fall as snow and become trapped in the ice cap.
- Changes in the ocean chemistry, moving to less favourable conditions.

The glaciation probably caused positive feedback, a runaway effect. Increased snowfall in a cooler period reflected more heat and light energy back into space due to the high albedo effect, so then the temperature would drop even further. This results in increased snowfall and so even more reflection, which could lead to the entire planet being iced over, known as 'Snowball Earth'.

The table below shows the reflectivity (albedo effect) of different surfaces.

Surface	Albedo/reflectivity (%)
Fresh snow	80–95
Ice	20–40
Desert sands	35–45
Forest	15–20
Water (solar elevation 30°)	6
Water (solar elevation 60°)	3

The falling sea level disrupted or eliminated habitats along the edges of the continental shelves. Reconstructions of the land masses over the South Pole around the time of the Ordovician-Silurian extinction shows the South Pole positioned in west Africa. The continents formed a large land mass called Gondwanaland, drifting across the South Pole. This meant that the continental shelves were already restricted to the periphery of Gondwanaland, as the ice caps formed on the land mass. A resultant fall in sea level would restrict the shelves even further, putting pressure on shelf dwellers in particular by eliminating their niches. This caused problems for brachiopods, bryzoans, bivalves, echinoderms and corals, the shallow benthonic organisms.

Planktonic and nektonic organisms were also seriously affected across all latitudes and in all marine environments. This saw the decline of graptolites and trilobites, as well as microfossils such as conodonts, chitinozoans and acritarchs.

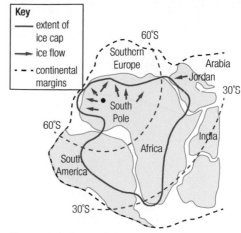

▲ **Figure 2** Reconstruction showing the position of the land masses and ice sheet around the time of the Ordovician-Silurian extinction

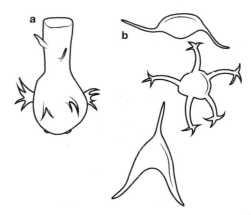

▲ **Figure 3** Diagrams showing microplankton **a** chitinozoan and **b** acritarchs. Both groups were affected by the Ordovician-Silurian extinction event.

CASE STUDY: MALFORMED PLANKTON

Microfossils called chitinozoans have been investigated from around the Ordovician-Silurian boundary and high concentrations of lead, iron and arsenic have been found. High levels of these elements are known to cause deformities in modern organisms, and so were the likely cause of deformity in ancient organisms. Researchers have suggested that oxygen depletion and toxic metals most likely caused the extinction of plankton out in the oceans, rather than the glacial episodes. High levels of metal suggest changes in the ocean chemistry.

QUESTIONS

1 Explain what is meant by the word 'extinction'.

2 Describe the patterns of extinction over geological time.

3 Research the morphologies and modes of life of the planktonic organisms that were most affected by the Ordovician-Silurian mass extinction.

KEY DEFINITIONS

Extinction where a species is no longer alive. It is usually judged as when the last member of a species dies and therefore cannot reproduce. Well over 99% of the species that have ever lived are now extinct.

Species groups of individuals with the ability to reproduce and to give rise to fertile offspring. Palaeontologists use morphology to define species as we have no way of knowing this information.

Genera (genus) a subdivision in taxonomy, a broader category than species.

Albedo effect the extent to which an object or medium reflects light. For example, snow, will reflect a high percentage of light energy, reducing the amount absorbed as heat energy.

Permo-Triassic mass extinction

This extinction event which occurred around 252 million years ago was the most devastating in Earth history. It marks the end of the Palaeozoic era, as around 95% of all marine species and 70% of terrestrial vertebrates became extinct.

In the seas, those affected included trilobites, tabulate and rugose corals, and brachiopods. Other groups, such as foraminifera and cephalopods, suffered great losses. Extinctions also happened on land, with a huge demise in the numbers of plant species, vertebrates and insects, indicating that this was a truly global event.

This opened up many niches to be filled, both on land and in the sea. This launched the recovery of life, and evolution of new forms, many of which are recognised today.

▲ **Figure 1** Palaeogeographic reconstruction at the Permo-Triassic boundary

Hypotheses to explain the Permo-Triassic extinction

The generally accepted hypothesis is that major volcanic activity was most likely to have been the cause for the extinction event. Other factors and events that may also have contributed, although less favourable, are discussed below.

Major volcanic activity (Siberian Traps)

The Siberian **Traps** are examples of a large igneous province (LIP), which are basaltic lavas (flood basalts) that erupted over a mantle plume. They form great thicknesses today and cover an area greater than the size of Europe, about two million km^2. The eruptions lasted for one to two million years. They are thought to cause global mass extinction by:

Areas in the vicinity of the eruption would be subjected to toxic gases, lava flows and pyroclastic flows. These would kill plants and animals nearby.

Initially ash particles would have been sent high up into the Earth's atmosphere, which blocked out sunlight, initially lowering the global temperatures. As this was such a large-scale eruption, the temperature change would have been significant. This in turn probably caused global glaciations and a fall in sea level. This cooling event would have lasted from hundreds to thousands of years.

Following the initial cooling, the continued emission of greenhouse gases such as CO_2 and SO_2 could have caused an increase in global temperatures. Greenhouse gases remained in the atmosphere for a very long time, and are believed to have caused an increase in global temperatures which lasted for thousands to millions of years. Acid rain would also have caused problems as gases react with water in the atmosphere.

Supercontinent formation and glaciation

Pangaea was formed when all the world's continents collided together. This formed one large land mass, called a supercontinent. The evidence to support this land mass comes from plate reconstructions and the stratigraphic record.

As a result of a supercontinent, there were fewer continental shelves so there was a lack of habitat for shallow marine dwellers, accounting for the massive decline in shallow marine species.

A single continent would have caused rapid fluctuations in climate, and unstable weather patterns. As parts of Pangaea moved over the northern and southern poles, there were widespread glaciations (evidenced in Australia, South Africa, South America and Antarctica). This caused the sea level to fall (regression), reducing shallow shelf environments.

KEY DEFINITIONS

Traps large-scale volcanism resulting in the formation of mafic igneous rocks.

Pangaea the name given to the supercontinent which formed at the end of the Permian.

Shocked quartz grains that have characteristics showing deformation under high pressure. It was first discovered at nuclear testing sites, and linked with meteorite impacts.

Iridium a transition element, rare on Earth, but found in meteorites. Luis Alvarez (1980) first proposed this as evidence that a giant object hit the Earth.

Tektites spheres or irregular shaped lumps of solidified molten rock, a few centimetres in diameter, black or green in colour, which are thought to have formed as a result of extremely high pressures and temperatures caused by a meteorite impact.

Tsunami a massive displacement of water, formed either due to an underwater earthquake or a meteorite impact.

Methane hydrates (methane ice)

This is a solid form of methane, believed to have formed within the sediments, which remains solid and stable up to around 18°C. The global increase in temperature at the end of the Permian is also thought to have triggered these methane hydrates in the seabed to become mobile and gaseous, causing them to be released.

Impact event

Evidence for this is quite flimsy, with questionable **shocked quartz** crystals found in Australia and Antarctica. Any impact craters found on the sea floor would have been destroyed (Wilson cycle), as there is no oceanic crust more than 200 million years old. Technically, a large impact event may have triggered the Siberian Traps. There is very little evidence for this hypothesis.

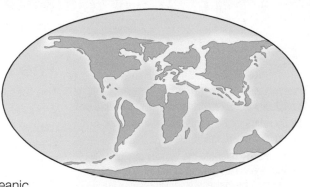

▲ **Figure 2** Palaeogeographic reconstruction at the Cretaceous-Tertiary boundary.

Cretaceous-Tertiary mass extinction

This is a large extinction event and 60 to 70% of species became extinct, marking the end of the Mesozoic Era around 66 million years ago. Once again, the event was gradual, showing a decline in species over several million years, leading finally to an abrupt extinction event.

Marine casualties included large reptiles (ichthyosaurs), brachiopods, ammonites, coccoliths, foraminifera, belemnites, fish and some bivalves. Terrestrial losses included dinosaurs, pterosaurs and plants. However, some groups were relatively unaffected by this extinction event. Animals such as crocodiles, turtles, lizards, mammals and birds all made it through without many casualties. The extinction of the dinosaurs did leave a large ecological niche, which allowed the radiation of mammals, eventually leading to the evolution of humans.

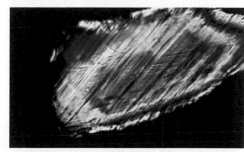

▲ **Figure 3** Shocked quartz grains

Hypotheses to explain the Cretaceous-Tertiary mass extinction

Impact of an asteroid or meteorite

This remains the most popular hypothesis. There is a lot of evidence to suggest that a large object did hit the Earth 66 million years ago. Evidence includes:

- There is a layer of **iridium** concentrated in clays at the boundary, and most iridium originates from space.
- Shocked quartz grains are found at the boundary as a thin layer within the sediments.
- The presence of **tektites**, usually found near impact craters. Tektites have a very low water content, suggesting they may not have formed on Earth, or they formed as the rock had melted very quickly on impact at the surface and were then thrown into the air.
- There is large-scale sedimentary evidence in Texas that there was a huge **tsunami** at this time.
- A large meteorite crater can be found on the Yucatàn Peninsula in Mexico, at Chicxulub. The crater is shown by gravity variations across the region, and shows a circular depression about 180 km in diameter. Although there are other contenders, it is the most likely impact site for the meteorite.

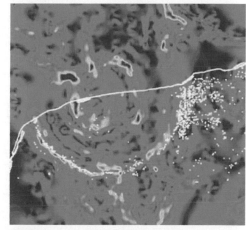

▲ **Figure 4** Gravity surveys onshore and offshore of the Yucatàn Peninsula, Mexico.

Major volcanic activity (Deccan Traps)

There was another enormous series of eruptions in India, covering an area of roughly 500 000 km². The eruptions took place over about 30 000 years. The effect of this volcanism was probably the same as with the Siberian Traps (see the section on the Permo-Triassic extinction).

CASE STUDY: DRILLING INTO THE IMPACT CRATER

in 2016 scientists published first results from a drilling expedition into the Chicxulub crater, off the Yucatàn Peninsula in Mexico. Scientists have brought up drill samples from 670m below the sea floor. They discovered shocked, granitic rocks from deep in the crust, described as being placed 'out of order' on top of the sedimentary rocks. This discovery shows a 'peak ring' crater, where there is a circular shaped ring inside of the crater, surrounding the centre. Chicxulub is the only well-preserved crater with a peak ring on Earth, evident on the gravity survey map.

QUESTIONS

1 What organisms were affected by the Cretaceous–Tertiary and the Permo-Triassic extinctions?

2 Explain how major volcanism and widespread glaciation could cause extinctions.

KEY DEFINITIONS

Biota refers to the plants and animals living in an area or environment.

Low-energy refers to the velocity of the water during deposition moving slowly or not at all. Examples of a low-energy environments are swamps, deep seas and lakes.

Anaerobic means without oxygen. This term is usually used to describe conditions within a sediment.

Anoxic means water without oxygen. Standard decomposing bacteria cannot live without oxygen, nor can scavengers.

The incomplete fossil record

The fossil record is biased, especially in the types of organisms preserved. When an organism dies, scavengers may fragment the remains and bacteria may act to decompose the soft parts. The only parts that usually survive are skeletal structures, if present, especially if they are mineralised skeletons ('hard parts') impregnated with crystals of calcium carbonate (calcite or aragonite), calcium phosphate or silica. Animals with mineralised skeletons include corals, brachiopods, molluscs, echinoderms and many chordates. It is no wonder that the fossil record is biased towards these groups, whilst those lacking a mineralised skeleton, such as worms and jellyfish, are poorly represented and can only be found in some, rare deposits as exceptionally preserved **biota**.

Lagerstätten deposits

The term '*Lagerstatt*' (German, *Lager* 'storage, lair' and *Statt* 'place'), with *Lagerstätten* being the plural, is a name used to describe sedimentary deposits

A summary of some of the world's major *Lagerstätten* deposits

Time division		Name of sedimentary unit	Time Ma	Location	Examples of dominant biota
Mesozoic	Cretaceous	Santana Formation	108–92	Brazil	Fossil fish, pterosaurs, reptiles, dinosaurs and amphibians. Insects.
		Yixian Formation	125–121	Liaoning, China	Huge numbers of vertebrates, including mammals, dinosaurs, pterosaurs, turtles, amphibians and fish. Insects and spiders.
	Jurassic	Solnhofen Limestone	140	Bavaria, Germany	See case study.
	Triassic	Madygen Formation	230	Kyrgystan	Fishes, amphibians, reptiles.
Upper Palaeozoic	Permian	Mangrullo Formation	c. 285	Uruguay	Petrified wood, conifers, seed ferns, clams and shrimps.
	Carboniferous	Mazon Creek	310	Illinois, USA	Many plants including those related to club mosses, horsetails and seed ferns. Sea scorpions, fish and jellyfish. Many insects including centipedes, shrimps, spiders.
		Bear Gulch Limestone	320	Montana, USA	Fishes and invertebrates.
	Devonian	Gogo Formation	380	Western Australia	Fish and arthropods such as eurypterids.
		Hunsrück Slates	390	Rheinland-Pfalz, Germany	Corals, trilobites, sponges, crinoids, sea cucumbers and echinoids.
		Rhynie Chert	400	Scotland, UK	Plants of various kinds preserved in 3D.
Lower Palaeozoic	Silurian	Wenlock Series	c.425	Herefordshire, England, UK	Worms, molluscs, sea spiders, ostracods, arthropods of unknown affinity.
	Ordovician	Soom Shale	450	South Africa	Microfossils, brachiopods, eurypterids and trilobites.
	Cambrian	Burgess Shale	508	British Columbia, Canada	See case study.
		Maotianshan Shale (Chengjiang)	515	Yunnan Province, China	See case study.
Precambrian	Precambrian	Ediacara Hills	550–545	South Australia	Enigmatic frond-like biota. (Also found in Leicestershire's Charnwood Forest.)
		Bitter Springs	1000–850	South Australia	Cyanobacteria and algae.

which have fossils that have been found in large numbers. Some of these deposits have organisms that have been exceptionally preserved.

There are two main types of *Lagerstätten:*

1 *Konzentrat-Lagerstätten* (concentation) where you can find many organisms in one bed, such as disarticulated bone beds and organic hard parts. It also includes high concentrations of fossils on reefs or bivalve oyster beds.

2 *Konservat-Lagerstätten* (conservation) are deposits known for exceptional preservation. These are crucial in providing evidence for the history of life on Earth, such as organisms found in the Burgess Shale and the Solnhofen Limestone. We will concentrate on this type of exceptional preservation in this book.

Exceptional preservation means that the fossils have very fine detail and the remains of soft tissues may be preserved. 'Soft tissue' may be the whole animal, such as a worm or jellyfish, or may be only part of the animal, such as muscle or gills, preserved along with other hard parts. It is usually only the hard parts that are preserved as fossils in most rocks, and so exceptional preservation is rare. Terrestrial animals have comparatively little chance of preservation unless they are washed into lakes, as the land surface is largely an area of erosion.

Konservat-Lagerstätten deposits have occurred throughout geological time, in many parts of the world and preserve a variety of different sedimentary environments, some of which are shown in the table.

Ideal conditions for exceptional preservation

For the few organisms that are exceptionally preserved, there is a sequence of stages that normally occur in the transition from dead body to a fossil:

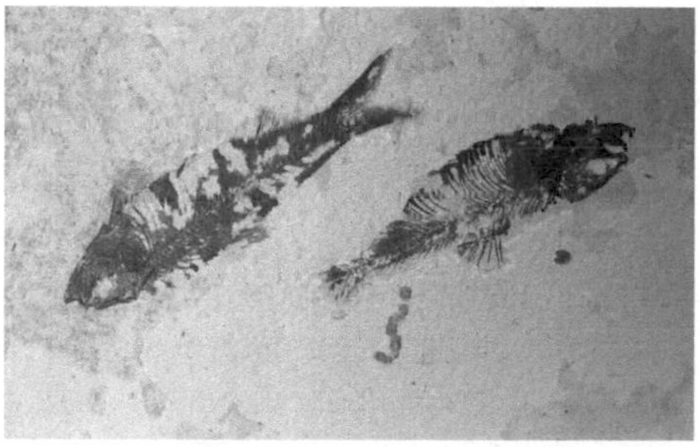

- Rapid burial in a protective medium such as soft, fine clay, to protect hard parts from the destructive action of scavengers and the weather.

- Burial in **low-energy** conditions for the best chance of fossilisation.

- Lack of oxygen, **anaerobic** conditions within the sediment or **anoxic** conditions in the water, to prevent decay.

- Low pH as decay is slowed down in conditions of high acidity, such as in peat swamps or altered ocean chemistry.

Figure 1 Exceptional preservation of fish from the Green River Formation, Wyoming, Eocene. Fish approximately 3 cm long.

Conditions that occur	Reasons
Quick burial in sediment, before or shortly after death	No time for breakage, scavenging or decay of material
Fine grained sediment	Fine particles preserve detail
Anaerobic conditions in the sediment and/or anoxic conditions in the water	No aerobic bacterial action, therefore no decay of soft tissue
No scavenging animals	Not fragmented, damaged or eaten
Original material replaced early in diagenesis	Less alteration of original material and so finer detail is preserved

1 Explain how methods of exceptional preservation allowed soft tissues to be preserved.

2 How can fossils in *Konservat-Lagerstätten* help us understand evolution of life on our planet?

These fossils are very old and pre-date the evolution of hard-shelled organisms. These are the oldest multicellular organisms on the planet, which means they had specialised tissues, presumably adapted for specific functions early in evolutionary history.

The most common resemble fronds and discs. *Charniodiscus*, a disc-like structure, which was first found in Charnwood Forest, Leicestershire, was named by Trevor D Ford in 1958. This was a holdfast which we now know anchored *Charnia*, a frond-like structure, to the sea bed. There are several species of *Charnia*, but all with the same basic morphology, but the fronds have slightly different shapes. *Charnia* and *Charniodiscus*, are the remains of soft-bodied, sessile, primitive animals, with an unknown affinity. Although they look like plants, they were animals.

The Ediacaran fossil *Dickinsonia* is a bilaterally symmetrical soft-bodied organism found as imprints in sandstone. They are of unknown affinity and have been interpreted as primitive worms or jellyfish.

The biota lived around 575 million years ago and are found on all continents, except Antarctica; in particular Australia, Siberia, Newfoundland and more locally in Charnwood Forest, Leicestershire, UK. The Ediacaran biota radiated in an event called the Avalon explosion, a period of rapid evolution following extensive glaciation (Snowball Earth). These were superseded by completely different biota in an event called the *Cambrian Explosion*.

Figure 2 Ediacaran fossils – *Charnia* and *Charniodiscus*. Frond approximately 15 cm long

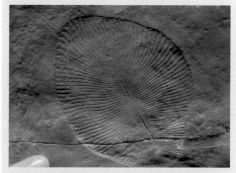

Figure 3 Ediacaran fossil – *Dickinsonia*. Finger for scale

CASE STUDY: CAMBRIAN EXPLOSION

This was an evolutionary landmark where large numbers of metazoan phyla appeared, seemingly quite suddenly, in the fossil record. Evolution over the following 20 million years resulted in the emergence of all the body plans that we recognise in extant species today. This event coincided with the evolution of 'hard' body parts, such as trilobite carapaces (exoskeletons), which increased the likelihood of preservation in the fossil record.

Other 'enigmatic' body plans from animals with unknown affinities also evolved which were less successful, perhaps as evolutionary experiments, but overall these organisms did not succeed beyond the Cambrian.

CASE STUDY: DISCOVERY OF THE BURGESS SHALE

This black shale was discovered in 1909 by a palaeontologist called Charles Walcott. The legend says that his horse stopped in front of a rock, which he then cracked open, discovering fossils. He tried to categorise the fossils using modern classification, with some success, but there were many that simply did not fit.

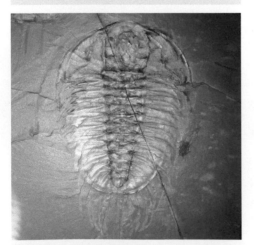

Figure 1 Trilobite *Olenoides*, showing preserved legs and gills outlined below the pygidium.

Rare and exceptionally preserved Cambrian biota provide us with critical insights into the evolution of life on this planet and the diversification of animals into those we know today. These exceptionally preserved organisms give us further evidence about the *Cambrian Explosion*.

Two important *Konservat-Lagerstätten* from the Cambrian are:

- *The Chengjiang Formation*, Lower Cambrian deposits from China (dated at around 515 Ma).
- *The Burgess Shale Formation*, Middle Cambrian deposits from Canada (dated at around 508 Ma).

Preservation of Cambrian biota

These deposits contain the remains of soft-bodied organisms, which are usually absent in the fossil record and provide a window into early **metazoan** evolution during the Cambrian. They contain preserved organisms that were completely soft bodied, which would have usually decayed in 'standard' conditions. It also includes the preservation of legs, gills, gut traces and muscle blocks in many different species. Some organisms appear unrelated to other known species, and are describes as having **unknown affinity**.

The soft parts are preserved as *aluminosilicate* films (type of clay), often associated with iron mineralisation. Iron pyrites within the sediments have enabled detailed fossilisation of some organic remains, as thin films. Iron pyrites preservation most probably resulted from inhibition of bacterial activity, possibly linked to low oxygen concentrations within the sediments (anaerobic). The bottom waters had sufficient oxygen for benthonic life (oxic) but the sediments lacked oxygen (anaerobic). Rapid entombment would result in organisms being 'sealed' into the sediments increasing the likelihood of exceptional preservation. There is also evidence that carbonate cements formed on the top of the sea bed, as a result of unique ocean alkalinity at the time of deposition. This would have also reduced the oxygen concentrations within the sediments and may have accelerated diagenesis.

Burgess Shale Formation

This has become the best known *Konservat-Lagerstätte* in the fossil record and is exposed in the Canadian Rockies.

This preserves hundreds of taxa as part of the Cambrian Explosion. This was important, as prior to the discovery of exceptionally preserved organisms, only hard parts remained in sediments. It helps us to understand that complex communities were in place early in evolutionary history.

Geological setting

It was initially thought that these deposits were the result of a catastrophic underwater landslide or turbidity current. It was postulated that the communities existed on a continental platform, which was then transported during a catastrophic event into an anoxic basin, over a distance of around 2 km. The assumption was that exceptional preservation was not possible if there was oxygen present on the sea floor, as decay would occur.

However, more recent studies have shown that the basin floor *was* suitable for life due to the following observations:

1 Most organisms were buried in their habitat, or have suffered only minor disturbance.
2 There were benthonic organisms preserved in life position, such as many species of sponge and some brachiopods.

3 Most decay processes took place prior to burial and resulted in some **disarticulation** when burial occurred.

4 There was evidence that variable amounts of disarticulation had occurred in some organisms, indicating transport, but only over very short distances.

5 Early mineralisation of body tissues occurred soon after burial (during early diagenesis).

The most likely explanation is these organisms actually lived in a relatively deep Cambrian basin, in tropical waters. There was an escarpment of an algal reef, several hundred metres high (Cathedral Escarpment) and a carbonate platform beyond, which probably extended 400 km towards the shoreline. The communities nestled at the base of the escarpment, in a deep water basin.

The sea floor was fed by periodic influxes of sediment, as a result of small turbidity currents or 'slurries', which then covered the organisms currently on the basin floor. At the point of any 'slurry', there were live and dead organisms on the sea floor. Some had already partly decayed, whilst others were buried alive (entombed).

Biota

Around 40% of the biota are **arthropods**; these possessed hard exoskeletons and include trilobites. The exoskeletons are often calcified during preservation. Limbs, antennae, gills, gut traces and other soft tissues are often preserved. There are many examples of the trilobite *Olenoides* which have preserved legs and gills.

Trace fossils are present in varying amounts, indicating that conditions on the sea floor were oxic and conducive to life at the time, whilst conditions within the sediments may have been anaerobic for at least some of the time. However, only a few infaunal organisms had evolved at this stage in the Cambrian, so scope for many infaunal organisms was limited.

⬆ **Figure 3** Reconstruction of *Marrella* (arthropod) is often called a lace crab. It has a strange head shield with four curving spines. Feathery filaments on the limbs were probably used for respiration.

⬆ **Figure 4** Reconstruction of *Aysheaia*, a velvet worm, was a limbed, soft-bodied, segmented animal. It had spines on its legs. The head had a pair of tapering limbs and small projections near its mouth.

⬆ **Figure 5** Reconstruction of *Hazelia*. A sessile filter feeder, part of the benthonic community.

Unknown affinity

There are many fossils that had many bizarre anatomical features, which are unknown in modern biology or elsewhere in the fossil record. There were many different and experimental body plans emerging in the Cambrian. Many of these forms died out during the Cambrian and only the successful lineages survived.

⬆ **Figure 6** Reconstruction of *Opabinia*. It had a long snout and paddle-like projections at the posterior end of the body, possibly acting as stabilisers in the water column.

⬆ **Figure 7** Reconstruction of *Hallucigenia*. It had a poorly defined head and spines (not bristles) on its back. It was named because of its 'bizarre and dream-like quality' and was originally reconstructed upside down.

⬆ **Figure 8** Reconstruction of *Pikaia*. It was an early representative of Phylum Chordata, as it had a primitive notochord, but it is not a vertebrate. It was a swimmer and its body was made up of muscle blocks known as myotomes, rather like a fish today.

CASE STUDY: MEASURING THE DECAY PROCESS

The states of preservation vary throughout the sequence and so through geological time. Some organisms were buried alive, whilst others had undergone various amounts of decay on the sea floor prior to burial. Studies looking at two different genera of polychaete (annelid) worms have identified index levels of decay in specific horizons. The two worms (*Canadia* and *Burgessochaeta*) were covered in bristles and show differences due to decay and the degree of fragmentation of the bristles.

a   b

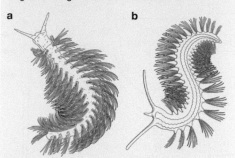

⬆ **Figure 2** Reconstructions showing bristled polychaete worms **a** *Canadia* and **b** *Burgessochaeta*

QUESTIONS

1 Describe the conditions that allowed exceptional preservation to occur during the Cambrian.

2 Compare and contrast the exceptional preservation shown in Cambrian with the Precambrian.

3 Research the fossil *Anomalocaris* and determine its place in a Cambrian ecosystem.

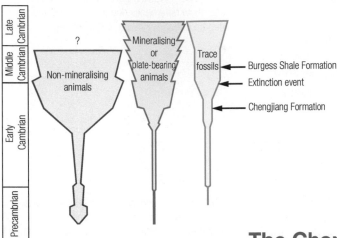

Figure 1 Record showing the evolution of non-mineralised (soft-bodied) organisms, mineralised organisms (hard parts) and trace fossils.

Evolution of 'hard parts'

The transition from the Precambrian to the Cambrian was noted by the evolution of multicellular marine animals which had evolved sufficiently to leave a fossil record. This included developing mineralised skeletons, such as those seen in arthropods. This was probably as a result of predation and the need for animals to protect themselves for survival. Thus, the more successful organisms with mineralised skeletons survived to pass on genes to their offspring.

The lowermost Lower Cambrian strata have small isolated and disarticulated shelly fossils and a few trace fossils. By the middle part of the Lower Cambrian there were more abundant invertebrates, such as many arthropods, molluscs, echinoderms, sponges and an increase in the number and type of trace fossils. This signified the evolution of life was rapidly occurring, both in and on the substrate.

There was an extinction event which occurred between the formation of the Chengjiang and the Burgess Shale formations, exemplifying the importance of the Chengjiang Formation in our understanding of evolution.

The Chengjiang Formation

The Chengjiang Formation, from the Lower Cambrian deposits from Yunnan Province in China, is dated at around 515 Ma. This formation is older than the Burgess Shale and provides us with an earlier window into evolution of life on Earth and the Cambrian Explosion.

Geological setting

The area which supported the Chengjiang biota was part of the South-west China Platform. Deposition took place in a relatively shallow, tropical shelf environment, which was in a restricted basin. The exceptionally preserved fossils usually occur within units which show cyclic sedimentation; alternating grey-green to yellow shales on a millimetre scale. There is evidence of sea level changes and tectonic activity which altered the conditions in the basin. The likelihood is that periodic turbidity currents or 'slurries' covered the fossils which showed little transport or disarticulation.

There may have been some periodic influxes of fresh water from the continents which may have contributed to the exceptional preservation by altering the seawater chemistry.

Biota

The fossils are preserved in a siltstone/mudstone deposit, which is only around 50 m thick, but is extensive laterally, covering thousands of square kilometres in Yunnan Province. Approximately 15% of the genera from the Burgess Shale are also found in the Chengjiang Formation. Echinoderms and molluscs are rarely found at Chengjiang but present in much larger numbers in the Burgess Shale, showing that an evolutionary step change occurred for these groups following the deposition of the Chengjiang Formation.

Soft tissues are preserved as aluminosilicate films, often with a high oxidised iron content and appear orange. The shales are heavily weathered.

Most of the Chengjiang animals were benthonic and some show evidence of short distance transportation following death. Sponges are sometimes found flattened on bedding planes, suggesting that they were uprooted prior to being buried, possibly aligned by a current. Burrowing organisms, such as worms, were also found on bedding planes. Benthonic, sessile organisms are seen to be smothered by sediment such as silt, indicating regular events or 'slurries' which caused burial and subsequent preservation.

The animals are diverse including annelid worms, arthropods (including trilobites), sponges, primitive chordates, echinoderms, algae and many more. Around one in eight animals are problematic forms with unknown affinities. Chengjiang is the richest deposit containing the unusual **lobopodians**, that is, a worm-like taxa with stubby legs, also called velvet worms.

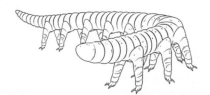

🔺 **Figure 2** Reconstruction of the velvet worm *Paucipodia* from the Chengjiang biota

QUESTIONS

1 Compare and contrast the preservation of fossils from the Chengjiang Formation and the Burgess Shale.

2 Research the size of the organisms shown in the photographs.

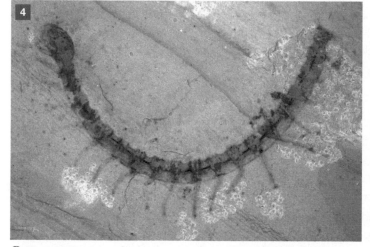

🔺 **Figure 3** *Lingulella*, a brachiopod with a very long pedicle.

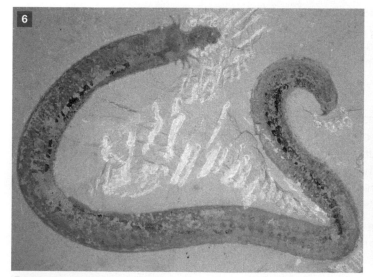

🔺 **Figure 4** *Cardiodictyon*, a lobopod showing numerous paired legs.

🔺 **Figure 5** *Eldonia*. This cluster shows a problematic fossil, whose true biological affinity is of debate. Perhaps a relative of an echinoderm or a hemichordate.

🔺 **Figure 6** *Cricocosmia*, an iconic worm of the Chengjiang Biota. The traces of the gut can be seen as dark areas.

🔺 **Figure 7** *Misszhouia*, an arthropod which lacked a mineralised skeleton. The slender antennae can be seen at the head of the animal. The digestive system is visible as an elongate dark patch running down the centre. Benthic scavenger or predator.

Images courtesy of Prof Derek J Siveter, University of Oxford

The Solnhofen Limestone

Geological setting

Around 155 million years ago, during the Jurassic, much of southern Germany was covered by an inland sea (shallow basin) which was cut off from the ocean. It had a limited input from terrestrial water sources. Small rises existed within the sea, which allowed colonisation by sponges and other tropical fauna. These rises formed reefs which separated the sea into individual lagoons. There was increased evaporation of water from these shallow lagoons, causing an increase in salinity within them to toxic levels for living organisms. Fine grained carbonate muds accumulated on the sea floor, precipitated from the seawater itself, along with other evaporite deposits, due to increased salinity.

Biota

Marine organisms

Fossils are generally very rare, but when present show unusual detail, including soft tissues or fragile elements, such as feathers. The fossils include jellyfish, dragonflies, beetles, crustaceans, echinoderms (sea lilies), horse shoe crabs, ammonites and the small vertebrates such as the small crocodile, *Alligatorellus,* turtles and fishes.

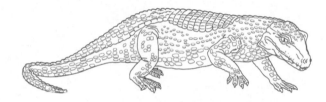

🔺 **Figure 1** Reconstruction of *Alligatorellus*. Specimens are approximately 20 cm long

There were also invertebrate trace fossils in some horizons, sometimes preserving the 'trace-maker', signifying adverse conditions and ultimately death. This would suggest that some organisms could tolerate high levels of salinity, at least for a short amount of time before death.

Terrestrial organisms

Terrestrial organisms are also preserved including: pterosaurs such as **Pterodactylus** and the dinosaur **Compsognathus**, as well as the 'bird-like' **Archaeopteryx**.

Archaeopteryx

Only a few specimens have ever been recovered, but they show preservation of feathers, bones and other soft tissues. They probably only had a mass of around 1 kg, and were about the size of a common raven. It had broad wings, with well-developed flight feathers, as well as tail feathers. It had body plumage around the legs, like trousers which was down-like and fluffy. Feathers have not been found on the neck or head.

Archaeopteryx had a primitive shoulder girdle which probably meant it had a limited range of movement. It probably glided or flew over short distances only.

Preservation

Nothing lived on the bottom of these lagoons due to the anoxic waters. It is likely that the water in the lagoons was stratified; oxic in the surface layers and anoxic at the bottom. If organisms fell into the lagoon or got trapped in the carbonate mud, they would be very quickly covered. The anoxic water at the bottom of the lagoon meant that no scavengers could live there and bacterial decay processes could not operate. Organisms lying in the mud would remain intact.

The soft tissue preservation suggests that there was rapid deposition. There are some fossils that show **necrolytic features**, which could have been due to the hypersaline environment. This included fish bent so much that their spinal columns have broken and the curved-back necks of flying vertebrates.

CASE STUDY: A NOTABLE LAGERSTATT FROM HEREFORDSHIRE

There are few Silurian *Konservat-Lagerstätten,* but a globally important site is found in the middle part of the Silurian (Wenlock Series) in the Welsh borderlands of Herefordshire.

These deposits were formed on the outer shelf or upper part of the continental slope of the Welsh sedimentary basin. The water depth is estimated to be around 150 metres. The fossils are preserved within nodules, entombed in volcanic ash that was deposited on the sea floor. There are fossils preserved here that are not found anywhere else in the world.

There are exceptionally preserved small invertebrates, such as worms, starfish, molluscs and brachiopods, many arthropods and several forms of unknown affinities.

The specimens were revealed by the use of micro-grinding and digital imaging. The specimens were ground down 20 microns at a time and then digitally photographed. There are hundreds of images for each specimen. Computer reconstruction techniques then produce a 3D 'virtual image' of the fossil. Of course, the specimens are destroyed during this process but the science is retained.

This is of global importance because the Silurian is the time period when the development of life on land was underway and the evolution of the first vascular plants (*Cooksonia*).

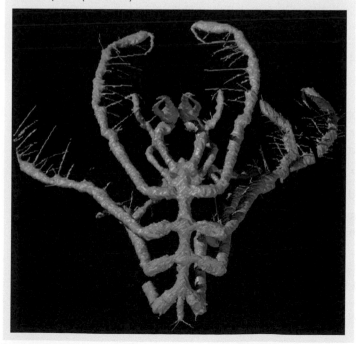

🔺 **Figure 2** Photograph of *Archaeopteryx*, showing a curved neck and wings

🔺 **Figure 3** Photograph of a single *Archaeopteryx* feather

QUESTIONS

1 Describe how an increase in salinity gave rise to the deposition of evaporite deposits in the Solnhofen Limestone and describe how this impacted on life in the lagoons.

2 Research the various marine and non-marine vertebrates that are found in the Solnhofen Limestone. Describe the likely food chains that existed at the time.

🔹 **Figure 4** A 'sea spider'. Extremely rare in the fossil record. Preserved in three dimensions.
Image courtesy of: Derek J Siveter, University of Oxford; Mark Sutton, Imperial College London; Derek Briggs, Yale University and David J Siveter, University of Leicester.

Whole basin facies analysis integrates concepts from sedimentary geology, environments and palaeontology, including exceptional preservation. It also assumes knowledge of oil, gas and structural geology. Earlier sections of this book should be revisited for a full understanding of this topic.

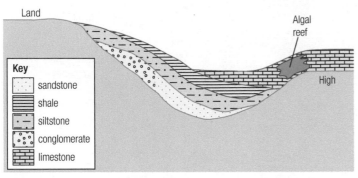

Figure 1 A cross-section showing a generalised sedimentary basin (not to scale and vertical axis exaggerated)

Key
- sandstone
- shale
- siltstone
- conglomerate
- limestone

Land

Algal reef

High

Formation of sedimentary basins

Sedimentary basins are subsiding areas of the Earth's crust where sediments have accumulated to greater thickness than in surrounding areas. Subsidence of the crust is most often caused by attenuation (stretching) at divergent plate margins, lithospheric contraction due to cooling at convergent margins, depression of the lithosphere by sediment loading or by faulting.

Sedimentary basins may form at plate boundaries, on continental shelves or in shallow seas. The mass of sediments accumulating on the crust causes it to sag downwards to accommodate new sediment, and it initiates new faulting along the basin margins or within the basin itself. For example, the formation of the Viking Graben in the North Sea.

What is sedimentary basin facies analysis?

It involves the study of vertical and lateral facies changes and can be a massive operation in a large basin. Data collection will include measuring stratigraphic sections and geological structures and also describing the depositional environment at the time of formation using sediments and sedimentary structures.

Identification of the depositional environment is usually based on the integration of all the evidence found in each rock unit when examined in detail in the field. The information is carefully recorded and used to construct a graphic log. Multiple graphic logs spread over an area allow reconstruction of the geometry of the basin, the positions of deep sea, shelves, reefs and shorelines in the geological past. They allow us to see changes in environment over time, shown vertically, and lateral changes by comparing a series of logs over a geographic area. Structural data helps establish the three-dimensional nature of the basin. A geological map is produced.

Type of sediment

Clastic sediments are deposited when energy levels fall. Coarse sediments represent high energy, and finer sediments low energy. Cyclic sedimentation provides information about changes in sea level, often beginning with a deepening event, shallowing upwards, and then repeating. Limestones form chemically, or by the accumulation of bioclasts in shallow basins that lack terrigenous input. Some originate in reefs. They indicate shallow, tropical water.

Sedimentary structures

Symmetrical ripple marks and small-scale cross-bedding indicate bidirectional flow, caused by waves in **marginal marine environments**. Channels can be formed by high-density flows, such as turbidites, flowing toward the edge of a continental shelf, forming a submarine canyon. Bouma sequences forming in deeper water thin away from a **land mass**. Flute casts are used as indicators of palaeocurrent directions. Deep water basins have parallel-bedded, low-energy shales and mudstones, deposited below the wave base.

KEY DEFINITIONS

Marginal marine environment neither marine nor terrestrial, but belongs to both categories, one at low and another at high tide. Often a beach environment.

Land mass means a continent or large body of land.

Graptolites extinct marine organisms that lived in colonies, most of which floated freely in the sea, whilst some may have been attached to the sea bed.

CASE STUDY: MICROFOSSILS AS ZONE FOSSILS

Microfossils are found in very large numbers in most sedimentary rocks. Their size means that although tricky to work with, they are preserved in the small chippings returned by drilling muds from boreholes and so are very useful for basin analysis. Many, such as foraminifera, are calcareous. Some microfossils, such as radiolaria, are composed of silica and can be extracted from limestone by dissolving the rock in HCl. Pollen, spores, acritarchs and dinoflagellates have resistant skeletons which can be extracted from clastic sediments using HF, which dissolves siliceous rock. Non-calcareous fossils will survive below the Carbonate Compensation Depth. From an economic geology viewpoint, microfossils are by far the most useful for correlating boreholes in basin analysis, although microscope techniques are needed to observe them.

Fossil evidence

Fossils found in their life position lived in the environment in which they were buried. So, coral reefs in life position formed in high-energy shallow water in the photic zone. Trilobite fossils and trace fossils within and on the substrate usually indicate that sea-floor conditions were suitable for life. Most trilobites (with some exceptions) were benthonic. **Graptolites** usually indicate low energy, deep sea deposition, as although they lived in the water column, they were only preserved in low energy environments. Fossils of terrestrial animals, such as flying vertebrates, indicate proximity to land. Pterosaurs may have become trapped in lakes or lagoons, indicating a marginal marine environment close to shore.

Energy levels impact on the proportion of whole organisms and the amount of fragmentation. High energy results in more collisions between fossils and other debris on the sea floor, increasing breakage. Whole fossils indicate lower energy. Shorelines normally have high energy conditions due to the influence of tides and waves.

Some fossils are specific to a certain horizon or period. These zone fossils can be used to correlate the same horizon across a basin. This is the basis of biostratigraphy. Trilobites, corals and graptolites are common zone fossils in the Lower Palaeozoic. Ammonites, bivalves and belemnites are common zone fossils in the Mesozoic. Microfossils, such as conodonts, acritarchs and ostracods are also important zone fossils.

Subsurface techniques

Seismic and borehole data, covered earlier in this book are summarised below. They are used to investigate inaccessible sedimentary basins where there are few surface outcrops.

Seismic data

Seismic reflection surveying uses a controlled source of seismic energy, such as dynamite, a seismic vibrator or air gun, and the reflections, like echoes, are recorded. They show two-way travel times, measured as arrivals reflected from subsurface interfaces, such as different rock types. Such surveys are most useful in prospecting for oil and gas. The images generated can be both 2D and 3D.

Well logs

This is the process of recording physical, chemical, electrical and other properties of rock by drilling a borehole. The downhole instruments measure and record the properties which are displayed as a graph as the borehole goes deeper. Fragments of rock retrieved can be processed to yield microfossils which can be used for biostratigraphic correlation.

CASE STUDY: ZONE FOSSILS

These are used to identify different geological time units, which are usually subdivisions of geological periods, known as stages. The ideal zone fossils should have the following qualities:

- Each species has a short stratigraphic range (evolved quickly) and is restricted to a small part of the stratigraphic column.
- Are abundant (that is common when it was alive *and* was easily preserved as a fossil).
- Have easily identified remains, with obvious differences between species.
- Are found in many different rock types, so they are free floating or free swimming in all (marine) waters.
- Have a wide geographical distribution, preferably worldwide regardless of climate.
- Have strong, hard shells or skeletons to enable them to be commonly preserved.

To hit all of these criteria would be remarkable. In fact all but a few of the tens of thousands of fossil species are not suitable as zone fossils. Often the mode of life of the organism is the deciding factor. Two good examples of zone fossils are the ammonites and the graptolites because they were widely distributed across the oceans and, when they died, fell into a variety of marine environments.

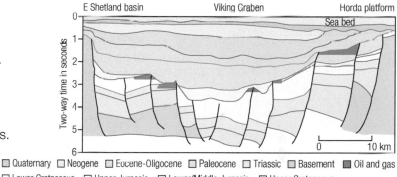

Figure 2 Cross-section of the Northern North Sea Basin, constructed from seismic reflection data

CASE STUDY: STRATIGRAPHIC RANGES

The range of fossil groups is often marked onto geological column diagrams as a vertical line. More information about the diversity of genera can be seen if the line is made into a 'kite' diagram – the thickest part of the line representing the maximum diversity.

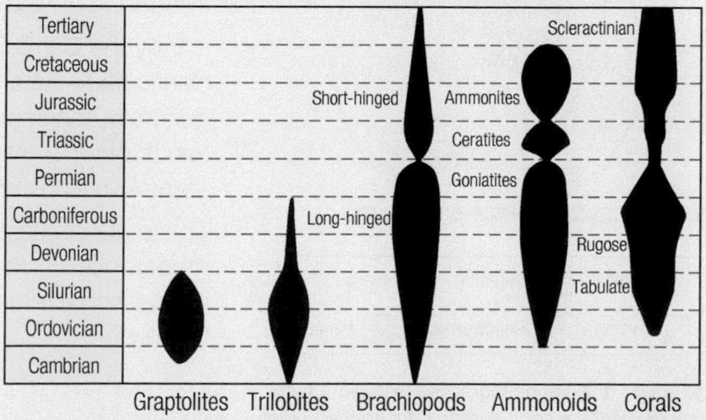

Figure 3 Diversity and range of common fossil groups. The wider the bar, the more organisms existed.

QUESTIONS

1 Describe the changes in sediments, sedimentary structures and fossils you would expect to see if you were to drill a series of boreholes moving away from a continental shoreline.

2 Explain why faulting can occur in sedimentary basins, regardless of their tectonic setting.

3 Using the British Geological Survey (BGS) website, research how the Weald and Hampshire basins have been investigated using 2D and 3D methods.

The Welsh Basin is a sedimentary basin which is located in the Welsh Borderlands and Central Wales and was formed during the Lower Palaeozoic. Here there is a great thickness of sediment which represents a zone which had regular subsidence events throughout the lower Palaeozoic.

Tectonic setting

At the start of the Cambrian, the supercontinent Pannotia, was breaking apart. Southern Britain lay on the margins of this supercontinent, on the Eastern edge of the fragment of crust called **Avalonia**. To the east lay the Midland Platform, an upland area of land at the time. Avalonia, **Baltica** and **Laurentia** microcontinents lay in the southern hemisphere.

The Lower Palaeozoic was a time where there was major plate tectonic activity in the Welsh Basin. There were divergent plates which caused the formation of a large ocean **Iapetus**. This was followed by convergence of plates and eventually resulted in the fusion of Avalonia, Baltica and Laurentia, as these plates collided together. This was also called the **Caledonian Orogeny**. It caused folding to occur in Wales and the Welsh Borderland and the development of a major mountain chain in Scotland, which was possibly as high as the present day Himalayas; the eroded remains of which we see today as mountain roots.

The Caledonian Orogeny also caused the low-grade metamorphism we see in Wales and the associated cleavages that are found within the Welsh Slate, still used today as roofing material.

In the Welsh Basin, the first Cambrian sediments are laid down unconformably on top of eroded Precambrian basement rocks. These Precambrian rocks are formed of earlier crustal blocks, called **terranes**. It is widely thought that these rocks formed as a result of island arc volcanism and the deposition of associated basin sediments, similar to Montserrat in the Caribbean island arc today.

The Cambrian

The palaeogeography at the time saw a land mass to the north-west, as the Irish Sea land mass and to the south-west, as the Midland Platform. These topographic highs provided eroded sediment which was transported into the basins. This was probably due to Laurentia rifting away from Gondwanaland as the Iapetus Ocean opened up. There was a shallow marine platform on its southern margins, with associated faulting, which allowed subsidence in the basin. The centre of the basin was deep marine, and remained so for most of the Lower Palaeozoic. The basin trended north-east to south-west.

Initially the Cambrian sediments are dominated by coarse to medium grained clastic sediments. They lie unconformably on top of Precambrian rocks and are generally conglomerates and quartzites. These are overlain by green, feldspar-rich sandstones or glauconitic sandstone, suggesting shallow water. There are also some thin limestones and sometimes red-coloured shales.

These are overlain unconformably by pebbly sandstones, sandstones and mudstones which are in turn followed by more dark shales and mudstones, suggesting a return to deeper water. Some of these layers contain trilobite zone fossils.

Precambrian, 600 Ma

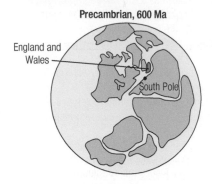

England and Wales

South Pole

Early Ordovician, 480 Ma

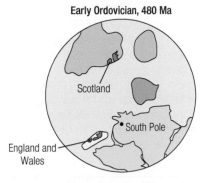

Scotland

South Pole

England and Wales

Late Ordovician, 440 Ma

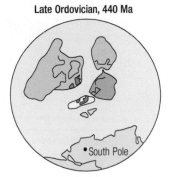

South Pole

Late Silurian, 420 Ma

South Pole

South Pole

	Baltica		Laurentia		Siberia
	Avalonia		Gondwanaland		Pannotia

🔺 **Figure 1** Palaeogeographic reconstructions of Avalonia, Baltica and Laurentia during the Lower Palaeozoic

The Ordovician

In the early Ordovician, subduction zones began to form along the margins of Avalonia and Laurentia and the Iapetus Ocean started to reduce in size. In North Wales and Pembrokeshire, the subduction fuelled the formation of volcanoes and associated volcanism, in an island arc setting, similar to the Caribbean today. Subsidence continued in the basin centre, and there was a marine transgression encroaching on the land masses.

At the end of the Ordovician, the volcanism was at its peak, with very large volcanoes in North Wales, Pembrokeshire and along the Welsh Borderland Fault System. Volcanism was also at its height in the Lake District, a little further north. At the end of the Ordovician the volcanism ceased because the Iapetus crust was no longer being subducted under this portion of the continental margin.

The Ordovician sees the real subdivision of black shales in the basin and clastic sedimentation on the shelf. It starts with the deposition of grit and quartzite, which is widespread across several areas. In marginal marine conditions there are conglomerates and sandstones. These are overlain by dark black shales which contain graptolite zone fossils. There are interbedded tuffs and rhyolitic lava flows near to the volcanic centres.

The Silurian

There were a number of transgressive and regressive cycles from the late Ordovician to the late Silurian. This may possibly be linked to climate change due to volcanism and linked to repeated glaciations. The fault systems bordering the Midland Platform were still operational, allowing subsidence in the basin centre.

There was a widespread break in sedimentation at the end of the Ordovician, corresponding to a mass extinction event. Following this the sedimentation followed the broad pattern established during the Ordovician, with dark black shales in the basin and other clastic sediments and thin limestones on the shelf.

The basin had interbedded turbidite deposits which thinned away from the shelf. Evidence of submarine channels can be seen on the shelf, where sediment was carried out into the basin. The deep part of the basin was anoxic.

The shelf or platform saw the development of reef systems, one of which is exposed today and forms a ridge, Wenlock Edge. The limestones are generally impure and examples include the Wenlock and Aymestry limestones. 'Silty limestones', 'limey siltstones' and shales dominate the shelf deposition from the middle to the late Silurian on the shelf. These are often highly fossiliferous and very variable.

At the end of the Silurian, a regression allowed the Old Red Sandstone terrestrial conditions to progress across the basin, marking the end of the Welsh Basin deposition. These sediments are closely associated with erosion of the Caledonian Mountain chain which formed as a result of collision between Baltica, Laurentia and Avalonia.

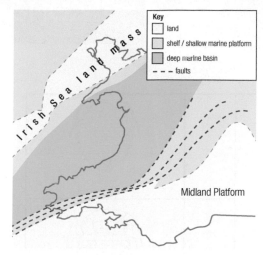

▲ **Figure 2** Palaeogeographic reconstruction of the Welsh Basin during the Cambrian Period

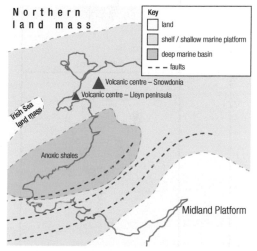

▲ **Figure 3** Palaeogeographic reconstruction of the Welsh Basin in the late Ordovician

KEY DEFINITIONS

Avalonia was a microcontinent during the Palaeozoic which contained Wales, part of Ireland, England and the southern part of Scotland, located in the southern hemisphere.

Baltica was a microcontinent during the Palaeozoic which contained the Baltic countries, located in the southern hemisphere.

Laurentia was a microcontinent during the Palaeozoic which contained northern Scotland and part of Ireland, and the North American continent, located in the southern hemisphere.

Iapetus was an ocean which existed in the Lower Palaeozoic, which closed due to the collision of Avalonia, Baltica and Laurentia microcontinents.

Caledonian Orogeny was a mountain building era, named after the Latin name for Scotland, *Caledonia*. The event took place over a considerable period of time, from the Ordovician to the early Devonian.

Terranes fragments of crustal material broken from one tectonic plate and accreted on another. The boundary, or suture, is usually a fault.

QUESTIONS

1 There are many volcanic centres during the Ordovician, mainly located in North Wales. Three large centres are Dolgellau, Bala and Snowdon. Research the types of volcanic rocks that can be found in the Welsh Basin, which were derived from these major centres.

2 Research and write an account of the Murchison-Sedgewick Controversy.

One fossil is usually chosen to be a biostratigraphic marker at a specific stratigraphic level, this is a *biozone* and the fossil would be known as a zone fossil. Sometimes more than one fossil can be used for the same stratigraphic level, but they are usually from two separate phyla. These biozones are established so that correlation between different parts of a basin and across the world is possible.

Zonation of the Welsh Basin

Trilobites

Trilobites are used to correlate the Cambrian, as they evolved quickly in the Cambrian seas. There were only a few competitors in their habitats allowing for diversification at a rapid rate. They had a good preservation potential as they had a mineralised exoskeleton. Common genera of trilobites found in the Welsh Basin, indicative of specific, but fairly wide time zones, are shown below.

Cambrian	Upper	*Olenus*
	Middle	*Paradoxides*
	Lower	*Olenellus*

🔺 **Figure 1** Simplified trilobite zonation of the Cambrian

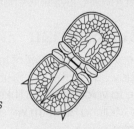

🔺 **Figure 2** *Glyptagnostus reticulatus*, a zone fossil from the Upper Cambrian

CASE STUDY: THE COMPLEXITIES OF ZONATION

Zonation of the Furongian (Upper Cambrian) using trilobites

The Cambrian is divided into zones based on the trilobites they contain. These zones are divided into subzones. The Furongian is the youngest of four internationally recognised units of the Cambrian, as described by the International Commission on Stratigraphy, 2016.

The lowermost unit of the Furongian biozone is based on two trilobites which existed at the same time; the agnositid, *Glyptagnostus reticulatus* and the olenid, *Olenus*. This zone is subdivided into four subzones based on different *Olenus* species; *O. gibbosus*, *O. truncatus*, *O. wahlenbergi* and *O. cataractes*.

The genus *Olenus* and *Glyptagnostus reticulatus* are easy to recognise. The differences between the four *Olenus* subzone species are very subtle, and so it can be difficult to identify fossils to subzone level.

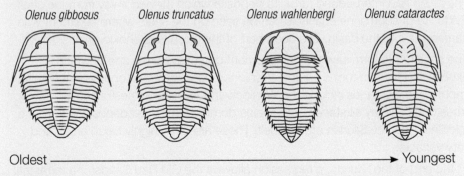

Oldest ————————————————→ Youngest

🔺 **Figure 3** Four different species of *Olenus*, marking four subzones in the Furongian, Upper Cambrian

Graptolites

Graptolite morphology

These were colonial organisms whose individuals laid down a scleroprotein skeleton in an orderly fashion. Based on similarities to modern hemichordates, it is thought that zooids lived in and built up tubes called **thecae**. The first zooid secreted a tube, called the **sicula**, which often had a long tapering point. Thecae were stacked vertically on top of each other and on top of the sicula. This may have been due to asexual budding from the first zooid. The thecae overlap and make up the **stipes**. True graptolites had varying numbers of stipes making up their entire skeleton, the rhabdosome.

Nema
A thin tube, an extension of the sicula, possibly used to attach to a floating object

Theca
Individual cup in which one zooid lived

Stipe
A branch of the rhabdosome

Aperture
Part where the zooid protruded through the skeleton, in order to filter feed

Sicula
Conical tube secreted by the first member of the colony

Virgella
Spine at the end of the sicula

Aperture of sicula
Where the first zooid protruded through the skeleton

🔺 **Figure 4** Graptolite morphology based on the example of *Monograptus*, a single-stiped form from the Silurian

Graptolites formed part of the micro-plankton in the oceans and were abundant, present in a large number of different sedimentary facies, although not well preserved where energy levels were high.

Zonation using graptolites

There are graptolite biozones identified throughout the Ordovician and Silurian. The changes in shape allow easy identification in the field, at least at generic level.

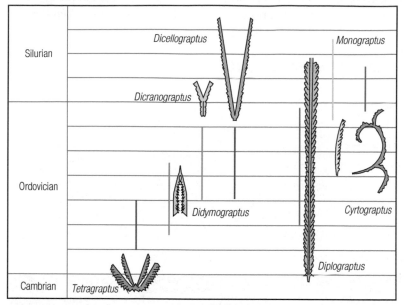

▲ **Figure 5** Simplified evolution of graptolite genera in the Ordovician and Silurian. Solid lines show the fossil ranges.

Corals (tabulate and rugose)

Corals are found in abundance in the extensive reef developments in Silurian rocks. They thrived when specific environmental conditions allowed and it is for this reason they are less useful for zonation than other fossils but are used in some instances, particularly in the Silurian.

CASE STUDY: GLOBAL BOUNDARY STRATOTYPE SECTION AND POINT (GSSP)

These are internationally agreed reference points that mark the lower boundary of a unit of time (called a stage) on the geological time scale. Most of these boundaries are marked by a change in palaeontology, whilst others are marked by global events, such as ash from volcanic eruptions. These 'stratotype sections' have been agreed by the International Commission on Stratigraphy (ICS) and have taken into account research from all over the globe. These sections are chosen to represent a standard for a particular time unit and are sometimes known as 'golden spikes'. The rocks in each stratotype section have to be representative of the same boundary across the world.

Global Boundary Stratotype Section and Point (GSSP) identified stratotype sections in the Silurian which are shown in Figure 7. It should be noted that six of the eight stratotype sections for the Silurian are from the Welsh depositional basin. This shows the global importance of this basin in geological terms. Of the other two stratotype sections, one is from Scotland, and the other from Czech Republic.

KEY DEFINITIONS

Theca a tube built onto the sicula, in which later members of the colony live.

Sicula a conical structure in which the first member of the colony lives.

Stipe the stack of thecae built up to form a colony.

CASE STUDY: BIOZONES

There are differences in the morphology of fossils found in Avalonia compared to those found from other microcontinents, such as; Baltica, Laurentia and Siberia, even though they lived at the same time! Even greater differences occur in parts of the world far from Avalonia during the Lower Palaeozoic, places such as Australia and China. Geographic isolation allowed organisms in each area to evolve separately. Each part of the world has a separate biozone chart, with their unique biozones noted. These have been carefully correlated between each other, so that areas can be compared.

CASE STUDY: CONODONTS

These are microscopic tooth-like structures from the feeding apparatus of an extinct primitive chordate, rather like a hag fish. This free-swimming animal enables the biozones identified throughout the Lower Palaeozoic to be used as a global tool. They are easily preserved in a variety of rock types, although best preserved in limestones or calcareous rocks.

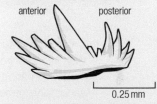

anterior   posterior

0.25 mm

▲ **Figure 6** Conodont element

QUESTIONS

1 Research the global distribution and stratigraphic ranges of the two corals, *Favosites* and *Ketophyllum*, from the Silurian.

2 Explain why it is more useful to use several different phyla to mark biozones.

Older ──→ Younger								
Global standard chronostratigraphy	**Llandovery Series**			**Wenlock Series**		**Ludlow Series**		**Přídoli Series**
	Rhuddanian Stage	**Aeronian Stage**	**Telychian Stage**	**Sheinwoodian Stage**	**Homerian Stage**	**Gorstian Stage**	**Ludfordian Stage**	**Not yet divided**
Stage basal boundary	Southern Uplands in Scotland Dob's Linn	Llandovery area Cwm-Coed-Aeron	Llandovery area Cefn Cerig section	Wenlock area Hughley Brook	Wenlock area Whitwell Coppice	Ludlow area Pitch Coppice	Ludlow area Sunnyhill Quarry	Czech Republic Požáry Section

▲ **Figure 7** The GSSP stratotype sections for each stage of the Silurian. Those shown in blue are stratotype sections found in the Welsh Basin and are of global significance. Key: Lower Silurian (pink); Middle Silurian (green) and Upper Silurian (orange).

Sediments forming the Jurassic system were deposited between 200 and 145 Ma. The geological history of Jurassic rocks is well known because palaeontologists have been able to make correlations between different localities using ammonites, which evolved most rapidly during this period. Zonation and correlation using fossils is covered in the next section and you will need to study it along with this one to get a more complete picture of Jurassic basin analysis.

Basin shape and structure

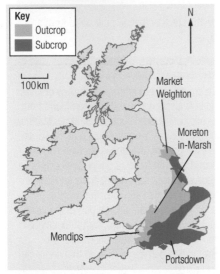

Figure 1 Position of outcrop and subcrop of Jurassic rocks in southern Britain

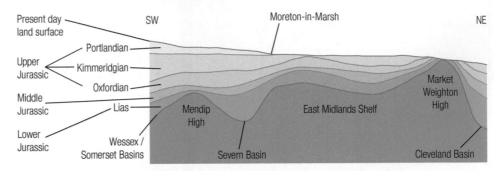

Figure 2 Simplified cross-section of the Jurassic basins in Great Britain

Simplified Summary of Jurassic Lithofacies

Series	Age / Location	Interpretation	Change in Sea Level
Upper Jurassic	*Portlandian*: Evaporites and limestones.	Lagoons became dry land. Freshwater limestones.	Marine regression.
	Kimmeridgian: Clays with high organic carbon content. Over 550m thickness in the south, less in Yorkshire.	No evidence of shallow water deposition. Parallel bedding. Few fossils.	Marine transgression. Deep water conditions.
	Oxfordian: Oxford Clay. Much thicker in the South than in Yorkshire. Then alternating limestone, clay and sandstone in the Corallian Beds.	Fine grained sediments in Oxford Clay deposited in deep water. Cross-bedding and shallow water features in Corallian Beds.	Marine transgression.
Middle Jurassic Ironstone found in Leicestershire is sandy and oolitic. Iron forms about 22% of the rock, giving it a rusty colour.	*The South*: Mudstone and limestone layers. Inferior Oolite, Great Oolite and Cornbrash.	Mainly carbonate but volcanic ash at base of Great Oolite.	Marine regression. Full marine conditions in the south only.
	East Midlands: Sands and ironstone, then dark siltstones and mudstones. Finally, alternating oolitic limestones and mudstones.	Ironstone with ooliths in higher energy shallow water.	Shallow marine followed by siltstones and mudstones in coastal lagoons.
	Yorkshire: Sandstones, shales, thin coals then shallow water limestones.	Coal deposits associated with coastal swamps.	Mainly continental facies, fluvial, deltaic and some coastal.
Lower Jurassic	*Lias*: Fine grained sediments, mainly mudstones. Oolitic ironstone in North Lincolnshire.	Part of succession shows cyclic sedimentation. Shale and mudstone alternate with limestones formed from planktonic remains when terrigenous sediment was absent.	Marine transgression at Triassic–Jurassic boundary. Deposits thin over highs which were land areas in the Permian. Ironstone ooliths formed in higher energy shallow water.

Ammonite

In Great Britain, Jurassic sedimentary rocks crop out from Dorset in the south west to East Yorkshire in the north-east. They often form a prominent escarpment with the dip slope on the south-east side. The major basins are shown in Figure 2. They are separated from one another by structural highs. The basins are grabens bounded by faults that are covered by younger sediments. The faults were detected by geophysical methods in the mid 1980s. Most basins are affected by tensional stresses as the crust subsides. This causes faulting rather than folding, as you can see for yourself by looking at the BGS 1:50 000 Moreton-in-Marsh map.

Facies changes

Analysis of sedimentary rocks allows depositional environments to be interpreted. The table shows a simplified summary for the Jurassic. The rocks provide evidence for cyclic sea level changes with some taking place over long periods of time, and others, such as those in the Corallian Beds, that were much more rapid. There is no evidence for polar ice caps in the Jurassic, so explanations involving glaciation are ruled out.

Cyclic sedimentation

A simplified section of a Corallian cycle is shown below.

Beds 1 to 3 are one cycle and bed 4 is the start of the next cycle.

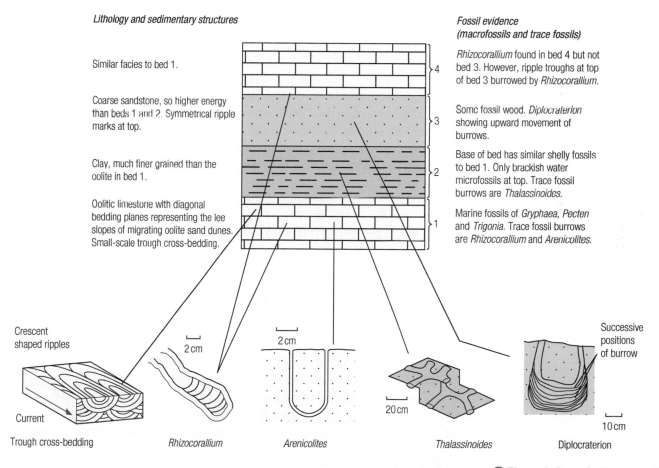

Lithology and sedimentary structures

Similar facies to bed 1.

Coarse sandstone, so higher energy than beds 1 and 2. Symmetrical ripple marks at top.

Clay, much finer grained than the oolite in bed 1.

Oolitic limestone with diagonal bedding planes representing the lee slopes of migrating oolite sand dunes. Small-scale trough cross-bedding.

Fossil evidence (macrofossils and trace fossils)

Rhizocorallium found in bed 4 but not bed 3. However, ripple troughs at top of bed 3 burrowed by *Rhizocorallium*.

Some fossil wood. *Diplocraterion* showing upward movement of burrows.

Base of bed has similar shelly fossils to bed 1. Only brackish water microfossils at top. Trace fossil burrows are *Thalassinoides*.

Marine fossils of *Gryphaea*, *Pecten* and *Trigonia*. Trace fossil burrows are *Rhizocorallium* and *Arenicolites*.

Crescent shaped ripples
2 cm
2 cm
Current
Successive positions of burrow
20 cm
10 cm

Trough cross-bedding   *Rhizocorallium*   *Arenicolites*   *Thalassinoides*   Diplocraterion

Figure 3 Cyclic Sedimentation in the Upper Jurassic of Dorset.

As you know from earlier in the course, ripple marks and cross-bedding indicate shallow water, as do the trace fossils. *Rhizocorallium* is almost parallel to the beds. *Thalassinoides* burrows also worked along the beds but moved up to different levels. *Arenicolites* and *Diplocraterion* were made by vertical burrowers. In *Dipocraterion*, the position of the burrow moves up in response to rapid deposition of sediment. The presence of *Rhizocorallium* at the top of bed 3, in which it is not normally found, is evidence that there was a rapid rise in sea level, allowing *Rhizocorallium* burrowers to move down from bed 4 into bed 3 whilst it was still soft.

An explanation for the rapid change in sea level shown in the Corallian Beds can be found in modern oolite sand banks found in tidal water <2m deep, in the Bahamas. Between the banks of oolite, which migrate in the currents, are troughs where clay-size muddy sediments are deposited. The process is shown below.

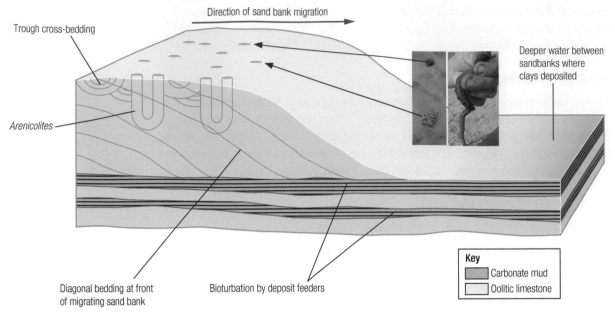

Direction of sand bank migration

Trough cross-bedding

Arenicolites

Deeper water between sandbanks where clays deposited

Diagonal bedding at front of migrating sand bank

Bioturbation by deposit feeders

Key
Carbonate mud
Oolitic limestone

△ **Figure 4** Deposition by lateral migration of an oolite sand bank

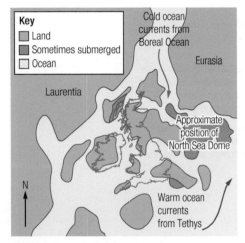

Key
Land
Sometimes submerged
Ocean

Cold ocean currents from Boreal Ocean

Eurasia

Laurentia

Approximate position of North Sea Dome

N

Warm ocean currents from Tethys

△ **Figure 5** Mid-Jurassic palaeogeography

As the sandbank moves, it covers the clay and sea level falls due to the extra thickness of sediment. In the Corallian Beds it seems that the sea became so shallow that grit was deposited in an estuary containing land plant remains, rather than marine fossils. These beds of terrigenous grit might also reflect local uplift and increased weathering, rather than rapid sea-level falls. The mass of sediment produced further subsidence of the basin and the sea returned, bringing back oolite deposition at the base of the next cycle.

Deep water sediments

Above the Corallian Beds is the fine grained Kimmeridge Clay containing no evidence of shallow water deposition. The lower beds are clay with bituminous shale, followed by oil shales with up to 40% organic carbon and no fauna living on the bottom. This suggests a lack of O_2 in bottom waters (anoxic). At the top is a layer of coccolith limestone, formed from planktonic remains. The Kimmeridge Clay extends from the South right up into Yorkshire and represents a major transgression forming a deep basin. It cannot be explained by local subsidence or lateral migration of sediments. Since glaciation is ruled out the only other variable that can affect sea levels is the volume of ocean basins, which is affected by the height or length of mid ocean ridges. This helps to explain changes in sea level affecting wide areas and taking place over longer time scales during the Jurassic.

Palaeogeography

There is evidence for a period of cooler climate beginning at the Middle Jurassic. Recent studies relate this to palaeogeography. Uplift of the area known as the North Sea Dome could have cut off what is now Britain from warm ocean currents moving north from Tethys, the ocean to the south. This would have allowed cold currents from the Boreal Ocean to move further south and reduce sea water temperatures by as much as 10°C. The evidence comes from oxygen and carbon isotope data collected from fossil samples.

QUESTIONS

1 What structural term describes the Mendip high?

2 Describe an example of how Walther's Law applies to parts of the Jurassic.

3 Describe the use of burrows in determining palaeoenvironments.

Reptiles

Marine reptiles, including ichthyosaurs and plesiosaurs were present throughout the Jurassic.

	Ichthyosaurs	Plesiosaurs
Largest known eyes of any vertebrate. An adaptation for deep diving in poor light		
Common characteristics	Breathed air, warm blooded and **viviparous**. Worldwide oceanic distribution.	
Top marine predator	Early Jurassic	Late Jurassic
Limbs	Many had dorsal fin and vertical tail fin	Four flippers
Propulsion	Used end of tail with vertical fin. Flippers used for direction like a rudder	Flippers made flying movement through water
Abundance and diversity	Reduced from Middle to Late Jurassic	More numerous from Middle Jurassic. Some species with longer necks
Length	1 to 10 m	5 to 10 m. Related pliosaurs reached up to 17m

Pterosaurs were common among the flying vertebrates.

- Pterosaurs were the earliest vertebrates to possess powered flight.
- Wings formed of membranes of skin attached to the long fourth digit.
- Wingspan of average adult *Pterodactylus* approximately one metre.
- Some pterosaurs had crests used as part of mating displays.
- Some had coats of hair-like filaments known as **pycnofibres**, which covered their bodies and parts of their wings.
- Fish was an important part of their diet.
- Able to swim and launch from the water.

Cephalopods and bivalves

Ammonites and belemnites are cephalopods. Ammonites can be distinguished from nautiloids and ammonoids by their **suture** pattern. The diversity of ammonites peaked in the Jurassic. Their rapid evolution and widespread distribution (they floated after death) make them extremely useful zone fossils worldwide. Belemnites are also used for zoning and correlation in many regions and are well represented in some Jurassic facies in Britain. Their guards have varying morphologies that can be recognised and used in zoning.

KEY DEFINITIONS

Viviparous means producing live young instead of eggs from within the body.

Pycnofibres filaments that compose the hair of pterosaurs.

Suture the line along which the septum and the shell fuse.

▲ **Figure 1** Pterodactylus a Jurassic Pterosaur

▲ **Figure 2** Ammonitic suture seen on the polished surface of the ammonite where the external shell has been removed.

▼ **Figure 3** Ammonite morphology

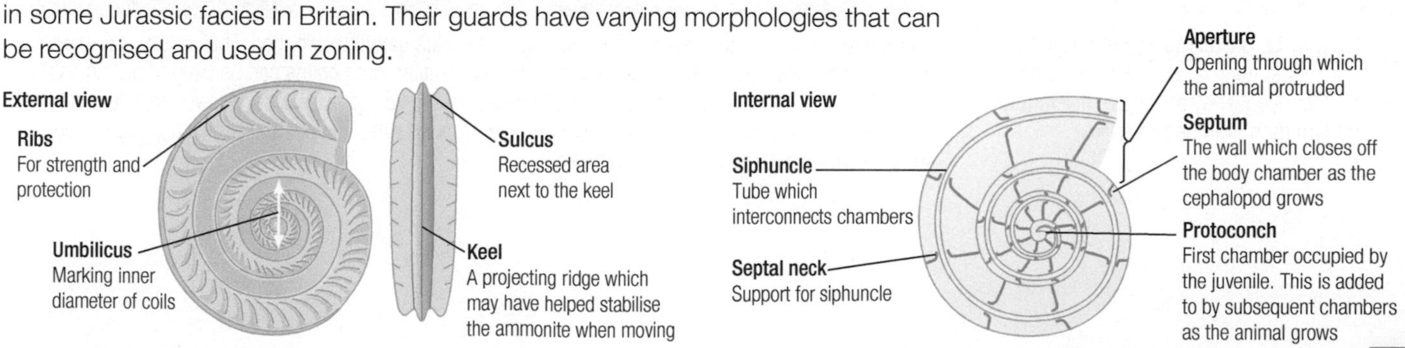

External view

Ribs
For strength and protection

Umbilicus
Marking inner diameter of coils

Sulcus
Recessed area next to the keel

Keel
A projecting ridge which may have helped stabilise the ammonite when moving

Internal view

Siphuncle
Tube which interconnects chambers

Septal neck
Support for siphuncle

Aperture
Opening through which the animal protruded

Septum
The wall which closes off the body chamber as the cephalopod grows

Protoconch
First chamber occupied by the juvenile. This is added to by subsequent chambers as the animal grows

Belemnite guard

Dactylioceras, a Lower Jurassic ammonite

Gryphaea, a Lower Jurassic bivalve

🔺 **Figure 5** Ammonite, belemnite and bivalve from the Lower Jurassic

0 ⊢——————⊣ 2
cm

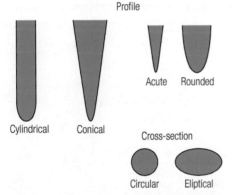

Profile

Cylindrical   Conical

Acute   Rounded

Cross-section

Circular   Eliptical

🔺 **Figure 6** Varying shapes of belemnite guards can be used to distinguish species

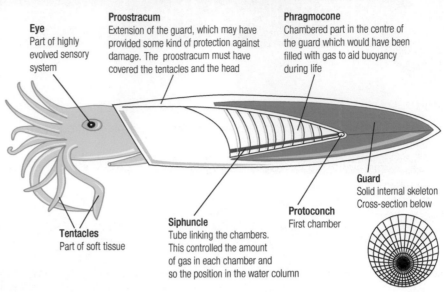

Eye
Part of highly evolved sensory system

Proostracum
Extension of the guard, which may have provided some kind of protection against damage. The proostracum must have covered the tentacles and the head

Phragmocone
Chambered part in the centre of the guard which would have been filled with gas to aid buoyancy during life

Tentacles
Part of soft tissue

Siphuncle
Tube linking the chambers. This controlled the amount of gas in each chamber and so the position in the water column

Protoconch
First chamber

Guard
Solid internal skeleton
Cross-section below

🔺 **Figure 4** Belemnite morphology

In some places in the Lower Jurassic, ancient sea floors were covered by assemblages of *Gryphaea* (bivalves). They are an extinct form of filter feeding oyster that lived on the sediment.

Gradually through time, the shell of gryphaea became larger, thinner and more flat. Ammonites, belemnites and *Gryphaea* are the most useful fossils for zoning and correlation in British Jurassic rocks.

Jurassic fossils and palaeoenvironments

Ammonites and belemnites

Both were nektonic marine cephalopods. Different species lived at different latitudes, due to variation in ocean water temperatures. This has been useful in determining the distribution of past climatic belts. The suture shows where the wall of the chamber, the septum, is fused to the inside of the shell. Due to its length, it made a very strong joint and allowed ammonites to withstand hydrostatic pressures at great depths in the oceans. Most of the soft tissue was in the last chamber. Ammonites moved vertically by using the siphuncle to vary the proportions of gas and water in the empty chambers. This, and their strong sutures, meant they were not limited to surface waters.

Like the ammonites, belemnites could move vertically as well as horizontally. They probably swam in the same way as modern squid. After death, they fell through the water column and their streamlined guards were sometimes aligned by currents. This alignment gives excellent palaeocurrent information.

Carbon and oxygen isotope ratios from ammonite and belemnite fossils have allowed the palaeotemperature and salinity of ocean water to be inferred.

> **CASE STUDY: AMMONITES AS ZONE FOSSILS**
>
> Ammonites swam in the oceans, ensuring they had a wide global distribution. Their strong calcareous exoskeletons provided an excellent prospect for preservation. Ammonites can be used to distinguish geological time intervals of less than 2×10^5 years. This is very precise. They evolved rapidly and were numerous, so they meet most of the requirements for good zone fossils! They are rare in high-energy seas, as they would probably be destroyed.

> **CASE STUDY:**
> **BELEMNITES AS ZONE FOSSILS**
>
> In the earliest Jurassic, belemnites were restricted to the European area but later they had a worldwide distribution. Jurassic rocks in cliffs near Charmouth on the Dorset coast include the Belemnite Marls. They are the best-preserved fossils in these rocks which were deposited during the Pliensbachian (part of the Lower Jurassic).
>
> Belemnites may be preserved in metamorphic rocks. Their robust guards survive where other fossils are destroyed. They are more resistant to metamorphism than ammonite fossils.

Trace fossils

Trace fossil assemblages are related to the environments in which the organisms that made them lived. They provide information about water depth, energy and other environmental variables.

Trace fossils	Palaeoenvironments
Diplocraterion	Beaches and sandy tidal flats, shallow water
Rhizocoralium, Arenicolites	Shallow, marginal marine
Thalassinoides	Below normal wave base, but may be above storm wave base

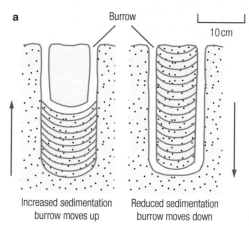

a   Burrow   10cm

Increased sedimentation burrow moves up   Reduced sedimentation burrow moves down

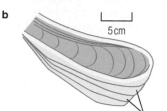

b   5cm

Evidence for former positions of tubes shows upward movement. Possibly made by crustaceans not worms because of scratch marks on inside of tubes

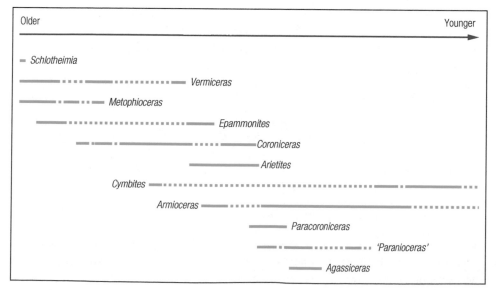

c

◄ **Figure 7 a** *Diplocraterion* and **b** *Rhizocorallium* burrows response to sedimentation; **c** *Thalassinoides* burrows form parallel to bedding in sediments but also moved vertically between beds

Zonation and correlation

If two widely separated rock units contain a sequence of identical zone fossils, then the rocks have the same relative age. Using ammonites, belemnites and some microfossils, Jurassic rocks can be correlated worldwide.

The range of a zone fossil can be very helpful when used with other fossils. Some fossils, such as *Agassiceras* below, have a short time range, others such as *Cymbites* have a longer time range, but where two or more fossil ranges overlap, for example, *Arietites* and *Paracoroniceras*, a biozone can be precisely determined. Zones are often named after an index fossil. The shorter the time range, the better the zone fossil.

Older   Younger

— *Schlotheimia*

— *Vermiceras*

— *Metophioceras*

— *Epammonites*

— *Coroniceras*

— *Arietites*

Cymbites ⋯⋯⋯⋯⋯⋯⋯⋯⋯⋯⋯⋯⋯⋯⋯⋯⋯⋯⋯⋯⋯⋯

Armioceras ⋯⋯⋯⋯⋯⋯⋯⋯⋯⋯⋯⋯⋯⋯⋯⋯⋯⋯

— *Paracoroniceras*

— *'Paranioceras'*

— *Agassiceras*

QUESTIONS

1 The dashed lines in Figure 8 are inferred 'ghost' ranges where the zone fossil has not been found. What evidence is there that the fossils existed during that part of the succession?

2 Why are fossils like *Gryphaea* less important than ammonites for zonation and correlation?

3 What palaeo-environmental information could be deduced from aligned belemnite guards and *Diplocraterion* trace fossils?

◄ **Figure 8** Fossil zones for part of the British Lower Jurassic

PRACTICAL ACTIVITIES IN GEOLOGY

Fall into two main areas:
- To assess skills in measurement, observation and recording of data, both in the field and in the laboratory.
- To assess skills in manipulation of data and other sources of information; such as interpretation of any collected results or interpretation of a geological map.

The geological data you will be working with will be both quantitative (using numerical measurements and results) and qualitative (using descriptive observations). The importance of collecting both types of data cannot be underestimated to enable a full analysis of a geological problem.

SAFETY IN THE LABORATORY

Carrying out simulations or other data-gathering exercises in the laboratory also has to be risk assessed. You should follow the code of conduct for your institution, school or college at all times.

General safety includes: taking care when dealing with water and electricity; using the Bunsen burner safely including the yellow flame when not in use; protection from any particulates that may cause harm when breathed in; wearing goggles to protect eyes when appropriate and taking care when dealing with glassware or sharp objects.

PRACTICAL SKILL AREAS

A minimum of twelve practical activities should be completed throughout your A level study. Six of these are of low-level demand and six are of medium to high-level demand. The content of the practical skills is specified and outlined as follows:

Lowest demand activities:
1 Minerals and rocks
2 Seismology
3 Crystalline processes
4 Sedimentary processes
5 Fossils
6 Geological sequences

Medium demand activities:
1 Orogenic processes
2 Fluid movement
3 Site investigation

High-level demand activities:
1 Geological resources
2 Investigation
3 Research skills.

It is expected that the skills acquired completing these tasks will be needed to attempt practically based examination questions.

Types of practical activity

For each laboratory-based assessment there will be a choice, to be determined by your teacher. These will cover a range of skills, but could involve:
- Setting up of practical equipment to simulate a geological condition or situation.
- The observation, measurement and recording of data to the appropriate level of accuracy.
- The assessment of risks or hazards involved with the equipment and consumables.
- The manipulation of data and display in suitable graphs or charts.
- An assessment of how close the practical activity is to the real geological condition or situation (is it a good model?).

Geology experiments are usually simulations of naturally occurring processes, as it is impossible to carry out real geology in the laboratory. This is because geological processes occur over long time periods and the conditions are difficult to replicate in the laboratory.

USE OF SCALES

Using the scales on photographs or thin section drawings is essential for making correct measurements. Some scales use ×1, meaning that the size shown is the same as the actual size.

A scale of ×0.5 means the feature shown is half size and ×3 means that the feature is three times larger.

Alternatively, a scale bar in mm, cm, m or km is used. An easy way to read a scale bar correctly is to mark the edge of a piece of paper to mark the length of the crystal or feature and then hold it against the scale bar. This is a method that saves calculations!

ANOMALOUS DATA

When individual readings fall outside the main set of the data these results are anomalous (not typical, or different from expected). There are various possible reasons for these anomalous points and explaining them is a key part of evaluating results.

Anomalies could be due to:
- Experimental error so that the points are wrong.
- Errors in measuring.
- Some morphologically similar measurements, for example dimensions of fossils, may lie outside of the natural range and may actually indicate a different species is present in the sample.

Anomalies can be seen on one of the graphs shown in Figure 2.

MEASURING FOSSILS

In some areas, large numbers of fossils can be found in the field. They can be measured in situ or collected and measured in the laboratory. Measurements taken in the laboratory tend to be more consistent as Vernier calipers or similar measuring devices can be used. Data can then be used to analyse the nature of the assemblages by observing the proportions of whole, broken or disarticulated fossils, and the range in sizes.

Where there are many fragmented fossils it is likely to be a death assemblage where fossils have been transported and then deposited. The graphs show two populations of fossils that can be analysed.

▲ **Figure 1** Brachiopod fossil showing where measurements were taken

▼ **Figure 2** Measurements of fossils in the laboratory **a** two different species of brachiopod **b** a single species of brachiopod, with individuals of all growth stages from small, young brachiopods to larger, older ones

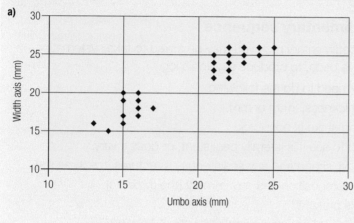

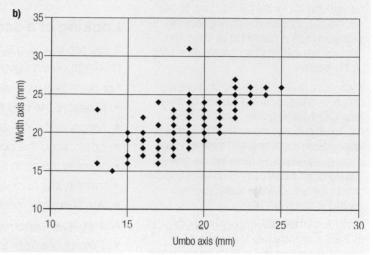

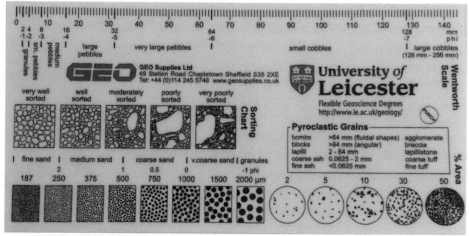

◀ **Figure 3** Grain size chart

Grain size charts

There are several different versions of this piece of equipment, but all do the same job. They allow you to classify rocks accurately using particle/grain size without needing to commit this information to memory. They can be used in the laboratory or field.

The grain size chart also gives you percentage cover diagrams, and so minor minerals such as mica, can be estimated using this part of the chart.

They may also have diagrams showing sorting, which serve as a reminder to describe the grain size distribution.

The advantage of using a grain size chart is that you do not have to memorise the numerical ranges for each type of clastic rock, as these are printed on the chart.

Grain size charts can be used in the laboratory or in the field. They are portable and light.

You will need to spend some time out of the laboratory or classroom examining real geology in *the field*.

The first thing to consider is the risk or hazards that may be present in the area that your teacher has chosen for your field trip. Although your teacher will have completed a risk assessment for the trip, you need to take responsibility for your actions and obey the geological code of conduct at all times. The risks and hazards vary enormously, depending on the locality.

If you visit a quarry or cliff-section, there is danger of falling rocks and you should always wear protective headwear, such as a hard hat. You should also make sure that if you are moving along a cliff face, that there is no-one below or above you, as rocks that become dislodged can cause injury. Slipping on unstable ground can also dislodge rocks, so care is needed at all times.

Similarly, coastal sections pose problems as there is an increased risk of slipping on lubricated surfaces and so extra care is needed, especially when walking on wave cut platforms. These areas may also be covered in seaweed which is also slippery. Ensure that you have tide timetables to hand, so that you do not get cut off by rising waters. In all cases, you should wear stout boots, such as walking boots and definitely not wellington boots, which are notoriously slippery and lack the rigidity needed for safety.

The main danger out in the field is twisting your ankle, so good ankle support is essential. In addition to this, you should also wear suitable clothing, including carrying a waterproof jacket. Jeans should not be worn as if these get wet, then they stay wet, and there may be a risk of hypothermia in cold weather. Always carry a spare layer in your rucksack, just in case.

It is not possible to use a hammer at many localities, so check before you arrive and wear goggles if you do have the opportunity to break off fresh samples.

Possible fieldwork exercises

It is impossible to give a comprehensive list of all the exercises that could be done in the field because the exercises will depend completely on the outcrop of rock you are studying and its geological history. For example, if field work is completed in Dorset, you could study sedimentology and palaeontology, in the form of cyclical sedimentation; whereas in Wales you could study sedimentology, as turbidites or shelf deposits; whilst in the Lake District you could look at metamorphism of slates or igneous products of volcanoes.

Some common activities are as follows.

Looking at a sedimentary sequence

If you are studying a sedimentary rock you may be asked to take systematic readings on successive beds, to produce a graphic log.

For each bed, you will need to do the following:

- Measure the bed thickness, in cm or mm.
- Measure the dip and strike [if possible].
- Look along the bed to see if is laterally persistent, or does it vary.
- Identify the grain size, shape and any sedimentary structures [the three s's].
- Identify the colour, composition and any cement [the three c's].
- Are there any fossils present?

A high-level response would also include observations of the following:

- Differences from the bottom to the top of a bed, such as pebbles at the base.
- The nature of the bedding planes, are they uneven, erosional or flat.
- The amounts of fossils present, sketches of them and possibly, identification.
- Sketches of any sedimentary structures that may be present.
- Trends or orientations of fossils or sedimentary structures to ascertain the flow direction (palaeocurrent direction) of the water at the time.

Remember, use the standard symbols to construct your log; and guidelines given in Chapter 3.

Identification of minerals and rocks in the field

Identification of some minerals found in the field, particularly vein minerals or those from igneous rocks, requires you to look at: the colour, cleavage, lustre, shape and hardness of each individual mineral. Using your finger nail, carrying a steel pin, a copper coin and a glass microscope slide will enable you to assess hardness fairly accurately in the field.

The common igneous rock-forming minerals are: quartz, K-feldspar, plagioclase feldspar, augite (pyroxene), hornblende (amphibole), biotite mica, muscovite mica and olivine. Common vein minerals include: calcite, quartz, fluorite and some metallic minerals such as galena and sphalerite. You should ensure that you can identify these in the laboratory before heading out into the field.

You should refer to the sections of this book which describe how to classify sedimentary, igneous and metamorphic rocks:

- Sedimentary (Chapter 3)
- Igneous (Chapter 2)
- Metamorphic (Chapter 4).

Gathering numeric data

Using a compass-clinometer

It is important to ensure that you know how to use the compass-clinometer to enable you to measure any structures in the field, primary or secondary. This is the key piece of field equipment in a geologist's toolkit. You may measure some of the following:

Primary structures include: bedding; dip; strike; palaeocurrent indicators (cross-bedding, flute marks, tool marks).

Secondary structures include: measurements on folds (angles of dip on limbs, measurements of interlimb angles); orientations/trends of any kind such as dykes and joints; lineations, crenulations and cleavage directions.

You may also need to be able to orientate yourself on a base map, OS map or geological map, using the compass.

Other measurements

You need to use a tape measure to record thicknesses of beds in a graphic log. You can also record: displacements across a fault; thicknesses of chilled or baked margins of intrusions; or thicknesses of veins. You should also include sizes and numbers of fossils that you may also find. These will need to be recorded in cm or mm.

Drawing sketches

Drawing sketches is a method that is used by geologists in the field. These can be drawn on varying scales. Large-scale sketches include: folding and faulting; or features visible in cliffs or quarry faces, such as reef systems. Small-scale sketches include: crenulations in schists; sedimentary structures such as ripples, cross-bedding and graded bedding; and fossils. These small-scale sketches can also be textures in rocks, such as: porphyritic, vesicular, schistose, porphyroblastic and oolitic.

Whatever, you sketch, please always use a suitable scale. If it is a sketch of a cliff or rock face, always record the compass direction or orientation.

🔺 **Figure 1** A compass clinometer used for measuring angles and directions of structures in the field

🔺 **Figure 2** Flute casts showing where an orientation could be taken to determine a palaeocurrent direction

🔺 **Figure 3** Recumbent fold and annotated sketch to show the main features. Note this includes measurements, a scale and an orientation.

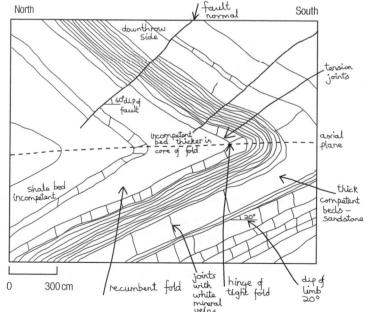

Maths examples

2.1 CALCULATING MASS % OF SILICA

We can calculate the mass % of silica in a rock using molecular masses. Most modern analyses are carried out using an X-ray fluorescence analyser (XRF). The mass % is given as oxides, e.g. of SiO_2, MgO, CaO, etc.

A similar exercise can be performed used the chemical formulae of the minerals present.

What is the silica mass % of a rock containing olivine and pyroxene in equal proportion, given:

Olivine: Mg_2SiO_4

Pyroxene: $CaMgSi_2O_6$

Oxide	Name	Molecular mass
CaO	Lime	56
MgO	Magnesia	40
SiO_2	Silica	60

Clue: $Mg_2SiO_4 = MgO + MgO + SiO_2$

Based on silica content would the rock be classed as ultramafic, mafic, intermediate or silicic?

5.5 CALCULATING AGE USING DECAY RATES

Those curves in the decay graphs can be plotted (by you) and used to find the age of the mineral or (sometimes) rock. Alternatively, you can apply the exponential equation. This may look alarming but can be tamed using the advice below:

$N = N_0 e^{-\lambda t}$

N_0 = the initial number of atoms in a sample, how many there are at time = 0

N = the number left after time t

λ = the decay constant. It is negative because the number will fall as time goes on.

It is easily found from the half-life $t_{\frac{1}{2}} = \dfrac{0.693}{\lambda}$

In realistic examples, there will be some powers of ten to attend to, so be careful!

E.g. suppose the half-life is 1250×10^6 years, the decay rate is

$\dfrac{0.693}{1250 \times 10^6} = 5.54 \times 10^{-10}$

In the dating of ocean floor basalts, you might be asked to find t or N and will have to use natural logarithms ('**ln**' on the calculator, not '**log**'). The trick is to take the natural logarithms of both sides of the equation which gives you:

$\ln N = \ln N_0 - \lambda t$

Then it's just a simple rearrangement of terms.

Try this on an example where you already know the answer. How many atoms are left after two half-lives if we started with 100?

So $N_0 = 100$, $\lambda = 5.54 \times 10^{-10}$ and $t = 2500 \times 10^6$ years

$\lambda t = 1.385$, $\ln (100) = 4.610$

$\ln (N) = 4.610 - 1.385 = 3.220$

Use the 'e' function key on your calculator to get $N = 25.03$, quite close to the 25 we were expecting!

9.10 VOLUME AND DENSITY

A seam has a working height of 2.2 m. If the longwall face is 400 m long and the mining starts 500 m from the shaft bottom, calculate:

a) The maximum volume of coal which can be mined. Give your answer in cubic metres (m^3)

b) The mass of coal which can be mined (the density of coal is $1\,300\,kg\,m^{-3}$).

a) Volume of a regular cuboid
= length × width × height
= $500 \times 400 \times 2.2 = 440\,000\,m^3$.

b) Density = mass/volume:
mass = density × volume
= $1300 \times 440\,000 = 572\,000\,000\,kg$
= 572 000 tonnes.

10.1 COMPARING EARTHQUAKES

1 The energy produced by the two largest known earthquakes is:

Chile (1960): 2.5×10^{23} joules   Alaska (1964): 7.5×10^{22} joules

$Mw = \frac{2}{3}logE - 6.1 = \frac{2}{3}log\ 2.5 \times 10^{23} - 6.1 = \frac{2}{3}\ 23.4 - 6.1 = 15.6 - 6.1 = 9.5$

$Mw = \frac{2}{3}logE - 6.1 = \frac{2}{3}log\ 7.5 \times 10^{22} - 6.1 = \frac{2}{3}\ 22.9 - 6.1 = 15.3 - 6.1 = 9.2$

2 What does one increment on the magnitude scale mean?

Compare a magnitude 8.0 and a magnitude 9.0 earthquake

$LogE = \frac{3}{2}Mw + \frac{3}{2}\ 6.1 = \frac{3}{2}\ 8 + \frac{2}{3}\ 6.1 = 12.0 + 9.2 = 21.2$ $E = 1.58 \times 10^{21}$

$LogE = \frac{3}{2}Mw + \frac{3}{2}\ 6.1 = \frac{3}{2}\ 9 + \frac{2}{3}\ 6.1 = 13.5 + 9.2 = 22.7$ $E = 5.01 \times 10^{22}$

The M 9.0 earthquake is 31.7 times stronger than the M 8.0 earthquake.
As an approximate check, the inverse log of $\frac{2}{3}$ or 1.5 is 31.6

3 Compare a magnitude 8.9 earthquake with magnitude 5.7 earthquake.

When comparing two magnitudes it is usual to approximate the above equation to

$M = \frac{2}{3}logE$ or $logE = \frac{2}{3}M = 1.5M$

This allows us to write:

$logE_{(8.9)} = 1.5\ M_{(8.9)}$

$logE_{(5.7)} = 1.5\ M_{(5.7)}$

a) Comparing magnitude, which is related to amplitude.
Using logarithms, the ratio of the magnitudes is found from the inverse log of $(8.9 - 5.7) = 3.2$.
The inverse log of $3.2 = 1585$.
The magnitude 8.9 earthquake is 1585 times bigger than the magnitude 5.7 earthquake.

b) Comparing the energy released from the earthquake, which actually does the damage.
Using logarithms, the ratio of the energies is found from the inverse log of $(8.9 - 5.7) \times 1.5 = 4.8$.
The inverse of $4.8 = 63\ 100$

The M 9.0 earthquake releases 63 100 times more energy than the M 8.0 earthquake.
The magnitude scale compares the amplitude of waves on a seismogram, not the energy released by the earthquake. So, the magnitude 8.9 earthquake is 1585 times 'bigger' as measured on seismograms, but about 63 100 times 'stronger' in terms of energy released! Magnitude, based on amplitude, is easier to explain, but the energy does the damage.

10.9 CALCULATING MASS MOVEMENT

Suppose a slope is a bedding plane of sandstone, with density $2270\ kg\ m^{-3}$ and has 15% porosity. What is the percentage increase in the force acting down the slope when the dry rock is saturated with water?

For simplicity, imagine a cubic metre of sandstone, dry weight 2270 kg; when saturated it contains 150 litres of water, another 150 kg, i.e. a 6.6% increase.

If a block of sandstone is 2 m thick with an area of $60\ 000\ m^2$, what is the force acting down the slope? The bedding plane dips at 24°.

If the angle of dip (θ) is measured as 24°.

The mass of the dry block is $2 \times 60\ 000 \times 2270\ kg. = 272\ 400\ 000\ kg$

The acceleration due to gravity is about $9.8\ m\ s^{-2}$

So the force acting down the plane is $m \times g \times sin\theta = 1.09 \times 10^9\ N$

Better hope that there's no clay on that bedding plane! This makes mass movement more likely.

Mass movement mechanism illustrated by a block of mass '**m**' on an inclined plane. The force down the slope **m g sinθ** is resisted by the frictional force '**f**'. The force acting into the plane (**m g cosθ**) and the reaction opposing it have been omitted.

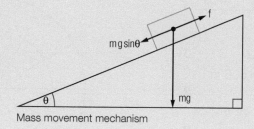

Mass movement mechanism

Scientific literacy questions

Practical skills in geology questions

1 a A student identified a sample of fossils from each of three beds in a disused quarry. She used the following null hypothesis (H_0);

There is no significant difference between the fossil assemblages in each bed.

She decided to use a chi squared test; $\chi^2 = \Sigma \dfrac{(O - E)^2}{E}$ to determine the significance of H_0 at the 0.05 confidence level.

Table 1.1 shows the data collection results.

		C1	C2	C3	
		Shale	Muddy limestone	Biosparite	Total
R1	Ammonties	30	16	12	58
R2	Worm burrows	24	52	15	91
R3	Bivalves	46	30	18	94
		100	98	45	243

🔺 **Table 1.1**

Table 1.2 shows some of the calculations carried out for the χ^2 test.

O	E	O – E	$(O - E)^2$	$\dfrac{(O - E)^2}{E}$
30	23.9	6.1	37.21	1.56
16	23.4	−7.4	54.76	2.34
12	10.7	1.3	1.69	0.16
24				4.80
52	36.7	15.3	234.09	6.38
15	16.9	−1.9	3.61	0.21
46	38.7	7.3	53.29	1.38
30	37.9	−7.9	62.41	1.65
18	17.5	0.6	00.36	0.02
			χ^2	18.50

🔺 **Table 1.2**

   i Complete the blank cells in **Table 1.2**. **3**

   ii Explain how you calculated the value of E in row 4. Show your working. **2**

   iii Explain how the degrees of freedom for this test are calculated. **2**

   iv The critical value for chi squared in this case is 9.49. Explain why H_0 is accepted or rejected at the 0.05 confidence level. **2**

   v Suggest reasons why the fossil assemblages are different in each of the beds. **3**

b Fossil belemnite guards were found concentrated at specific horizons. Orientations of 15 specimens were recorded from a single bedding plane. The results are shown in **Table 1.3** below.

specimen number	orientation
1	021°
2	065°
3	212°
4	035°
5	220°
6	330°
7	040°
8	055°
9	040°
10	005°
11	060°
12	103°
13	132°
14	235°
15	056°

🔺 **Table 1.3**

   i Complete the tally chart in **Figure 1.1** below using the data supplied. Plot the rose diagram. **3**

orientation		tallied number of specimens
000–030°	181–210°	
031–060°	211–240°	
061–090°	241–270°	
091–120°	271–300°	
121–150°	301–330°	
151–180°	331–360°	

🔺 **Figure 1.1**

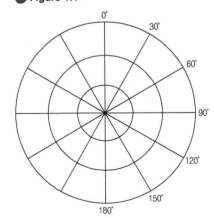

   ii Describe and explain the distribution of the data you have plotted. **2**

2 a A number of structures are formed by compressive and tensional stresses.

The cross-section in **Figure 2.1** below shows faulted strata.

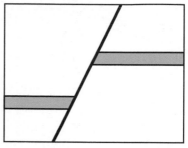

▲ Figure 2.1

i On the cross-section clearly label the following features:

- Fault plane
- Hanging wall
- Fault dip
- Downthrow side. **4**

ii Identify the type of fault **1**

The diagram in **Figure 2.2** below shows folded strata.

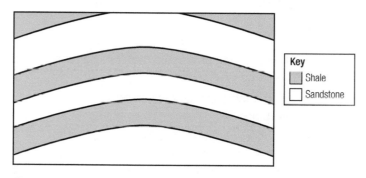

Key
- Shale
- Sandstone

▲ Figure 2.2

iii On the diagram, draw joints where they are most likely to form. **1**

iv Explain your choice of location for the joints. **1**

b Analyse the differences between a North Sea oil reserve and onshore fields where fracking is used. **4**

c **Figure 2.3** shows onshore oil and gas fields in the USA.

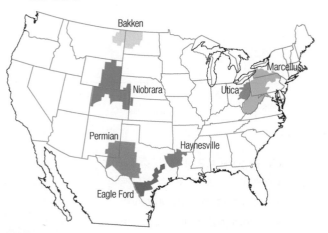

▲ Figure 2.3

Table 2.1 shows oil production per fracking rig in barrels per day.

Region	June 2016	July 2016
Bakken	832	850
Eagle Ford	1067	1097
Haynesville	30	31
Marcellus	68	69
Niobrara	925	947
Permian	493	508
Utica	345	356

▲ Table 2.1

i Calculate the percentage change in mean oil production per rig in the seven onshore oil fields shown in **Table 2.1** between June and July 2016.

Show your working. **2**

ii Describe the facies to which the source rocks belong. **2**

3 Read the text below, then answer the questions that follow. **2**

Bardarbunga is a strato-volcano beneath the Vatnajökull ice cap. It is cut by a swarm of fissure dykes which extend 115 km in the direction of 240° from North and 55 km in the direction of 030° from North. In August 2014, approximately 1600 earthquakes occurred in 48-hours, with magnitudes up to 4.5. Monitoring by a seismic network provided data showing that activity was confined to the region close to the front of the propagating dykes. Coverage is less good to the south of the volcano due to difficult field conditions on the ice cap. Once a pathway has formed and remains open, the flow of magma seems to be aseismic. After 13 days, the dyke erupted at Holuhraun. The eruption took place over a six-month period, and was the largest in Iceland since 1783. The ice filled caldera of Bardarbunga collapsed during the eruption. Large volumes of sulfur dioxide were emitted. There have been 32 confirmed eruptions of Bardarbunga between 1702 and 2014, of which 23 occurred in the eighteenth century.

a **Figure 3.1** is a map of Iceland.

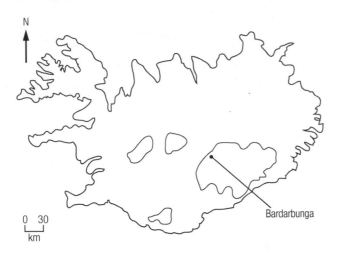

Figure 3.1

On the map:

i Label the Vatnajökull ice cap. **1**

ii Draw and label two lines to show the trends of the dyke swarms. **2**

b i Describe how the magnitude of seismic waves is measured. **3**

c i Draw a fully labelled diagram to show the features of a strato-volcano. Ensure that you label at least six features. **3**

ii Describe the process of caldera collapse. **3**

d i Calculate the return period, to the nearest year, for eruptions of Bardarbunga. Show your working. **2**

ii Calculate the expected frequency of eruptions of Bardarbunga. Include units in your answer. **1**

e Probability or prediction? Analyse the usefulness of both approaches in the context of the eruption of Bardarbunga. **6**

4 The map, **Figure 4.1**, shows an igneous intrusion and its surrounding country rocks.

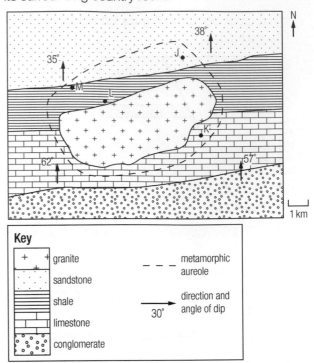

Figure 4.1

a i Describe the processes that take place in a metamorphic aureole. **2**

ii Name the rocks that would be found at **J, K, L** and **M**. **4**

iii Suggest two reasons why the width of the metamorphic aureole is greater in the north than in the south. **2**

b

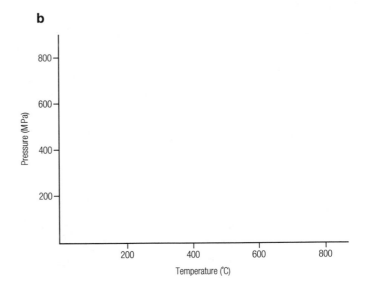

Figure 4.2

On **Figure 4.2** above, shade and label where the following fields would plot:

• Contact metamorphism

• Regional metamorphism **2**

c i Describe how the rock shown in **Figure 4.3** below is produced by metamorphism. **2**

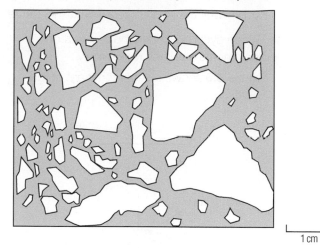

⬤ Figure 4.3

ii **Figure 4.4 a** and **b** below show thin section diagrams of metamorphic fabrics.

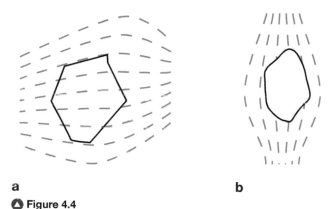

a   **b**

⬤ Figure 4.4

Analyse the diagrams and describe the information that each fabric contains. **4**

5 **Figure 5.1** below shows the geology of a small area.

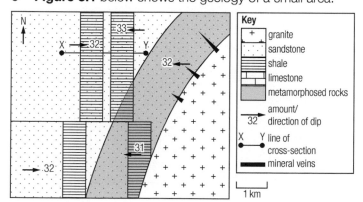

⬤ Figure 5.1

a i Fully describe the fault and the movement on the fault. **2**

ii Draw a clearly labelled cross-section along the line **X–Y** shown on the map, to show the structure. **2**

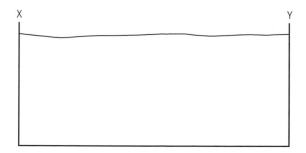

b Fossils **C** and **D** have been found in the area and are shown in **Figure 5.2** below.

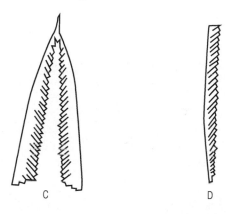

⬤ Figure 5.2

i Name the group to which both these fossils belong. **1**

ii Label **two** different morphological features on fossil **C**. **2**

iii Clearly label on **Figure 5.1** where fossil **C** may be found. **1**

iv Why are fossils **C** and **D** **not** found in the same rock unit? **1**

c There are mineral veins in and around the granite.

i Describe the process that formed these mineral veins. **3**

ii The minerals galena, sphalerite and fluorite are found in the veins.
Give two diagnostic physical characteristics of each mineral. **3**

6   **Figure 6.1** is a diagram showing polar wandering curves for two continents.

A	250 Ma
B	175 Ma
C	100 Ma
D	75 Ma
E	50 Ma

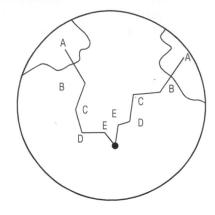

⬤ Figure 6.1

a   i   When did the continents begin to move apart? **1**

   ii   Explain how polar wandering curves are constructed.   **2**

b   **Figure 6.2** shows photographs of palaeoclimatic indicators.

a

b

c

⬤ Figure 6.2

i   Describe the geological features shown in each photograph.   **3**

ii   Explain why they are palaeoclimatic indicators. **4**

c   The Wilson cycle describes the breakup of supercontinents and their reassembly.

   i   **Figure 6.3** shows a supercontinent at the Equator.

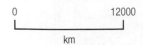

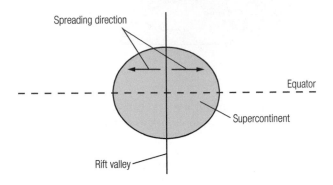

⬤ Figure 6.3

   Assume that Earth's equatorial circumference is 40 000 km and that the average spreading rate at the rift valley is 2.8 cm yr^{-1}.

   Calculate the time needed for the two halves of the supercontinent to collide on the other side of the Earth.

   Show your working.   **3**

   ii   Evaluate the Wilson cycle model as a framework for understanding long-term global climate change.   **6**

7 The diagram shows a silicate mineral structure in which all the oxygens are shared. Which of the following minerals has this structure?

- **A** Quartz
- **B** Pyroxene
- **C** Olivine
- **D** Mica   Your answer ☐

8 A rock is found to have crystals with an average diameter of 5.4 mm and a composition of approximately 60% Plagioclase Feldspar and 40% Pyroxene. Which of the following is it most likely to be?

- **A** Diorite
- **B** Dolerite
- **C** Gabbro
- **D** Peridotite   Your answer ☐

9 A sedimentary rock is found to be made up of 23% fossil fragments supported in a lime mud. What would it be classed as under the Dunham scheme?

- **A** Mudstone
- **B** Wackestone
- **C** Packstone
- **D** Grainstone   Your answer ☐

10 Which of the following statements is/are correct?

1. Sea-level changes due to the formation of ice are isostatic and can be correlated world-wide.
2. Scandinavia is currently experiencing isostatic rebound after the melting of ice in the last 12 000 years.
3. Transgressions due to the melting of ice are evident in the raised beaches around Scotland.
4. Changes in the rate of spreading and the lengths of ocean ridges cause eustatic changes in sea-level.
5. Regressions result from global warming and can be seen in seismic sections as 'onlap'.

- **A** 1 and 2
- **B** 2 and 5
- **C** Only 4
- **D** 2 and 4   Your answer ☐

11 Which of the following resources is usually detected using electromagnetic (EM) survey methods:

1. Oil and gas   2 Coal
3. Sulfide ore deposits   4 Salt deposits

- **A** All of these
- **B** 2 and 3
- **C** 3 and 4
- **D** Only 3   Your answer ☐

12 Describe and explain the hydrogeological impact of dams. Suggest how engineering geology can be used to mitigate adverse impacts.   **6**

13 The map shows Dalradian metamorphic rocks, the ornamentation indicates the grade of the metamorphism based on Barrovian zones.

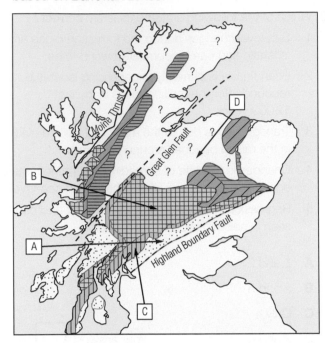

a i If A indicates the index mineral Chlorite and C is the zone in which Biotite is found; what is the index mineral found in zone B? **1**

ii Areas marked '?' have rock types that do not show the index minerals used in Barrovian zones. If they did, which index mineral would you expect at D? **1**

b i Given the minerals found in zones A and C, what was the original rock type? **1**

ii What type of metamorphism is being mapped in this diagram? **1**

iii Suggest the cause of this metamorphic event. **1**

c i What do the metamorphic zones suggest about the timing and type of movement on the Great Glen Fault? **2**

ii Given that the Dalradian rocks were deposited between 730 and 570 Ma, comment on the plate tectonic setting that could give rise to the metamorphism. **2**

14 a In the boxes provided, draw and label external views of a brachiopod and a bivalve so that the differences in symmetry can be made clear. **2**

b i In the boxes labelled 'internal'; draw and label sections through a brachiopod and bivalve to show how the opening and closing of the valves was achieved. **2**

ii Explain the differences in mechanisms for opening and closing the valves for each organism. **2**

Brachiopod	Bivalve
external	external
internal	internal

c Describe and explain the differences between brachiopods and bivalves in their feeding and respiration. **4**

15 Explain the principles of radiometric dating using K-Ar as an example. Describe a different possible difficulty in obtaining an accurate age for igneous, sedimentary and metamorphic rocks. **6**

1 a i

O	E	O – E	$(O – E)^2$	$\dfrac{(O – E)^2}{E}$
30	23.9	6.1	37.21	1.56
16	23.4	–7.4	54.76	2.34
12	10.7	1.3	1.69	0.16
24	37.4	–13.4	179.56	4.80
52	36.7	15.3	234.09	6.38
15	16.9	–1.9	3.61	0.21
46	38.7	7.3	53.29	1.38
30	37.9	–7.9	62.41	1.65
18	17.5	0.6	00.36	0.02
			χ^2	18.50

1 mark for each correct number highlighted in the grid above. **3**

ii $E = \dfrac{\text{row total} \times \text{column total}}{\text{grand total}} = \dfrac{91 \times 100}{243} = 37.4$

1 mark for explanation; 1 mark for correct answer. **2**

iii degrees of freedom = number of rows – 1 × number of columns –1

(3 – 1) (3 – 1) = 2 × 2 = 4

1 mark for explanation; 1 mark for correct answer. **2**

iv The value of chi squared is greater than the critical value.

Therefore, H_0 can be rejected at the 0.05 confidence level. **2**

v Credit any sensible suggestions. 1 mark for each bed.

Examples:

Ammonites were pelagic so more are found in deeper water deposits indicated by shale.

More worm burrows in muddy limestone due to soft substrate and higher organic content.

Biosparite probably the bed with coarser (sand size) grains so less suitable for burrowing organisms. **3**

b i Correct numbers for the tally chart are: 2,9,1,1,2,0

Correct plotting of rose diagram **3**

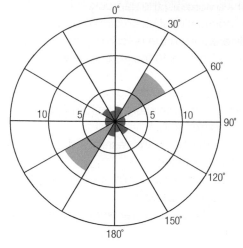

3

ii Belemnite guards are aligned with a predominantly NE-SW trend.

They were aligned by currents (either from NE or SW) when they fell to the sea floor. **2**

2 a i

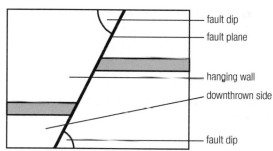

Fault dip must be the acute angle. 1 mark for each correct label. **4**

ii Normal fault. **1**

iii Joints are in the sandstone (white beds). Need to be some joints at the hinge and at least 3 correct joints. If in shale as well, or only at base of bed then = 0. **1**

Key
- Shale
- Sandstone

iv In the sandstone as it is a competent bed / brittle at the hinge / crest where tension or stretching is occurring at the top of each bed and compression is occurring at the bottom of each bed. **1**

b Conventional reservoirs composed of porous / permeable rocks, e.g. sandstone / limestone *but* the reservoir in this basin is impermeable shale / shale normally the caprock.

In conventional reservoirs, the source rock is different from the reservoir rock/ *but* in this basin the source rock is also the reservoir rock.

In conventional setting gas often migrates to the top of a structure / anticlinal/salt dome/fault/ stratigraphic trap. *but* in this basin the reservoir does not have any structural highs.

Conventional reservoirs are normally found around the edge of a basin *but* in this area the reservoir is in the middle of the basin.

(1 mark for statement about North Sea **and** corresponding unconventional reservoirs

Allow 1 mark for any 2 points that are not matching pairs to max 2) **4**

c i June mean = 537 barrels per day
July mean 551 barrels per day **1**

percentage change was $\dfrac{551 - 537}{537} \times 100 =$ 2.6% **1**

ii shale facies; offshore marine; containing planktonic remains as source of petroleum; low energy; oxygen poor; fine grained / muddy. Any 2 **2**

3

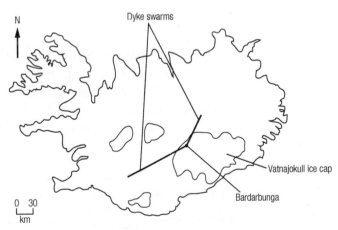

Dyke swarms
N
Vatnajokull ice cap
Bardarbunga
0 30
km

a i Label to correct area. **1**

ii 1 mark for each line drawn to scale, labelled and in correct direction. **2**

b Recordings made using seismometers; seismic moment (M_W) provides data on physical scale of earthquake; size of fault rupture / slip displacement; amount of energy released. Any 3 **3**

c i

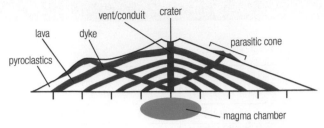

vent/conduit crater
lava dyke
pyroclastics
parasitic cone
magma chamber

Lava and pyroclastics labelled in alternate layers = 2
Any additional two features = 1 **3**

ii Volcano supported by magma in chamber beneath.
Violent eruption / level of magma in the chamber drops.
Magma chamber not full / there is a void / space / cavity.
Unsupported top / cone / volcano top or crater falls into space below.
Any 3. **3**

d i $\dfrac{n+1}{m}$ where n is number of years on record and m is number of recorded occurrences.
Return period is 312 + 1 ÷ 32 = (9.78) 10 to nearest year. **2**

ii Expected frequency is $\frac{1}{10} = 0.1\,\text{yr}^{-1}$ **1**

e Return period does not mean earthquakes of given magnitude will occur in that period.

Forecasts based on sets of data which may not be entirely accurate or reliable.

Data are generally not values from an idealised distribution.

Return period is useful in risk analysis for natural events and the safety of structures.

Forecasts for Bardarbunga depend on time scale used for data / 18th century pattern different from modern pattern in Bardarbunga data.

Any 3

Difficulty of predicting exact place/date/magnitude.

Public reaction to imprecise warnings.

Unnecessary damage to economy/ loss of earnings / income/ social functioning of settlements if prediction not precise.

False negatives can lead to lack of preparedness.

Successful prediction could greatly reduce loss of life / hospitals and rescue teams to be prepared/ on standby.

Use of micro-seismics for activity at front of dyke useful for directional trend at Bardarbunga.

Any 3

Real solution may not be prediction but may be forecasting coupled with infrastructure modification.

Prediction can often only give very short advance warning but forecasting allows preparedness.

Any 1

Max 6 marks

Max 4 marks if no reference to Bardarbunga. **6**

4 a i Alteration / change / recrystallisation / in country rock.

Due to heat from igneous intrusion / baking of country rock by contact metamorphism. **2**

ii J = metaquartzite; K = marble; L = hornfels; M = spotted rock / spotted slate / andalusite slate. **4**

iii In the north, the edge of the intrusion is dipping at a shallow angle.

In the south, the dip of the granite is steep.

May be more eroded in north if topography sloping.

May be more jointing / fracturing in north.

Any 2 **2**

b

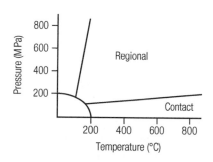

2

c i Rocks are crushed and ground into angular fragments;
by the mechanical process of cataclasis;
dynamic metamorphic zones.

Any 2 **2**

ii Diagram (a)
Crystal grew at the same time as a period of deformation.
Compression / pressure was perpendicular to the foliation.
Flattening of foliation less within the mineral, which offered resistance.
Any 2

Diagram (b)
Garnet / crystal formed during a period of metamorphism.
Later metamorphism produced the foliation.
Foliation wrapped around crystal so occurred later.
Any 2 **4**

5 a i Normal fault;
downthrow to the south. **2**

ii

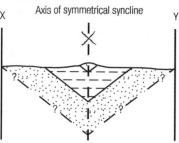

Sketch cross-section showing syncline

Labelling **2**

b i Graptolites **1**

ii

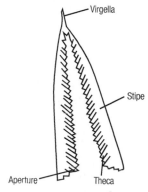

Any 2 **2**

iii In unmetamorphosed shale. **1**

iv C is Ordovician; D is Silurian / didn't live at the same time. **1**

c i Gases / fluids escaping from intrusion react with country rocks and replace minerals within them.

Composition of rock at contact is altered.

Elements added to contact / country rock may result in ore deposits.

Silicic magmas contain more water than mafic magmas and so develop greater volume of reactive gas.

Contact metasomatism.

Any 3 **3**

ii galena
SG 7.5 / high density; lead grey colour; hardness 2.5; cubic; metallic lustre; grey streak.

sphalerite
SG 3.9 – 4.1; hardness 3.5 – 4.0; streak white to reddish brown.

fluorite
SG 3.18; hardness 4; streak white; cubic; vitreous lustre.

[Allow rounding for the hardness of each mineral]

Any 2 for each mineral = 1 mark **3**

6 a i At C on a map OR 100 Ma **1**

ii At the time of formation, some rocks preserve the direction of the magnetic field.

Preservation of magnetic field occurs when rocks cool below the Curie temperature / point.

The inclination (Im) and declination vectors (Dm) are preserved in some rocks; paleolatitude (λp) and paleolongitude (ϕp) of the pole can be found.

Any 1

Samples of magnetised rocks collected on land and in oceans.

Ocean floor rocks only date back 200Ma / land-based rocks needed for earlier

Any 1 **2**

b i a Red coloured / desert sandstone with dune cross-bedding

b Very poorly sorted / mineralogically and texturally immature sediment

c Fossilised colonial coral **3**

ii Wind-blown sands with iron cement are formed in arid desert conditions.
Cross-bedding directions can be used to determine palaeowind directions.
Large, angular fragments in a clay matrix are typical of deposits from glacier ice.
Boulder clay / till formed in arctic climates.
Corals live in warm shallow tropical waters.
Water temperatures need to be between 25 and 29°C. **4**

1 point for each feature + any 1 other

c i The ocean spreading rate (28mm yr^{-1}) = 2 8 km per Ma

For continents to meet on other side of Earth the distance travelled = 20 000 km (half the Earth's circumference) – 6000 km (half width of supercontinent) = 14 000 km

$$\text{time} = \frac{\text{distance}}{\text{velocity}} = \frac{14000}{28\text{km Ma}^{-1}} = 500\,\text{Ma}$$

1 mark for each correct figure in the calculation; 1 mark for correct answer. **3**

ii Indicative points:

- Embryonic, young, mature, subduction, terminal and end stages of model.
- Different thermal properties of land and ocean.
- Mountain belts modify local and global climate patterns.
- Global climate controlled by ocean currents.
- Ocean currents depend on shape of ocean basins.
- Movement of plates and continents affects distribution of land masses.
- Mountain ranges and connectivity of the oceans affected.
- Phanerozoic climate change occurred over periods of 100 Ma.
- Similar timescale to Wilson cycle.

Level 3 (5–6 marks)
Clear recognition of value Wilson cycles. Uses a wide range of evidence and draws appropriate conclusions. *There is a well-developed line of reasoning which is clear and logically strucured.* *The information presented is relevant and substantiated.*
Level 2 (3–4 marks)
Sound interpretation of some evidence. Realises that Wilson cycles are of value but may not be able to make clear links to examples in support. *There is a line of reasoning presented with some stucture. The information presented is relevant and supported by some evidence.*
Level 1 (1–2 marks)
Limited interpretation in terms of evidence / examples. May make statements that are not part of a consistent argument. Lacks direct link to concepts included in Wilson cycle. *The information is basic and communicated in an unstructured way. The information is supported by limited evidence and the relationship to the evidence may not be clear.*
0 marks
No response or no response worthy of credit.

Answers to practice questions

7 The diagram shows a framework silicate and the information provided would give a formula of SiO_2. Thus Quartz (**A**) is the only correct answer. **1**

8 The mineral composition makes the rock mafic and the mean crystal size is coarse. Thus Gabbro (**C**) is the correct answer. **1**

9 Mud supported is either a mudstone or wackestone. More than 10% grains means that it must be (**B**) a wackestone. **1**

10 Statement 1 is incorrect as only eustatic changes can be correlated worldwide. Statement 2 is correct. Statement 3 is incorrect as raised beaches show uplift of the land and therefore regression. Statement 4 is correct. Statement 5 is incorrect; regressions would show as offlap and global warming would result in transgressions. **1**

Thus (**D**) 2 and 4 correct is the expected answer.

11 Electromagnetic survey is generally used for shallow subsurface investigations and therefore not for oil and gas exploration. Coal is just another sediment in a succession and would not show a significant change in electrical properties. Metal ore deposits would show anomalous conductivity and could be detected by EM survey. This would be the expected answer except that salt is such a good insulator that it can be detected by the way in which currents flow round or over salt domes.

Best answer (**C**). **1**

12 One mark if ANY of the following points are described **AND** explained:

Impounding water in the reservoir raises hydrostatic pressure which can act as de-stabilising uplift if water has access to the base of concrete dams.

Raised hydrostatic pressure acts on joints/ bedding planes/ foliation and can de-stabilise slopes above the reservoir.

Raised hydrostatic pressure may 'lubricate' faults causing reactivation especially with the added load of impounded water.

Raised hydrostatic pressure may find weaknesses in earth dams leading to leakage, erosion and failure.

Water can be prevented from accessing the base of the dam by a grout curtain/cut off curtain – a series of boreholes injected with high-pressure cement or silicone or similar compounds.

Water can be prevented from accessing the base of the dam by laying down impermeable compressed clay or a geomembrane liner during construction.

Geology should be mapped in detail to avoid EITHER dipping clay layers OR dips into reservoir OR faults OR strata dipping downstream.

Strata liable to be destabilised by increased hydrostatic pressure could be grouted/rock anchored/rock bolted/ drained to prevent failure. **6**

13 a i In the Barrow zones the third index mineral encountered would be Garnet. **1**

    ii Assuming the grade is increasing then Kyanite/ Sillimanite. **1**

  b i These index minerals are only present if the original rock contains a high proportion of clay minerals. **1**

      ANY rock with high clay minerals, e.g. shale/ mudstone/pelitic.

    ii This is regional metamorphism OR allow contact/thermal. **1**

    iii Continental collision/thickening of the crust OR major igneous intrusion(s)/granite intrusions/ gabbro intrusions. **1**

  c i Timing: that the metamorphism occurred before faulting on the GGF. **1**

      Fault movement is sinistral tear/wrench/strike-slip. **1**

    ii Any two points from:
Crustal thickening in mountain-building event.
Magma rising above a subduction zone.
Caledonian orogeny.
Closure of the Iapetus Ocean.
Collision of an island arc with Laurentia/ Dalradian margin. **2**

14

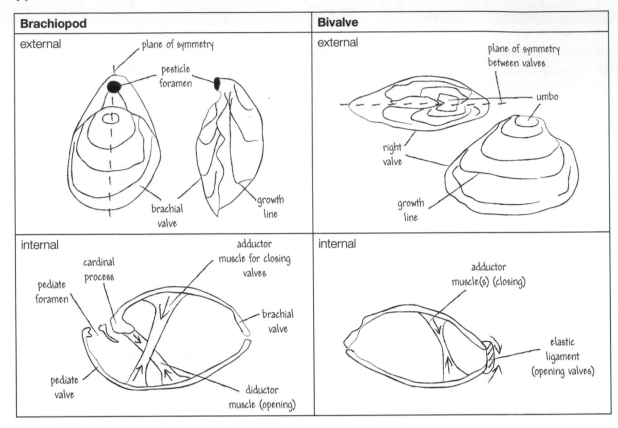

Brachiopod	Bivalve
external — plane of symmetry, pesticle foramen, brachial valve, growth line	external — plane of symmetry between valves, umbo, right valve, growth line
internal — cardinal process, pediate foramen, pediate valve, adductor muscle for closing valves, brachial valve, diductor muscle (opening)	internal — adductor muscle(s) (closing), elastic ligament (opening valves)

a Both diagrams should have at least 3 correct labels.   1
Both diagrams should contain a clear, labelled plane of symmetry.   1

b i Both diagrams should have at least 3 correct labels.   1

Brachiopod should have both diductor and adductor muscles labelled, Bivalve must have adductor and elastic ligament labelled.   1

ii Explanation that brachiopod diductor muscle acts on a lever on the inside of the brachial valve/cardinal process to open valves.
Closed by contraction of adductor muscle(s).   1

Explanation that bivalve opens due to elastic ligament and shuts as adductor muscle contracts.   1

c Brachiopod uses a <u>lophophore,</u> a coiled organ covered in filaments which generate currents within the mantle to bring in oxygenated water.   1

It traps nutrient particles/filter feeds with the same organ passing particles back to the mouth.   1

Bivalves have <u>siphons.</u> The inhalant brings in fresh oxygenated water and passes it over the gills which generate the current. The exhalent siphon removes water and waste from the anus.   1

The nutrient particles are trapped by the gills/ filter feeds and passed to the mouth.   1

15 Any 3 of the following points for 3 marks.

Radioactive parent atoms decay at a statistically constant rate.

To form stable daughter atoms.

50% of the parent will decay during one half-life.

^{40}K-^{40}Ar has a half-life of 1.26 Ga.

Measuring the ratio of parent to daughter atoms reveals the time since decay commenced.

Loss of daughter atoms such as $^{40}Argon$ in ^{40}K-^{40}Ar decay will change the ratio making the rock appear younger.   3

Any 3 of the examples for 3 marks.

Major igneous intrusions will cool slowly such that different minerals and different parts of the intrusion would become closed systems at different ages.

Sedimentary rocks are porous and weather more easily and are prone to lose Argon.

Sedimentary rocks are composed of a variety of clasts older than the rock.

Metamorphism can 're-set' the decay clock as heat drives off Argon.

Different minerals have different closure temperatures and show different ages.   3

Glossary/Index

Aa lava flows have a rough, blocky, jagged surface. 39–41

Abrasion the wearing away of the Earth's surface by the action of wind, water or ice moving sediment over a surface. 54, 74, 100, 103, 192

Absolute dating gives specific dates for rock units or events given in millions of years before present (Ma). 108, 114

Abstraction the removal of water from any source. 184, 186–187, 235

Abyssal plain is an almost level area of the deep ocean floor covered by deep sea sediments. 68, 84–85, 156

Acid mine drainage caused when water flows over or through sulfur-bearing materials, for example rocks containing iron pyrites. This causes acidification of the water by iron pyrites reacting with air and water to form sulfuric acid and dissolved iron ions. 187, 200–201

Adaptive radiation a process when organisms diversify rapidly into many different forms, usually as a response to a change in the environment or to exploit a new ecological niche. 252, 255, 258

Adiabatic cooling occurs when crust or mantle material rises, undergoes expansion and the temperature falls; with no loss or gain of thermal energy. 28, 161

Adiabatic (decompression) melting rising magma experiences a decrease in pressure. Expansion causes a reduction in temperature with the loss of heat as molecules use energy to move further apart. Melting occurs because the melting point also decreases as the pressure decreases. 160

Adiabatic heating occurs when crust or mantle material descends and temperature rises as it contracts; with no loss or gain of thermal energy. 28

Adiabatic process a thermodynamic process in which no heat enters or leaves the system during expansion or compression. 28

Adsorption the process by which particles, such as heavy metal ions, become attached, semi-permanently, to the surface of a clay mineral. 210–211, 232

Advection a process by which thermal energy is transferred through a medium by a fluid. It is an important process close to mid-ocean ridges where sea water is drawn down into the crust and replaces rising hydrothermal fluids. 135, 156, 158

Albedo the fraction of solar energy (shortwave radiation) reflected from the Earth back into space. It is a measure of the reflectivity of the Earth's surface. Ice, especially with snow on top of it, has a high albedo: most sunlight hitting the surface bounces back towards space. 244–245, 247, 267

Amniotic eggs types of eggs with shells produced by birds and reptiles. 261–263

Amphibian a creature with two life styles, one in water, the other on land. 251, 260–261, 266, 270

Amplitude the maximum extent of an oscillation, measured from the position of rest. 125, 127, 220, 222–225, 230–231, 237, 295

Amygdaloidal where there are large vesicles that have been filled with a secondary mineral. 23, 26–27, 49

Anaerobic means without oxygen. This term is usually used to describe conditions within a sediment. 62, 86–88, 101, 103, 201, 212, 244, 270–273

Anhedral crystals show poorly formed crystal faces. 24–25

Anions atoms that have gained electrons leaving them with a net negative charge. 16–17, 211

Anoxic means water without oxygen. Standard decomposing bacteria cannot live without oxygen, nor can scavengers. 101, 103, 119, 190, 212–213, 215, 270–272, 277, 281, 286

Anthropocene the suggested name for the current geological epoch. The time during which human activity has been the dominant influence on climate and the environment. From the Greek *anthropos* meaning 'human'. 250–251

Anticline an upright fold with the oldest rocks in the core. 114, 146, 170–172, 177, 179, 202, 205–206, 213, 215–216

Antiform an upward closing fold. 170

Apparent dip a dip that is measured to be less than the maximum inclination. 115, 164

Apparent polar wandering curve a line on a map which joins up the apparent positions of the magnetic north pole over time. 144–145

Apparent thickness is the thickness of the bed measured as seen at the outcrop but not at right angles to the bedding plane. 164

Aquiclude an impermeable rock that does not transmit water. 184–185

Aquifer a body of porous and permeable rock capable of storing and yielding significant amounts of water.

Aquitard a rock which has very low permeability and only allows the transmission of water at very low rates. 185

Archaeopteryx a genus of a small, bird-like dinosaur, of a similar size to a raven. 264–265, 276–277

Archean an era within the Precambrian, covering a period from 4.0 billion to 2.5 billion years ago. 88, 156

Archosaurs a group of reptiles dominating the Triassic. These were the ancestors of dinosaurs, birds and crocodiles. 262–263

Arenaceous sedimentary rocks are those in which the grain size of clasts is 0.0625 to 2 mm. 58–59

Argillaceous sedimentary rocks are those in which the grain size of clasts is less than 0.0625 mm. 58–59, 232

Artesian basin a large, synclinal confined aquifer under hydrostatic pressure. 184

Artesian wells hold water under hydrostatic pressure, which rises up the well to the piezometric surface on release. 184

Arthropods invertebrate organisms with jointed appendages, segmented body and an exoskeleton. 105, 111, 250, 254, 270, 272–275, 277

Articulated organisms found whole or connected, as in life. 103, 111

Asperity the term mostly used to describe the roughness of the surface of a discontinuity. 240

Assemblage the combination of different fossils identified in a rock unit. 73, 99, 102–103, 108, 116, 118, 288–289, 291

Assimilation the melting process that incorporates blocks of country rock, freed by stoping, into the magma. 35, 48–49

Asteroids rocky objects which failed to form a planet. 107, 120–123, 251, 269

Asthenosphere a layer of the mantle below the lithosphere. 28–29, 128–129, 131–133, 140–141, 149–150, 153, 155–157

Astrophic means the hinge line is not straight, may show a shallow or highly curved pattern. 258–259

Attenuation the loss of energy experienced by a wave shown as a reduction in amplitude as it propagates through a material. 220, 278

Attrition the wearing down of sedimentary grains due to collisions with other grains during transport. 54, 66, 74, 76–78, 192

Augite belongs to a group of minerals called pyroxene. 12–13, 19–21, 23, 25, 27, 46, 162, 292

Aureole a region surrounding an igneous intrusion in which the country rocks have undergone contact metamorphism. 34, 90, 92–93, 114–115

Avalonia was a microcontinent during the Palaeozoic which contained Wales, part of Ireland, England and the southern part of Scotland, located in the southern hemisphere. 245, 255, 280–281, 283

Average crustal abundance describes the amount of metal in 'average' continental crust. 188–189

Axial plane a plane that joins the hinges of all the beds. It bisects the fold. 94, 114, 170–171, 175–179

Axial plane trace the outcrop of the axial plane at the Earth's surface. 114, 170–171, 175, 178–179

Baked margin in the country rock where it was heated by the intrusion and altered. 32–33, 38–39, 50, 115, 293

Lithification the process of changing unconsolidated sediment into rock. 62, 106–107, 212

Lithofacies include all the physical and chemical characteristics of a rock. 64, 284

Lithosphere the upper rigid layer of the crust and upper mantle. 28–29, 128–129, 131–135, 140–141, 148–161, 278

Lithostatic pressure the vertical pressure due to the mass of the rock only. It is also referred to as the overburden pressure. 238–239

Littoral zone is the area between the extreme low and extreme high water of the spring tides. 66–67, 78–79, 103

Lobe-finned fish are fish that possessed both lungs and gills and had four fleshy fins supported by bones in a similar structure to a hand, e.g. lungfish. 260–261

Lobopodians are a group of worm-like taxa with stubby legs, often called velvet worms. 274–275

Longshore drift the combination of littoral drift and beach drift. 78, 193

Low-energy refers to the velocity of the water during deposition moving slowly or not at all. Examples of a low-energy environments are swamps, deep seas and lakes. 81, 84, 87, 103–104, 116–117, 212–213, 258, 270–271, 278–279

Low velocity zone (LVZ) is characterised by low seismic velocities. 28, 128–129, 158–159

Low viscosity where magma or lava are fluid and flow freely. 23, 25, 30, 41–44

Lustre the surface appearance of a mineral, as it interacts with light. 8–10, 12, 20–21, 63, 188, 292

Macrofossils fossils large enough to be visible with the naked eye. 80–81, 285

Mafic igneous rocks have a silica content of 52 to 45% and are dark (melanocratic). 20, 23, 49–50, 130, 162, 268

Mafic minerals dark coloured, silica poor and rich in magnesium and iron. 20–22, 46, 50–51, 91, 95

Magma molten rock, which is beneath the Earth's surface. 9, 18–25, 27–51, 92, 94, 107, 112–113, 115, 130, 134–137, 140, 144–145, 148–156, 158–161, 191, 194, 213

Magnetic inclination the angle of dip of the lines of a magnetic field. It is the dip angle made with the horizontal and the Earth's magnetic field lines, measured with a compass. 138–139

Magnetometer an instrument which detects the strength and direction of the magnetic field. 138–139, 194

Magnitude a measure of the amount of strain energy released by an earthquake. 37, 45, 107, 109, 126–127, 139, 147, 150–151, 220–224, 226–231, 237, 295

Major intrusions are plutonic and cool deep below the surface and include batholiths. 21, 24, 34–35, 109

Mantle plume a stationary area of high heat flow in the mantle, which rises from great depths and produces magma that feeds hot spot volcanoes. 28, 135, 137, 154–155, 159, 161, 244, 268

Marginal marine environment neither marine nor terrestrial, but belongs to both categories, one at low and another at high tide. Often a beach environment. 278–279

Marker horizon shows a bed or bedding plane with a change of lithology easily distinguished which covers a wide geographical extent. 116, 251

Matrix the background material of small grains in which larger grains occur in a sedimentary rock. 58–61, 64, 67, 72–73, 84, 96, 142, 164, 181, 193, 217

Maturation process converts plankton into petroleum by the effects of increasing temperature and pressure during burial. 212–213

Mechanically formed sedimentary rocks result from the processes of erosion, transport and deposition of clasts. 58–59

Melt the name given to magma or lava, in the liquid phase. 39–41, 46–48, 50, 137, 159–161

Mercalli, Guiseppi (after whom the Mercalli scale is named). 127

Mercalli scale measures the intensity of an earthquake and is based on the effects that are felt in the area. 126, 222–223, 226

Mesozoic 110–111, 214, 246, 262–263, 269–270, 279

Metallic meteorites high density, composed of iron and nickel with a similar composition to the Earth's core. 122

Metamorphic aureole a large area around a batholith where the rocks have been metamorphosed. 34, 90, 92–93, 114–115

Metamorphic grade a measure of the intensity of metamorphism. Although increases in temperature only result in increasing grade in contact metamorphism, grade is also used to describe regional metamorphism where both temperature and pressure vary. 34, 91–93, 98–99

Metamorphic zone the name given to an area between two isograds. The zone is named after the lower grade isograd. All locations within a metamorphic zone experienced the same metamorphic grade. 90, 98–99

Metasomatism the hydrothermal alteration of rock. 162–163

Metazoans multicellular, eukaryotic organisms of the animal kingdom. 272

Meteorites rock fragments which fall to Earth from space. 116, 121–123, 127–129, 231, 268–269

Micrite a microcrystalline calcite, a depositional matrix of lime mud. 60–61, 80–81

Microfossils fossils that are usually less than 1 mm and can only be seen with a lens or microscope. 80–81, 85, 100, 103, 119, 130, 148, 160, 216–217, 249, 267, 270, 278–279, 285, 289

Migmatite a metamorphic rock, such as gneiss, that melts to form an igneous rock. 93, 153

Migration the movement of petroleum from a source rock to a reservoir rock. 212–213

Mineral a naturally occurring chemical substance having a definite composition and crystalline structure. 8–29, 34–35, 39–40, 46–56, 58–66, 72, 76–77, 79, 88, 90–101, 108–109, 116, 121–123, 128, 130–131, 134, 136, 138–139, 142, 145, 149, 152, 162–164, 166, 169, 176–177, 180, 186–198, 200–201, 203, 206, 210–211, 216, 230, 232–233, 236, 238–241, 270, 272–275, 282, 290–292, 294

Mineralogical maturity a measure of the extent to which minerals have been destroyed by weathering and attrition. 54–55

Mineral resources can be metallic and non-metallic or industrial minerals. 188–189, 196–197

Minor intrusions cool at hypabyssal depth below the surface and include sills and dykes. 21, 24, 32–33, 38, 109, 113, 115

Mohorovičic´, Andrija (after whom the Moho is named). 126–127

Mohorovičic´ Discontinuity; Moho is the boundary broadly between the crust and the mantle where P and S wave velocities decrease. 126–127, 131, 133, 149, 150–151, 153, 159

Mohs, Friedrich (Mohs' scale of hardness) a non-linear scale which compares the hardness's of common minerals and can be used as a diagnostic tool to identify those that are unknown. 11, 15

Moment the turning effect of a force around a pivot. 209, 220

Moment magnitude scale measures the magnitude and leverage on two sides of the fault in an earthquake. 126–127, 150, 222

Monomictic conglomerate a coarse grained sedimentary rock containing clasts of a single rock type. 66

Morgan, W Jason (proposed the theory of mantle plumes). 154

Moon or natural satellite is a body that orbits a planet. 120–123, 130, 132, 134, 143

Morphology describes the shape of the organism. 102, 104, 110–111, 254, 256, 260–261, 265, 267, 271, 282–283, 287–288

Mould the impression of the outside or inside of a fossil. 100–101

Mouth bars crescent-shaped deposits of sand and silt forming below sea level, where distributaries enter the sea. 87

Multi-beam echo-sounder sends out up to 100 beams of sound covering an arc up to 10 km wide, allowing wide areas of the sea floor to be covered in one transect of the ship. The profiles are only 100 m apart. 158–159

Mural pores connections between adjacent corallites of coral, perhaps for communication. 256